# New Concepts of a Blood–Brain Barrier

# New Concepts of a Blood–Brain Barrier

Edited by

John Greenwood
*Institute of Ophthalmology*
*University College London*
*London, England*

David J. Begley
*King's College*
*University of London*
*London, England*

and

Malcolm B. Segal
*United Medical and Dental Schools*
*St. Thomas' Hospital*
*London, England*

**Springer Science+Business Media, LLC**

Library of Congress Cataloging-in-Publication Data

On file

Proceedings of a symposium, held in honor of Professor Michael Bradbury, on New Concepts of a Blood–Brain Barrier, held July 4–6, 1994, in London, England

ISBN 978-1-4899-1056-1 ISBN 978-1-4899-1054-7 (eBook)
DOI 10.1007/978-1-4899-1054-7


Originally published by Plenum Press, New York in 1995
Softcover reprint of the hardcover 1st edition 1995

10 9 8 7 6 5 4 3 2 1

**Michael Bradbury, M.D.**

To whom this volume is dedicated. The photograph was taken at the Symposium Dinner during the symposium "New Concepts of a Blood–Brain Barrier" held in Professor Bradbury's honor to mark his contribution to blood–brain barrier studies. (Photograph by Dr. Mark Habgood.)

## PREFACE

The scientific contribution of Mike Bradbury to the study of the blood-brain barrier is considerable and wide-ranging, starting with his M.D. thesis in 1962 and still continuing today. The varied spectrum of topics relating to the blood-brain barrier presented in this volume and the many geographical locations from which both speakers and participants gathered to attend the symposium are a fitting testament both to Mike's wide-spread influence in the field and to the esteem in which he is held.

When we first had the idea of a *festschrift* to mark Mike's retirement as Professor of Physiology at King's College London and the beginning of a new *vie libre* as scientist and yachtsman the plan was initially for a gathering of all of the many colleagues, collaborators and students who had worked with him over the years. However what we had not taken into account was the closeness of the international community of workers in the field, the widespread influence that Mike had wielded and the speed with which word would spread. We should have anticipated all three. The final outcome was an excellent symposium with the majority of the world's key workers either contributing or attending. We hope that this volume presents an adequate record of the meeting.

In deciding on a title for both the symposium and for this volume we felt that we could not do better than to adapt the title of Mike's 1979 book "The Concept of a Blood-Brain Barrier" which is a key work in the subject and widely quoted in current publications. We would like to thank Mike for allowing us this privilege.

John Greenwood
David Begley
Malcolm Segal
Susan Lightman

*Symposium Organising Committee*

The organising committee would like to thank the following for their generous sponsorship of the meeting.

The Biomedical Sciences Division, King's College London
Allergan Ltd. UK
Glaxo Ltd. UK
Zeneca Ltd. UK

CONTENTS

# DEVELOPING VIEWS OF THE BLOOD-BRAIN BARRIER

Michael W.B. Bradbury

Physiology Group
Biomedical Sciences Division
King's College London
Strand
London WC2R 2LS

## HISTORY

It is helpful sometimes to look at our work in the context of both past history and of current developments. Just as plants and animals develop and evolve so do the ideas of science. One may detect earliest mention, mainstreams of development, defunct issues and growing points - the latter being very dependent on practicable techniques as well as on concepts.

Though previous experimenters had found that certain chemicals with potential biological activity were effective when administered into cerebrospinal fluid but not into blood (e.g. Biedl and Kraus, 1898), Lewandowsky (1900) appears to have been the first to attribute special properties to the brain capillaries. He stated that they excluded certain substances in blood from entering brain, whilst allowing others to pass through.

We are more familiar with the name of Goldmann whose colorful experiments showed that trypan blue administered into the bloodstream left the brain white and the CSF colorless (1909). Conversely, trypan blue administered into cerebrospinal fluid (CSF) stained adjacent central nervous system (CNS) tissue avidly (1913). Interestingly, Ehrlich (1902), though frequently cited as the discoverer of the blood-brain barrier, argued firmly against the cerebral blood vessels having special properties. He attributed absence of intravital staining of brain by acid dyes to a lack of affinity.

### A Glial Barrier?

Goldmann ascribed a general impermeability of cerebral capillaries to 'the glial limiting membrane'. Controversy continued as to whether the site of the barrier was endothelial or glial until distribution of transport markers could be observed in electron micrographs (EM) (Reese and Karnovsky, 1965), see below. It is now known that an

*New Concepts of a Blood–Brain Barrier*
Edited by J. Greenwood *et al.*, Plenum Press, New York, 1995

endothelial barrier is near universal in vertebrates, including cyclostomes, but not in most elasmobranchs (reviewed by Abbott, 1992). The holocephalan Chimaera thought to be related to the ancestors of elasmobranchs has an endothelial barrier (Bundgaard, 1982). Sharks, dogfish and skates have a tight glial investment beyond the leaky capillary endothelium. Similarly, amongst invertebrates, the cuttlefish Sepia, and presumably other cephalopods, has a respectable glial barrier arround its capillary endothelium (Abbott and Bundgaard, 1992). The important adaptation would seem to be separation of the interstitial fluid of brain from blood irrespective of the particular cells involved.

To return to our main story, notable insight based on careful reading of the existing literature was shown by Krogh in his Croonian lecture published in 1946, having been postponed by the war. In relation to the blood-brain barrier he stated - 'On the whole the permeability of the vessels in the CNS is very reminiscent of that generally found or assumed for the cellular membrane, and it would appear that in the search for drugs which act on this system one should be guided mainly by studies of their solubility in lipoid and not as very generally supposed, of their electrical charge. It must be admitted that there is no proof that the endothelium of the vessels in the CNS have no secretory function.' Note the subtle use of the double negative in the last sentence.

## An Electron-Microscopist's Heresy

An interesting heresy concerning the blood-brain barrier arose from early electron micrographs of the brain (Wyckoff and Young, 1954; Schultz *et al.*, 1957). These and other authors noted a lack of extracellular space in brain. The large apparent spaces between cells seen under the light microscope, appeared under the EM to be filled with cellular processes, the neuropil, which included the watery processes of astrocytes. Maynard *et al.* (1957) argued that the blood-brain barrier was an illusion. The phenomenon of slow entry of hydrophilic solutes into brain could be attributed to slow transport across the plasma membranes of brain cells, mainly glia. Slow entry into cerebral extracellular fluid did not exist because there was effectively no extracelular fluid. This hypothesis gained very wide acceptance not least amongst physiologists. When I made my first experiments on this 'illusion' in 1961, the new concept gave practitioners in the field an

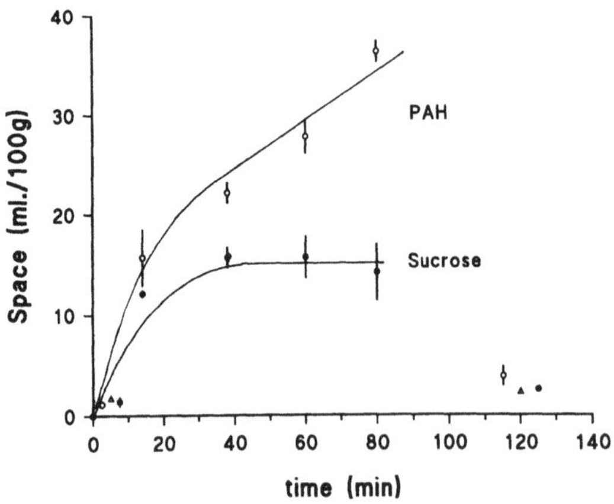


**Figure 1.** Penetration of PAH and sucrose into pieces of cerebral hemisphere incubated at 40°C - lines. Points at bottom of figure represent penetration of PAH, sucrose and $^{131}$I-iodide into brain of rabbits infused intravenously. From Davson and Spaziani (1959).

aura of unacceptability and even naughtiness! The hypothesis totally failed to account for the early findings of Biedl and Kraus, Lewandowsky and Goldmann.

### The Right Conclusion

One who held fast to the original, rational explanation was Hugh Davson. I remember clearly reading his paper with Spaziani of 1959 in the science library of Oxford University. Its clarity of thinking and simplicity of experimentation impresses me to this day. Using radio-labelled extracellular markers, e.g. iodide, sucrose and PAH, Davson and Spaziani showed that these tracers barely penetrated from blood to brain in vivo, but were readily taken up into fragments of brain in vitro and indicated an extracellular space of 14-20 % (Figure 1). Similar results were later shown by extracellular markers administered into CSF.

For those who would not believe and demanded visual evidence, the final proof of a blood-vessel barrier, sited at the level of the endothelium, came from the experiments of Reese and Karnovsky (1967) and of Brightman and Reese (1969). It was shown in EM's that horseradish peroxidase (HRP) and colloidal lanthanum hydroxide were prevented from entering cerebral interstitial fluid by the endothelial cells joined by a continuous seal of occluding junctions whereas after injection into CSF these compounds penetrated very readily into interstitial fluid even reaching the pericapillary regions from the brain side. These were truly 'Goldmann' experiments at a fine structural level. From this time the study of uptake of substances from blood into brain became that of inward transport across the blood-brain barrier.

## METHODS OF STUDY

Figure 2 indicates several techniques which have contributed and continue to contribute to the analysis of transport at the blood-brain barrier. The reductionist goal must be to elucidate mechanisms at the level of the capillary wall and at the endothelial cell. The last two of the four general methods of the list below allow this objective to be approached.

### Intravenous Infusion or Injection

Intravenous infusion or bolus injection of a radiotracer followed by regional brain analysis at a set time after. If it is assumed that the blood-brain barrier is rate- limiting, as is generally valid, such methods allow the normal flux of a solute into brain from blood to be measured, but allow little control of the fluid perfusing cerebral microvessels. Exceptionally, the transport kinetics of a molecule such as glucose can be analysed because glucose has no competitors in blood plasma.

### Short Exposure Methods

A method which allowed repetition in the same animal was the indicator diffusion technique (Crone, 1965), but greater flexibility and consistency in small laboratory animals, such as the rat, was Oldendorf's BUI Method (1970). This allowed at least semi-quantitative characterization of a number of transport mechanisms at the blood-brain barrier and liberated the field from phenomenology. A lineal descendant of the Oldendorf Method

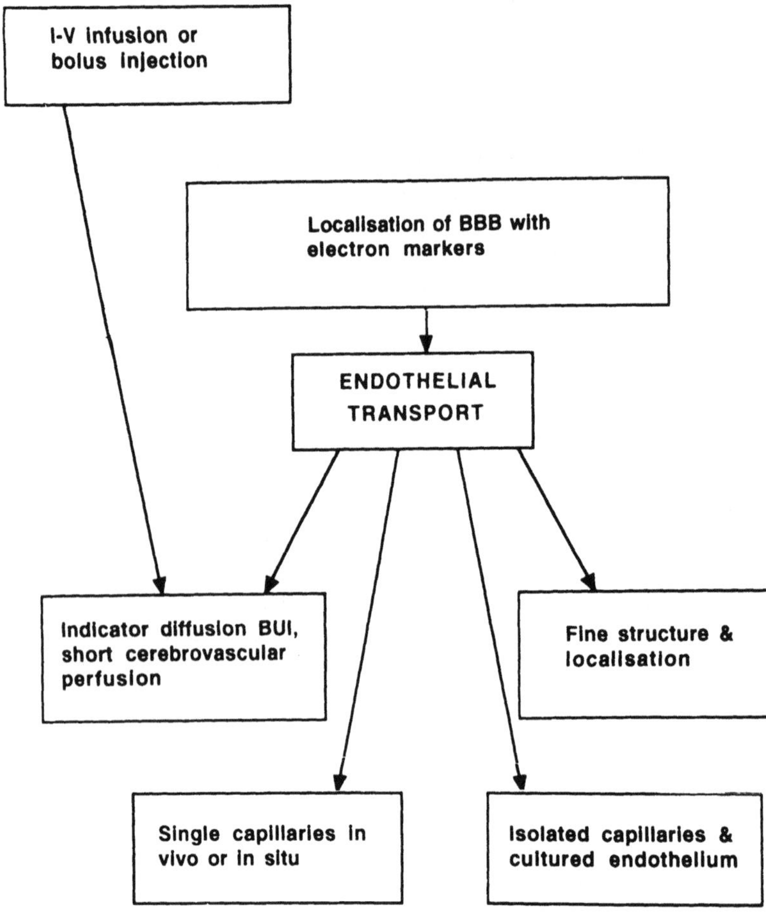


**Figure 2.** Phylogeny of BBB methods. Some techniques which have contributed and continue to contribute to the analysis of transport at the blood-brain barrier.

is the short duration perfusion technique of Takasato *et al.* (1984). This permitted extension of the duration of the Oldendorf Method beyond the time of a single capillary pass and allowed complete cotrol of the composition and pressure of the fluid perfusing the capillaries of one cerebral hemisphere.

## Study of Single Capillaries

Crone and Olesen (1982) measured the resistance of the wall of single microvessels on the surface of frog brain by means of current-delivering and voltage-recording microelectrodes. The filtration coefficient of such capillaries has been measured by a development of the Landis method (Fraser and Dallas, 1990), and it is now possible to measure the actual permeability of single microvessels in the rat brain to fluorescent compounds (Easton and Fraser, 1994).

### Isolated Capillaries and Cultured Endothelium

Studies of isolated capillaries have had limited usefulness. Purification and viability have been problems. Experiments on transport have been limited to those of uptake and efflux from the tissue, and even in these the relative roles of the abluminal and luminal plasma membranes have been uncertain.

Separation of isolated microvessels from brain has been most useful as a preliminary to culture of the endothelium. This, as you know, has become something of a small industry. Both primary cultures from the brains of different species and immortal cell lines have their proponents. The technique allows study of human material, electrophysiology and measurements of intracellular ionic activities with fluorescent markers. An objective must be the culture of an endothelial monolayer which in terms of its tightness and asymmetry is a true blood-brain barrier *in vitro*. By astrocyte coculture and by other means, this goal is being approached with persistence but has not yet been achieved.

## TRANSPORT OF TRACE METALS

### Background and Techniques

Both circumstances and my background have led to investigation of the transport of some trace metals at the blood-brain barrier. The necessity of control of the fluid within the capillaries of a blood-brain barrier *in vivo* has led me to make considerable use of a method involving short vascular exposure. The rest of this chapter will be devoted to considering the basis of such studies and to presenting some indicative results.

In marked contrast to the cations of the monovalent alkali metals, all the trace metals in blood plasma are heavily bound with the concentration of free ion for each metal forming a very small fraction of its total. The Table shows the general distribution of species for four trace metals in plasma. In many cases, the concentrations of species of a metal in plasma have been computed from the known formation constants of each putative complex together with the known concentration of each ligand and the total concentration of the metal (May *et al*, 1977).

**Table 1.** Chemical state and associated ligands of some trace metals in blood plasma.

| Dynamic equilibrium | | | High affinity, slow |
|---|---|---|---|
| Free | Low MW | High MW | or non-exchangeable |
| $Zn^{2+}$ | Histidine | Albumin | $\alpha_2$-macroglobulin |
| $Pb^{2+}$ | Cysteine | Albumin | $\alpha_1$-intertrypsin inhbitor |
| $Fe^{3+}$ | (Citrate) | (Albumin) | Transferrin |
| $Al^{3+}$ | Citrate | (Albumin) | Transferrin |

Ligands in parenthesis only form complexes when saturation of transferrin is high.

It is likely that the transported form or forms of the metal will be one or more of those actually present in plasma. In general, we have found it convenient to initially examine the rate of uptake of the metal by infusing intravenously the metal ion as a radiotracer. The radioactive cation is given free of carrier and is of high specific activity so that the total concentration of the metal in plasma is not substantially raised. The infusion is given at a diminishing rate so that the radiotracer concentration in arterial plasma is maintained fairly constant. If the rate of uptake from blood is fast enough to suggest that brain entry might be measured during short cerbrovascular perfusion, then the latter method is attempted. The perfusion fluid is generally an oxygenated simple salt solution with the pH buffered by 10 mM HEPES. The ability of the latter to complex trace metals is very limited. Perfusions may be made in the presence and absence of a ligand which can complex the metal under study. Ideally, the concentration of the free metal ion is measured with an ion-selective electrode. Alternatively, its concentration and that of any complexes may be computed as discussed above.

## Lead

The amino acid cysteine and the protein albumin are the main ligands for lead in plasma. Figure 3 shows that 5% albumin or 200 $\mu$M cysteine effectively abolished entry of $^{203}$Pb into several brain regions during 1 min of perfusion of a single cerebral hemisphere. Hence, it is unlikely that complexes of lead with either albumin or cysteine are appreciably transported into brain. However, the free Pb ion or a simple inorganic

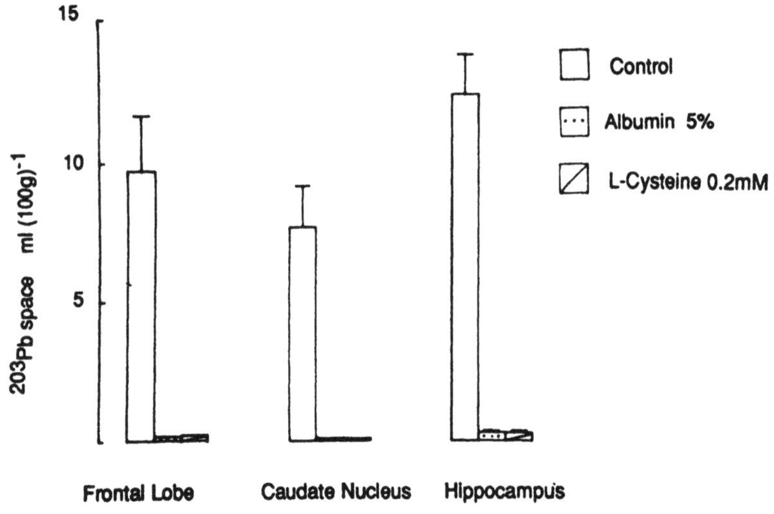


**Figure 3.** Effect of albumin and of cysteine on $^{203}$Pb uptake into 3 regions of brain after 1 min of carotid perfusion of $^{203}$Pb (1 $\mu$M total lead) in saline buffered to pH 7.4 with 20 mM HEPES. From Deane and Bradbury (1990).

complex of it must penetrate very rapidly. In separate experiments the lead flux varied linearly with lead concentrations between 0.1 and 4 $\mu$M. The presence or absence of $HCO_3$ did not influence lead transport into brain. Influx did, however, vary with $[H^+]$ in a manner which suggested that $PbOH^+$ is the transported species (Deane and Bradbury, 1990).

## Zinc

In contrast to cysteine and lead, histidine is known to enhance zinc transport into brain and other tissues. Initially it was found possible to buffer the free zinc, $[Zn^{2+}]$, to concentrations approaching those in plasma by use of bovine albumin (constants are available for binding of zinc at the site of highest affinity). Zinc flux into brain followed Michaelis- Menten kinetics in relation to $[Zn^{2+}]$ with a $K_m$ of 16 nM.

If $[Zn^{2+}]$ was buffered with histidine, the flux was approximately linearly enhanced at constant $[Zn^{2+}]$ by either L- or D-histidine. Zinc flux into brain with histidine was unaffected by the presence of either 500 $\mu$M L-phenylalanine or 500 $\mu$M L-arginine. This, together with the lack of stereospecificity of the histidine effect, suggested that a Zn-histidine complex was not being carried via either the large neutral amino acid (L) transporter or by the basic amino acid ($y^+$) transporter. The approximately linear relation of zinc transport to histidine concentration suggested a role for the complex containing one histidine molecule, $ZnHis^+$, rather than for that containing 2 histidines, $Zn(His)_2$. For a range of free zinc and histidine concentrations, various models for a relation of Zn flux to $[Zn^{2+}]$, $[ZnHis^+]$ or $[Zn(His)_2]$ were tested for goodness of fit. The better models contained $[Zn^{2+}]$ or $[ZnHis^+]$ or both but not $[Zn(His)_2]$. The minimal value for the sum of the squared difference between the observed and predicted values of the flux was given by a saturable and linear relation of flux to the concentration of $ZnHis^+$ (Figure 4).

It appeared to us that the important influence of $ZnHis^+$ might be due to its ability to carry zinc during diffusion across unstirred water layers containing negatively-charged macromolecules, the layers not being accessible to Zn-albumin. Diffusional flux of the free ion is likely to be limited by its very low concentration (probably $10^{-9}$ to $10^{-10}$ M in extracellular fluids and $10^{-10}$ to $10^{-11}$ M inside cells). Such unstirred layers might occur at the luminal glycocalyx of the endothelium, inside endothelial or other cells and within the

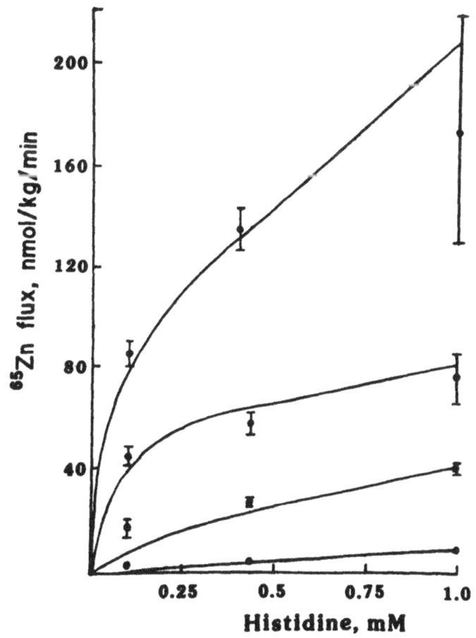


**Figure 4.** Effect of 3 concentrations of L-histidine on $^{65}$Zn flux into parietal cortex at 4 values of $[Zn^{2+}]$: 1, 10, 100 nM and 1 $\mu$M (from bottom up). Curves are of flux = 64.4 $[ZnHis^+]$/(390 + $[ZnHis^+]$ + 0.00378 $[ZnHis^+]$, concentrations being in nM. From Buxani-Rice *et al.* (1994).

interstitial fluid of brain. Transport across the membranes of the endothelial cells is considered to be due to the $Zn^{2+}$-transporter characterized in the absence of histidine (Buxani-Rice *et al.*, 1994).

## Iron

I will complete this brief survey of work in my laboratory on transport of trace metals across the blood-brain barrier by adding some fuel to the controversy concerning iron transport across this endothelium. It is clear that the bulk of Fe transport from blood into endothelium is via receptor-mediated endocytosis of Fe-transferrin. Indeed, we have

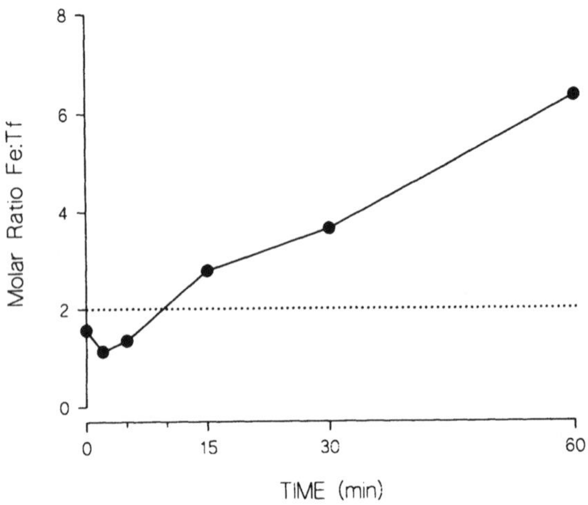


**Figure 5.** The molar ratio of $^{59}Fe$ to $^{125}I$ (i.e. transferrin) uptake into brain endothelial cells from $^{59}Fe$-$^{125}I$-transferrin. An increase in the ratio, to a level greater than 2, indicates that iron accumulates in concentrations above that present in the transferrin. From Taylor (1993).

been able to show that most of this transport may be blocked by appropriate antibodies in both the mouse and the rat *in vivo*. Studies with cultured endothelium from rat brain permit further analysis. Figure 5 shows uptake of $^{59}Fe$ and of $^{125}I$ from $^{59}Fe$-$^{125}I$- diferric transferrin into a layer of such endothelial cells. The $^{125}I$ representing the protein of transferrin remained constant after 15 min (not shown). The molar ratio of $^{59}Fe$ to $^{125}I$ increased over 60 min to beyond 6 (Taylor, 1993). This is very clear evidence that, as in all cells investigated, iron is being removed from the transferrin, since transferrin in the cells cannot bind more than two atoms of iron per molecule. The next question is how the iron released intracellularly crosses the abluminal membrane of the endothelium to reach the cerebral interstitial fluid. This is unknown but we do have evidence that under the conditions of perfusion, $Fe^{2+}$ ions can cross both membranes of the endothelium (Bradbury, 1994).

In conclusion, the study of transport of trace metals at the blood-brain barrier has revealed intricacies of behaviour at this interface, inaccessible to the techniques of its early investigation. Mechanisms available relate very much to the chemical species of the metals present in plasma. There is a place for transport of the free ions, of simple inorganic complexes, of organic complexes and of endothelial uptake of a metal with its specific binding protein.

## REFERENCES


Abbott, N.J., 1992, Comparative Physiology of the blood-brain barrier, in "Handbook of Experimental Pharmacology Volume 103 Physiology and Pharmacology of the Blood-Brain Barrier", M.W.B. Bradbury, ed., Springer-Verlag, Berlin, Heidelberg, etc., pp.371.

Abbott, N.J. and Bundgaard, M., 1992, Electron-dense tracer evidence for a blood-brain barrier in the cuttlefish *Sepia officinalis*, J. Neurocytol., 21:276.

Biedl, A. and Kraus, R., 1898, Uber einer bisher unbekannte toxische Wirkung der Gallensäuren auf das Zentralnervensystem, *Zentralblatt inn. Med.*, 19:1185.

Bradbury, M.W.B., 1994, Transport of $Fe^{2+}$ into brain during cerebrovascular perfusion in the anaesthetized rat, *J. Physiol. (Lond.)*, 479:37P.

Brightman, M.W. and Reese, T.S., 1969, Junctions between intimately opposed cell membranes in the vertebrate brain, *J. Cell. Biol.*, 40:648.

Bundgaard, M., 1982, The ultrastructure of cerebral blood capillaries in the ratfish *Chimaera monstrosa*, Cell Tissue Res., 226:145.

Buxani-Rice, S., Ueda, F. and Bradbury, M.W.B., 1994, Transport of zinc-65 at the blood-brain barrier during short cerebrovascular perfusion in the rat: its enhancement by histidine, J. Neurochem., 62:665.

Crone, C., 1965, The permeability of brain capillaries to non-electrolytes, *Acta Physiol. Scand.*, 64:407.

Davson, H. and Spaziani, E., 1959, The blood-brain barrier, *J. Physiol. (Lond.)*, 149:135.

Deane, R. and Bradbury, M.W.B., 1990, Transport of lead-203 at the blood-brain barrier during short cerebrovascular perfusion with saline in the rat, *J. Neurochem.*, 54:905.

Easton, A.S. and Fraser, P.A., 1994, Variable restriction of albumin diffusion across inflamed cerebral microvessels of the anaesthetized rat, *J. Physiol. (Lond.)*, 475:147.

Ehrlich, P., 1902, Uber die Beziehungen von chemischer Constitution, Vertheilung und pharmakologischer Wirkung, reprinted and translated in 'Collected Studies in Immunity', 1906, John Wiley, New York, pp.567.

Fraser, P.A. and Dallas, A.D., 1990, Measurement of filtration coefficient in single cerebral microvessels in the frog, *J. Physiol. (Lond.)*, 423:343.

Goldmann, E.E., 1909, Die äussere und innere Sekretion des gesunden und kranken Organismus im Lichte der 'vitalen Farbung', *Beitr. Klin. Chirurg.*, 64:192.

Goldmann, E.E., 1913, Vitalfarbung am Zentral-nervensystem, *Abh. Preuss. Akad. Wiss., Phys.-Math.* Kl. I:1.

Krogh, A., 1946, The active and passive exchanges of inorganic ions through the surfaces of living cells and through living membranes generally, *Proc. Roy. Soc. B.*, 133:140.

Lewandowsky, M., 1900, Zur Lehre der Cerebrospinal flüssigkeit, *Z. Klin. Med.*, 40:480.

May, P.M., Linder, P.W. and Williams, D.R., 1977, Computer simulation of metal-ion equilibria in biofluids: models for the low-molecular-weight complex distribution of calcium (II), magnesium (II), manganese (II), iron (III), copper (II), zinc (II) and lead (II) ions in human blood plasma, *J. Chem. Soc. Dalton:* 588.

Maynard, E.A., Schultz, R.L. and Pease, D.C., 1957, Electron microscopy of the vascular bed of the rat cerebral cortex, *Am. J. Anat.*, 100:409.

Oldendorf, W.H., 1970, Measurement of brain uptake of radio-labelled substances using a tritiated water standard, *Brain Res.*, 24:372.

Reese, T.S. and Karnovsky, M.J., 1967, Fine structural localization of a blood-brain barrier to exogenous peroxidase, *J. Cell Biol.*, 34:207.

Schultz, R.L., Maynard, E.A. and Pease, D.C., 1957, Electron microscopy of neurons and neuroglia of cerebral cortex and corpus callosum, *Am. J. Anat.*, 100:369.

Takasato, Y., Rapoport, S.I. and Smith, Q.R., 1982, An *in situ* brain perfusion technique to study cerebrovascular transport in the rat, *Am. J. Physiol.* 247:H484.

Taylor, E., 1993, Transfer of iron across cellular barriers, Ph.D. thesis, University of London.

Wycoff, R.W.G. and Young, J.Z., 1956, The motoneuron surface, *Proc. Roy. Soc. B.*, 144:440.

# DEVELOPMENT OF THE BLOOD-BRAIN BARRIER

Britta Engelhardt and Werner Risau

Max-Planck-Institut für physiologische und klinische Forschung,
W. G. Kerckhoff-Institut, Abteilung Molekulare Zellbiologie,
61231 Bad Nauheim, Germany

## INTRODUCTION

The experiments leading to the concept of a vascular blood-brain barrier date back more than 100 years, when Paul Ehrlich and his collaborators discovered in a series of experiments that certain dyes, when injected into the vascular system, were rapidly taken up by all organs with the exception of the brain (Ehrlich,1904). Ehrlich himself interpreted these findings as a lack of affinity of nervous tissue for these dyes and found it hard to believe that the cerebral vascular endothelium might be selective. However, shortly afterwards, Goldmann (1913), an associate of Ehrlich, could show that the very same dyes, when injected into the cerebrospinal fluid, readily stained neural tissue but were incapable of staining all the other organs. Thus it became clear that these dyes were indeed prevented from getting access to the blood circulation. The concept of a vascular blood-brain barrier, which also functions as a brain-blood barrier was born (Bradbury, 1984; Bradbury, 1979). It remained controversial until 70 years later Reese and Karnovsky (1967) and Brightman and Reese (1969) were able to ultrastructurally locate the barrier. They showed that in mammals horseradish peroxidase, a small molecule of 40 kDa was prevented from passing between endothelial cells by tight junctions connecting most brain endothelial cells.

There are some specialized tissues located at strategic positions in the midline of the ventricular system that lack a blood-brain barrier; these are collectively referred to as circumventricular organs. Included under this designation are the pineal body, the median eminence, the subfornical organ, the organum vasculosum of the lamina terminalis, the neurohypophysis and the area postrema. Within all these regions capillaries are fenestrated providing specific sites for the transfer of proteins and solutes. Exposure of neurons to the blood milieu is of functional importance in these areas as they mostly serve as neuroendocrine organs (e.g. neurohypophysis) or rather as brain monitors where neurons with receptor properties sense a variety of substances in the blood (e.g. area postrema). Another exception where endothelial cells do not form a barrier is the choroid plexus. The choroid plexus is a villous structure extending from the ventricular surface into the lumen of the ventricles; its major known function is the secretion of cerebrospinal fluid. The choroid plexus consists of an extensive capillary network enclosed by a single layer of cuboidal epithelium. The capillaries are fenestrated as in the circumventricular organs, while tight junctions surround the apical regions of choroid plexus epithelium. Similarly, a complex network of tight junctions connecting specialized ependymal cells seal off the central nervous system (CNS) parenchyma from the circumventricular organs (Leonhardt, 1980; Bouchoud and Bosler, 1986).

The structural barrier formed by complex tight junctions between the brain endothelial cells along with the low abundance of pinocytotic vesicles represents only one distinctive feature of brain endothelial cells related to their function as a barrier, which protects the CNS microenvironment from changes in the blood milieu. In order to preserve homeostasis in the nervous system, necessary metabolites should have ready access to the CNS whereas unnecessary metabolites or toxic substances need to be removed. Indeed, there are specific carrier molecules in the plasma membrane of brain endothelial cells that ensure directed transport of metabolites into and out of the CNS. Finally, blood- or brain-borne substances can be metabolized or modified by specific enzymes specifically localized in blood-brain barrier capillaries.

The cellular and molecular mechanisms that control the development and maintenace of the highly specialized endothelium forming the blood-brain barrier are not well understood. This review will give an overview on our understanding of how the unique vascular system of the central nervous system develops, specializes to and maintains the blood-brain barrier.

## DEVELOPMENT

The formation of the vascular system of the body occurs very early during embryonic development. The first morphologically distinguishable structures are so called blood islands, which differentiate from the splanchnopleuric mesoderm (Risau, 1991b). Blood islands are cyst-like structures where blood cell precursors appear as a morphologically distinct central cell cluster surrounded by a peripheral cell layer. The predetermined endothelial precursors, located at the periphery of blood islands, are called angioblasts. Angioblasts proliferate, migrate and finally join to form endothelial strands which connect blood islands, and thus give rise to a primary capillary plexus. The in situ differentiation of a primitive vascular plexus derived from migrating angioblasts, is called vasculogenesis (Risau and Lemmon, 1988; Risau et al., 1988; Risau, 1991b). The primary capillary plexus differentiates further later during embryogenesis through a complex series of regressions, sproutings, fusions and enlargements to form an arteriovenous vascular network (Evans, 1909; Romanoff, 1960).

The development of the brain vascular system begins when angioblasts invade the head region to form the perineural vascular plexus which in a 2-day chicken embryo or in a 10 day rodent embryo covers the entire surface of the neural tube (Bär, 1980; Risau and Wolburg, 1990; Dermietzel and Krause, 1991). None of the precursors that form the perineural vascular plexus nor the intraneural vessels are derived from the neuroectoderm (Stewart and Wiley, 1981b; Noden, 1991). Thus the vascular system within the CNS does not develop by vasculogenesis. Rather, vascular sprouts, originating from the perineural plexus penetrate the proliferating neuroectoderm at 4 days of embryonic development in the chicken and 11 days in rodents (Bär, 1980). This mechanism, where new vessels form by sprouting from pre-existing vessels, is called angiogenesis (Risau and Wolburg, 1990; Noden, 1991; Feinberg and Noden, 1991). Additionally, transplantation studies, in which embryonic brain tissue of the chicken was transplanted into ectopic sites of quail embryos, confirmed that the transplanted brain tissue of the chicken could induce ingrowth of vascular sprouts (angiogenesis) in the quail (Stewart and Wiley, 1981b; Risau et al., 1986b). Moreover, the grafted neural tissue induced blood-brain barrier characteristics like upregulation of specific marker molecules in the host-derived endothelial cells (Stewart and Wiley, 1981b; Risau et al., 1986b). During brain angiogenesis newly forming vessels radially invade into the neuroectodermal tissue. Vessels elongate, give rise to manifold branches and finally anastomose with adjacent sprouts to form a plexus of undifferentiated capillaries in the ventricular zone of the developing brain (Bär, 1980; Roncali et al., 1986). It is striking that the onset of angiogenesis in the brain occurs at precisely reproducible stages of embryonic development (i.e. on day 4 during chicken embryonic development and day 11 in rat and mouse). The rate of angiogenesis is maximal in the early postnatal period and down-regulated in the adult brain (Robertson et al., 1985). Besides the timing of brain angiogenesis, the process of sprouting into the neuroectoderm occurs at specific locations thus ensuring a reproducible pattern of cortical vascularization in mammals (Bär, 1980; Kuban and Gilles, 1985). The molecular and/or cellular mechanisms establishing this pattern still remain unknown. One possibility is that an endothelial specific chemotactic factor, produced locally within the neuroectoderm, might guide the sprouting capillaries to their branching points in the brain. The diffusion distance or local concentration of such a factor could determine the spacing of sprouts.

The precise regulation of the timing and pattern of vascular development in the brain prompted us to investigate whether the embryonic brain produces factors, which are chemotactic for endothelial cells and/or stimulate the growth of blood vessels. We hypothesized a paracrine mechanism, in which angiogenic factors produced by the brain could bind to specific receptors expressed on endothelial cells thus leading to migration and proliferation of these cells.

Growth factor and angiogenic activities for endothelial cells have been isolated from embryonic day 3.5 chicken brain. This is the time of the onset of angiogenesis (Risau, 1986). Acidic and basic FGF were identified in these preparations and have been shown to be potent endothelial cell mitogens *in vitro* and potent angiogenic factors *in vivo*. Therefore they seemed to be good candidates for mediators of angiogenesis in the brain (Vlodavsky et al., 1987; Ingber et al., 1987; Folkman and Klagsbrun, 1987; Risau et al., 1988a). Subsequent studies, however, showed that the pattern of aFGF and bFGF gene expression did not correlate with endothelial cell proliferation and angiogenesis in the brain (Schnürch and Risau, 1991) as these factors are present in neurons in the adult brain, in which angiogenesis does not occur and in which endothelial cells have a low turnover rate unter physiological conditions (Emoto et al., 1989; Schnürch and Risau, 1991). No other cell type could be detected to express aFGF in the brain *in vivo*. An additional problem concerning the physiological role of aFGF and bFGF is their lack of a hydrophobic signal sequence required for secretion via the classical secretory pathway. In addition, although four distinct receptors for the different members of the FGF family have been characterized, there is still no evidence that FGF receptors are expressed by brain endothelial cells *in vivo* (Heuer et al., 1990; Wanaka et al., 1991). In fact, *in situ* hybridization analyses have indicated that there is no expression of FGF receptors 1 and 2 in capillary endothelial cells during embryonic development (Heuer et al., 1990; Umemori et al., 1992; Peters et al., 1992; Tannahill et al., 1992).

In 1989, a new angiogenic factor was reported in the vascular biology literature named vascular endothelial growth factor (VEGF). This molecule was derived from a variety of sources including follicular pituitary cells, aorta and some tumors, and was capable of exerting mitogenic effects on cultured endothelial cells (Ferrara and Henzel, 1989; Leung et al., 1989; Globus et al., 1989; Gospodarowicz, 1989; Mergia et al., 1989; Plouet et al., 1989; Gospodarowicz and Lau, 1989; Tischer et al., 1989; Conn et al., 1990a; Conn et al., 1990b). In contrast to other endothelial growth factors, which are mitogens for a large number of different cell types, its mitogenic activity seems to be less powerful however exclusively specific for endothelial cells (Ferrara and Henzel, 1989; Leung et al., 1989b). It was soon recognized that VEGF was identical to the independently isolated vascular permeability factor (VPF), which induces vascular leakage *in vivo* (Senger et al., 1983; Keck et al., 1989). VEGF/VPF is a 46 kD glycosylated dimeric protein with structural homology to platelet-derived growth factor (PDGF) (Keck et al., 1989). By alternative splicing of messenger RNA (mRNA) VEGF may exist as any of four homodimers composed of monomers of 121, 165, 189 and 206 amino acids in man (Mergia et al., 1989; Tischer et al., 1989). In the mouse, three different VEGF isoforms, namely $VEGF_{120}$, $VEGF_{164}$ and $VEGF_{188}$, which also arise by alternative splicing of mRNA, have been described so far (Figure 1 and Breier et al., 1992; Plate et al., 1992). Homodimers of murine $VEGF_{120}$ and $VEGF_{164}$ are readily secreted into conditioned medium by COS-cells transfected with murine VEGF cDNA *in vitro*. Both isoforms have been shown to stimulate the proliferation of bovine aortic endothelial cells to similar levels (Breier et al., 1992; Plate 1992). In contrast, murine $VEGF_{188}$ polypeptide is retained by the transfected cells. The mitogenic activities of the high molecular weight VEGF isoforms remain to be determined.

Strong evidence has accumulated suggesting that VEGF is a key regulator of brain angiogenesis. We have shown that the spatial and temporal expression pattern of VEGF mRNA correlates well with angiogenesis during embryonic development in the mouse brain (Breier et al., 1992; Plate et al., 1992). By *in situ* hybridization we have shown that in day 17 mouse embryos VEGF mRNA is expressed in the ventricular layer of the developing neuroectoderm (Figure 2). In adult brain, where vascularization is complete and the rate of endothelial cell proliferation is very low (Robertson et al., 1985), VEGF mRNA could not be detected any longer in the ependymal cells, which form the lining of the ventricles. Therefore, it is conceivable that during brain development VEGF is released by cells of the ventricular layer, thus promoting angiogenesis by initiating migratory and mitogenic responses by endothelial cells from the perineural vascular plexus. A concentration gradient of the diffusible VEGF-isoforms declining from the ventricular layer towards the perineural vascular plexus may cause the radial ingrowth of capillaries from the vascular plexus towards the angiogenic stimulus provided by VEGF secreting ventricular epithelial cells (Breier et al., 1992; Plate et al., 1992). The hypothesis that VEGF may function as an angiogenic growth factor for fetal blood vessels, is further substantiated by the presence of high levels of VEGF mRNA in the early embryonic kidney, in the highly vascularized placenta, and by the correlation between VEGF expression and vascularization in rat corpus luteum (Phillips et al., 1990).

Interestingly, in contrast to the transient expression in ependymal cells, VEGF expression persisted in the choroid plexus epithelium of the brain on adult mice, when vascularization of these structures is complete (Breier et al., 1992). VEGF might therefore have an additional function which is different from a role as mediator of neovascularization. In the choroid plexus it might be stabilizing the fenestrated capillary phenotype.

The key role of VEGF as a regulator of embryonic angiogenesis in the brain is consistent with the expression-pattern of its two known receptors. Both VEGF-receptors belong to the large family of transmembrane receptor tyrosine kinases. Receptor tyrosine kinase activity is stimulated upon ligand binding, which results in phosphorylation of the receptor itself and multiple cellular substrates and, subsequently, in a variety of cellular responses (Ullrich and Schlessinger, 1990). Various receptor tyrosine kinases have been shown to be key regulators of developmental processes, which is reflected in a high level of protein tyrosine phosphorylation in the early embryo, that decreases in the later stages of embryogenesis and is relatively low in adult animal tissues (Maher and Pasquale, 1988). Recently, the fms-like tyrosine kinase flt-1 (Shibuya et al., 1990) was shown to have binding affinity for VEGF (Figure 1 and De Vries et al., 1992). We isolated a second high affinity VEGF-receptor from proliferating endothelial cells of postnatal day 4 - 10 mouse brain, when there is maximal endothelial cell proliferation (Millauer et al., 1993). This second VEGF-

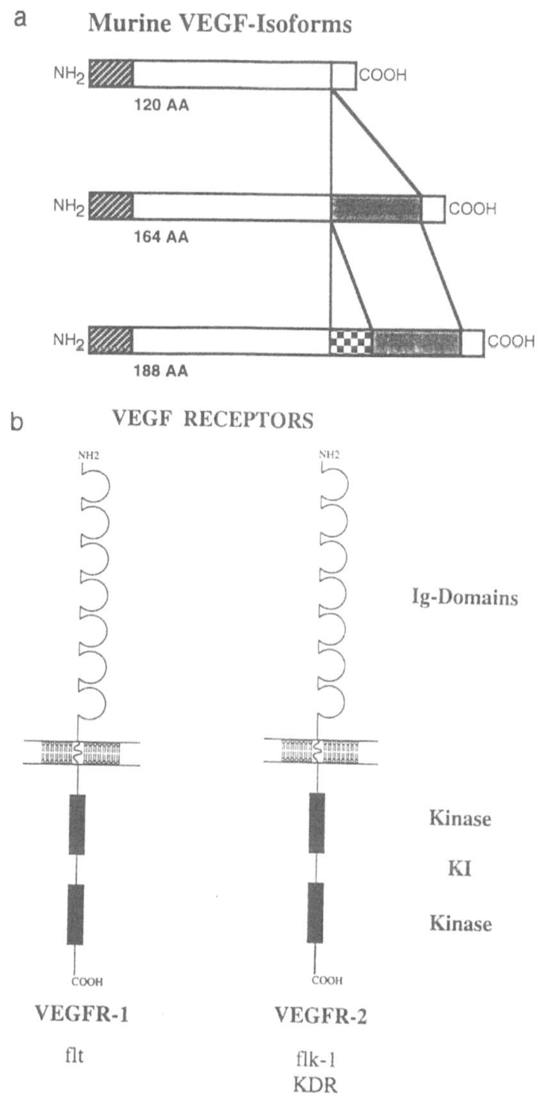


**Figure 1: VEGF isoforms and VEGF receptors.** (a) Structure of murine VEGF proteins as deduced from cDNA sequences. The cross-hatched box at the amino terminus represents the putative 26-amino acid signal peptide for secretion. Additional domains in $VEGF_{164}$ and $VEGF_{188}$ are represented as shaded boxes. The number of amino acids (AA) of the mature proteins are indicated. (b) Structure of the two transmembrane tyrosine kinase receptors for VEGF: flt-1 and flk-1.

receptor also belongs to the family of receptor tyrosine kinases and had been previously designated fetal liver kinase 1 (flk-1; see Figure 1; Matthews et al., 1991). In situ hybridization in the embryonic day 14.5 mouse embryo revealed that flk-1 is restricted to capillaries and blood vessels (Figure 2). At day 11.5 in the mouse embryo, when the first vascular sprouts begin to radially invade the neuroectoderm from the perineural plexus, expression of flk-1 was high in the perineural vascular plexus and in structures resembling invading vascular sprouts (Millauer et al., 1993). These in situ hybridization analyses indicated that the proliferating endothelial cells of an angiogenic sprout express the flk-1 mRNA. At embryonic day 14.5, when the neuroectoderm is already highly vascularized, numerous radial vessels as well as branching vessels of the intraneural plexus contained large amounts of flk-1 mRNA (Millauer et al., 1993). Finally, in the adult brain, when angiogenesis has ceased, flk-1 expression was very low and appeared to be mainly restricted to the choroid plexus endothelial cells.

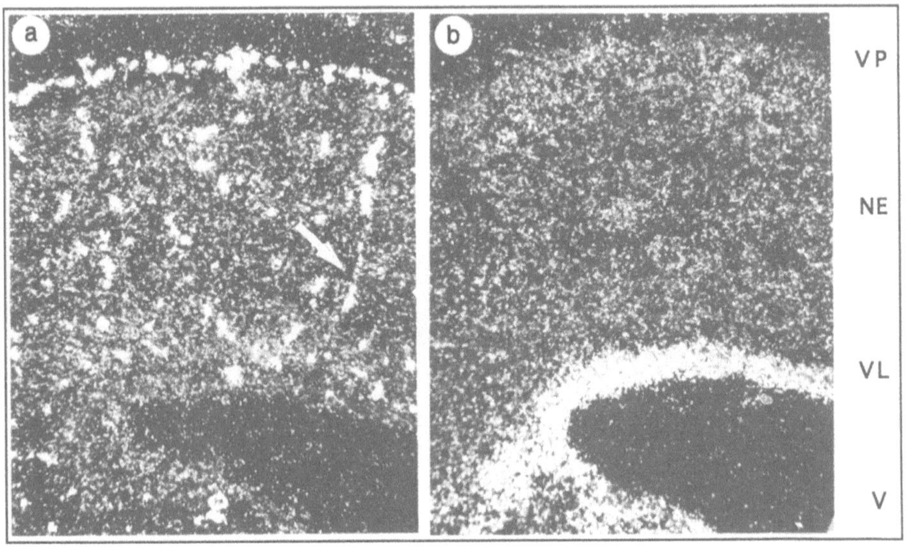


**Figure 2: Complementary expression of VEGF and flk-1 in developing mouse brain.** In situ hybridization of a 17.5 day mouse embryo; dark field image. (a) flk-1 transcripts were detected in the perineural vascular plexus (VP) and in the capillaries of the neuroectoderm (arrow). Capillaries sprouting from the vascular plexus grow towards the ventricular layer (VL) where VEGF-transcripts can be detected (b). V = ventricle, NE = neuroectoderm.

Taken together these findings support the hypothesis that flk-1 expressing capillary sprouts, originating from the perineural vascular plexus, migrate towards an angiogenic stimulus provided by VEGF, which is produced by cells in the ventricular layer (Figure 2). Consistent with this, receptors for VEGF are highly expressed on endothelial cells throughout brain angiogenesis and are downregulated in the adult brain, when angiogenesis has ceased (Millauer et al., 1993; Plate et al. 1992). A pivotal role of the VEGF-flk-1 ligand-receptor system is suggested by these findings, however further studies with animal models will be necessary to establish the overall role of flk-1 in brain angiogenesis.

It is unlikely that the multiple aspects of endothelial cell physiology are controlled by a single receptor. It is therefore conceivable that different combinations of receptors, whose composition changes in a time- and tissue dependent manner, specifies endothelial cell behaviour during blood vessel development and maintenance. Recently, two additional endothelial cell-specific receptor tyrosine kinases, tie-1 (tie) and tie-2 (tek) have been described that are expressed in developing vasculature of the brain (Sato et al., 1993; Schnürch and Risau, 1993; Partanen et al., 1992; Korhonen et al., 1992). High mRNA levels are observed during vascular development in the brain. Consistent with a possible role in brain angiogenesis, tie-1 and tie-2 mRNA is expressed specifically by endothelial cells and down-regulated in the adult organism (Schürch and Risau, 1993; Maisonpierre et al., 1993). Their expression patterns resemble the expression of the high affinity VEGF receptor flk-1 mRNA. However, both tie-1 and tie-2 differ from flk-1 especially in their extracellular domains. Tie-1 and tie-2 are unusual receptor tyrosine kinases, which combine a peculiar mosaic of immunoglobulin-, EGF-like- and fibronectin type III - domains in their extracellular sequence. The functional significance of this sequence motif is not obvious. However, taken together they have significant similarities to extracellular matrix proteins and adhesion molecules. Whether these homologies indicate a role for these receptor tyrosine kinases in cell-cell or cell-matrix interactions, or whether the ligands for tie-1 and tie-2 are diffusible molecules remains to be determined. Identifications of ligands may help to define which aspects of endothelial cell growth, differentiation and function are regulated by tie-1 and tie-2 as compared to flk-1 and flt-1.

Finally, in order to identify signal transduction proteins specifically expressed in developing blood-brain barrier endothelium, we searched for proteins containing src-homology 3 (SH3) domains. SH3-domains are structural elements typically found in signal transduction proteins. Unexpectedly, we found that all murine and bovine brain endothelial cells encoded the SH3-domain of one protein namely the non-receptor tyrosine kinase Lyn (Achen et al., 1995). Lyn has been shown to be involved in signal transduction in B lymphocytes (Yamanashi et al., 1993) and in cell-cell adherence junctions in hepatocytes (Tsukita et al., 1993). The mRNA for lyn is localized in embryonic and early post-natal mouse brain endothelium but not in endothelium outside the brain (Achen et al., 1995). The relevance for Lyn expression during brain angiogenesis remains to be determined, however its specific expression suggests that Lyn could be involved in transduction of endothelial growth or differentiation signals critical for blood-brain barrier development.

## DIFFERENTIATION

The differentiation of the blood-brain barrier, namely the development of the selective permeability of brain capillaries is usually investigated by measuring the decreasing permeability for vascular tracers like dyes, horseradish peroxidase or lanthanum. Blood-brain barrier differentiation appears to be a gradual process which is independent of vascular proliferation in the brain, because e.g. rat brain endothelial cells *in vivo* commence barrier formation at about embryonic day 16, while they still proliferate (Robertson et al., 1985). Blood-brain barrier differentiation occurs at different time points in different locations in the brain. This means that not all brain capillaries become impermeable at the same time. Rather, the actual time is dependent on the anatomical location. In the mouse the barrier to proteins forms first in the spinal cord and last in the telencephalon; within the telencephalon there is an ependymal-cortical gradient of barrier differentiation, with injected protein still visible in the subependymal layer at embryonic day 16 (Risau et al., 1986a). There is also a remarkable difference if one compares different tracer molecules. While macromolecules like horseradish peroxidase are excluded from the brain parenchyma around embryonic day 14 the barrier to ions as measured by electrical resistance in pial vessels only matures postnatally (Butt et al., 1990). Therefore, despite some controversial results obtained by using different techniques (Wakai and Hirokawa, 1978; Risau et al., 1986a; Saunders and Mollgard, 1991; Risau and Wolburg, 1991) it is clear that all the capillaries that are present in the brain before embryonic day 13 in the chick and embryonic day 15 in the rat and mouse lack a mature barrier.

The transition from a leaky capillary, e.g. perineural capillaries which are fenestrated early in development (Yoshida et al., 1988) to capillaries with features of a blood-brain barrier is characterized by several changes in endothelial cell morpholopgy, biochemistry and function that make these endothelial cells distinct from every other endothelial cell in the body.

Tight junctions between the endothelial cells of the brain microvessels have been shown by qualitative freeze-fracture and ultrathin section electron microscopic studies to be more complex than those in other endothelial cells in the body (Nagy et al., 1984; Nico et al., 1992). Tight junctions are thought to function as a seal only if they are continuous and branched (Claude and Goodenough, 1973; Cereijido et al., 1989). The complex network of tight junctions between brain endothelial cells is therefore primarily responsible for the paracellular impermeability and unlike simple tight juctions provides a high electrical resistance (about 2000 Ohm x $cm^2$; Crone and Olesen, 1982). During brain angiogenesis it has been shown in the rat that the decrease in vessel permeability to protein correlates with a conformational change of the tight junctions between endothelial cells lining the brain vessels leading to an increase in the length of "fused" junctional contacts during embryonic development (Stewart and Hayakawa, 1987; Schulze and Firth, 1992; Schulze and Firth, 1993). We have performed a quantitative analysis of the structure and function of tight junctions in primary cultures of bovine brain endothelial cells using quantitative freeze-fracture electron microscopy and ion and inulin permeability (Wolburg et al., 1994). The complexity of tight junctions, defined as the number of branch points per unit length of tight junctional strands, decreased 5 hours after culture but thereafter remained almost constant. In contrast, the association of tight junction particles with the cytoplasmic leaflet of the endothelial membrane bilayer (P-face) decreased continuously with a major drop between 16 and 24 hours of culture (Figure 3). The P-face association of tight junctions could be restored to a certain extent by co-culture of endothelial cells with astrocytes or astrocyte conditioned medium. Co-culture with fibroblasts had no effect on P-face association. These data suggest that the association of tight junction particles with the P-face rather than the branching frequencies of tight junctions correlate with blood-brain barrier function (Wolburg et al., 1994). Since the molecular nature of the tight junction is still enigmatic (see discussion in Citi, 1993; Fleming et al., 1993; Javed et al., 1993) - it is difficult to speculate about the molecular mechanisms that might be involved in the loss of P-face association of tight junctions. However, there is evidence that tight junctions are associated with the cytoskeleton, which seems to be important for the regulation of paracellular permeability (Madara, 1987; Madara and Pappenheimer, 1987; Drenckhahn and Dermietzel, 1988). In addition, adherens junctions which are strongly associated with the cytoskeleton, e.g. via cadherin/catenin interactions seem to be intermingled with tight junctions in blood-brain barrier endothelial cells (Schulze and Firth, 1993). Lyn expression may be important in this context considering its presence in cell-cell adherence junctions in hepatocytes. The notion that P-face association may be crucial for tight junction barrier function is also consistent with the observation that tight junctions in peripheral, non-barrier endothelial cells are E-face associated *in vivo* (Simionescu et al., 1988; Mühleisen et al., 1989). It is tempting to speculate that tight junction proteins such as ZO-1, cingulin, occludin and other yet unknown molecules may play a role in this process (Citi et al., 1988., Risau, 1991a and 1991b; Furuse et al., 1993). To date, neither expression of ZO-1, cingulin nor occludin has been proven to be sufficient to establish "tightness" of junctions as measured by electrical resistance over

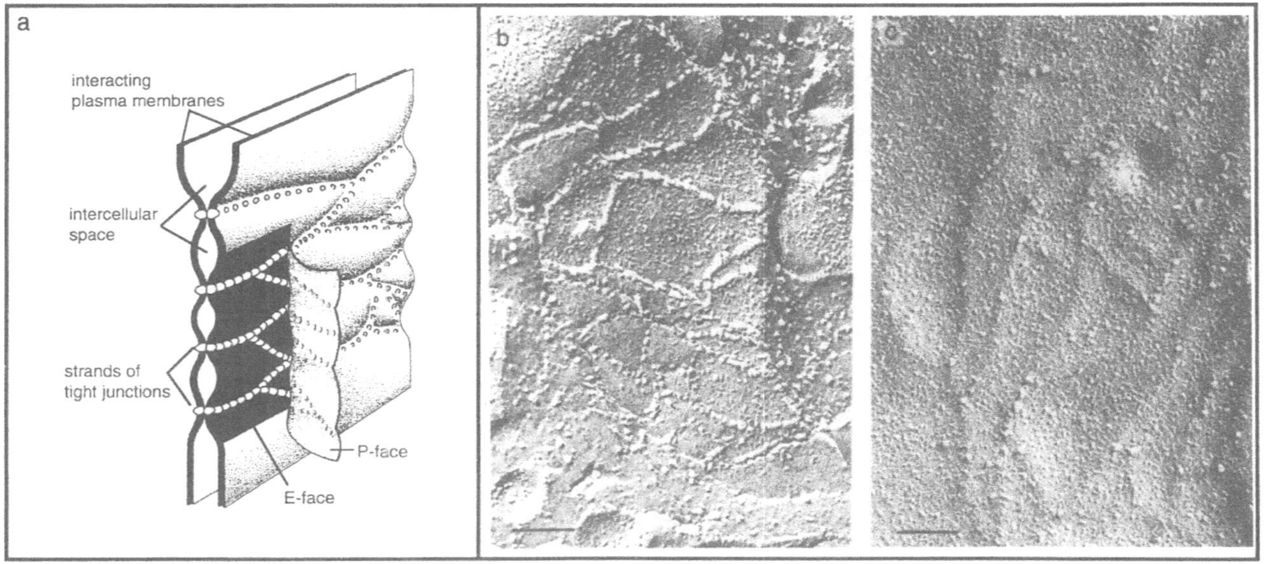


**Figure 3: P-face association of tight junctions.** (a) shows a schematic drawing of tight junctions between two adjacent cells. In freeze-fracture electron microscopy of freshly isolated cerebral capillary fragments the tight-junction particles are P-face associated (b). P-face association of tight junctions is lost in cultured brain endothelial cells (c). Bars, 0.1 mm.

epithelial cell monolayers *in vitro* (Stevenson et al., 1988; Anderson et al., 1988; Krause et al., 1991; Dermietzel and Krause, 1991). Nevertheless, our data (Wolburg et al., 1994) suggest an important functional role for the P-face association and thus the cytoplasmic anchoring of the tight junction particles for brain endothelial barrier function in particular and probably paracellular permeability in general.

Since there are virtually no pinocytotic vesicles found in brain capillary endothelium, unspecific transport of molecules across the endothelium is prohibited (Coomber et al., 1987; Stewart and Wiley, 1981). In order to facilitate transport of nutrients from the blood into the brain specific transport systems are present in brain endothelial cell membranes, which carry nutriens, ions and hormones into the brain (Betz and Goldstein, 1986). In addition, there are transport molecules that pump substances out of the brain to maintain a constant neuronal environment (Cordon-Cardo et al., 1989; Schinkel et al., 1994). Furthermore, brain capillaries provide a "metabolic" barrier by expression of specific enzymes that catabolize molecules taken up from the blood-stream (e.g. dopamine by monoaminoxidase in endothelial cells). Table I gives an overview over these molecules, namely transporters, enzymes and antigens that have so far - in comparison to other vessels - been specifically localized to brain capillaries (see also other articles in this volume). Because the correlation of their expression in brain capillaries with the development of the permeability barrier and/or their expression in the differentiated blood-brain barrier (Risau et al., 1986a; Vorbrodt et al., 1986a; Vorbrodt et al., 1986b; Vorbrodt et al., 1986c; Vorbrodt, 1986; Lossinsky et al., 1986), they have been proposed to be specific blood-brain barrier markers and as such important for barrier function. However, in most cases the specificity of expression and hence their exact role for the blood-brain barrier has to be further elucidated.

**Table I. Molecules associated with the blood-brain barrier**

| Molecule | Name | Reference |
|---|---|---|
| **TRANSPORTERS & RECEPTORS** | Glucose transporter | Dwyer and Pardridge, 1993 |
| | Transferrin receptor | Jefferies et al., 1984 |
| | LDL receptor | Méresse et al., 1989 |
| | acLDL-Receptor | Gaffney et al., 1985 |
| | $Na^+$-$K^+$-ATPase | Schielke et al., 1990 |
| | Multi-Drug Resistance | Cordon-Cardo et al., 1989 |
| | Retinol-binding Protein | MacDonald et al., 1990 |
| | Insulin receptor | Duffy and Pardridge, 1987 |
| | Aminoacid Transporters | Betz and Goldstein, 1978 |
| | Ion channels | Betz et al., 1986 |
| **ENZYMES** | γ-glutamyl-transpeptidase | Risau et al., 1992 |
| | Aminopeptidase N | Kunz et al., 1994 |
| | Alkaline phosphatase | Mori and Nagano, 1985 |
| | Butyryl cholinesterase | Trancard et al., 1989 |
| | Monoamine oxidase | Kalaria et al., 1987 |
| **ANTIGENS** | EBA | Rosenstein et al., 1992 |
| | HT7/OX-47/Basigin | Seulberger et al., 1992 |
| | PC-1 | Harahap and Goding, 1988 |
| | ApoA1 | Weilerguttler et al., 1990 |
| | Thrombomodulin | Ishii et al., 1986 |
| | Meca 32 | Engelhardt et al., 1994; Hallmann et al., (submitted) |

For instance, γ-glutamyltranspeptidase has been widely used as a marker for brain capillaries and differentiated properties of brain endothelial cells (Orlowski et al., 1974; Meyer et al., 1990; Puumala et al., 1990). The differentiated blood-brain barrier, however, represents a complex cellular system consisting of endothelial cells, pericytes, perivascular microglial cells and astrocytes. When studying blood-brain barrier endothelial properties *in vitro* these cellular contaminants must be considered. In order to gain further insight into whether it is the endothelial cell expressing the specialized property associated with the blood-brain barrier *in vivo* highly purified endothelial cultures are required for *in vitro* studies (Risau et al., 1990; Risau and Wolburg, 1990a; Risau and Wolburg, 1990b; Seulberger et al., 1990). Using rat brain and bovine brain endothelial cell cultures, contaminated mainly by pericytes, we localized the major amount of γ-glutamyltranspeptidase activity to pericytes and not to endothelial cells *in vitro*. It is interesting that an induction of γ-glutamyltranspeptidase activity could be observed in endothelial cells, when cocultured with astrocytes in a transfilter system (Risau et al., 1992). Consequently, whatever the function of γ-glutamyltranspeptidase in brain capillaries might be, any consideration of its significance for the blood-brain barrier has to take into account the presence of this enzyme in pericytes.

We have characterized a monoclonal antibody called HT7 that recognizes a highly glycosylated 48 kDa protein, which belongs to the immunoglobulin-superfamily (Seulberger et al., 1990; Seulberger and Risau, 1990; Albrecht et al., 1990). The HT7 antigen is specifically expressed on chicken blood-brain barrier endothelium but not on other endothelial cells (Risau 1986b; Albrecht et al., 1990). Interestingly, it is absent from the fenestrated choroid plexus endothelium but present on choroid plexus epithelium, the site of the blood-cerebrospinal fluid barrier (Figure 4). Several homologues in different species have been described (Seulberger et al., 1992). To date, HT7 and its homologues serve as markers for differentated blood-brain barrier endothelium but the function of the protein remains unknown.

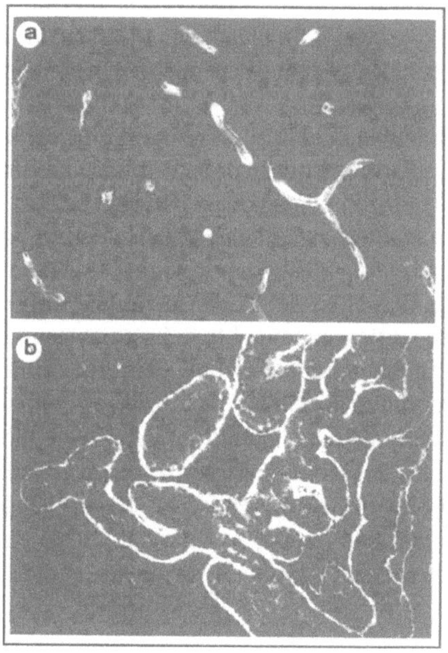


**Figure 4: Expression of the HT7 antigen in chicken brain.** Immunofluorescence staining of chicken brain with HT7 shows specific expression on brain endothelium (a) and on choroid plexus epithelium (b). Fenestrated capillaries of the choroid plexus show no staining with HT7 (b).

Another endothelial membrane protein in the mouse is recognized by the mAb Meca-32 (Engelhardt et al., 1994; Hallmann et al., 1995). In contrast to the HT7 protein, which commences expression exclusively on brain endothelium at a time in embryonic development when the blood-brain barrier begins to develop, expression of the Meca-32 antigen is suppressed during differentiation of the blood-brain barrier (Hallmann et al., 1995). The importance of downregulation of the Meca-32 antigen during the differentiation of the blood-brain barrier is as yet unknown.

Two recent accomplishments in the field of blood-brain barrier transport deserve to be highlighted. First, disruption of the mouse mdr1a P-glycoprotein gene was observed to lead to a deficiency in the blood-brain barrier and to increased sensitivity to certain drugs thus providing the first direct evidence for a role of this transporter in blood-brain barrier function (Schinkel et al., 1994). Second, the low-density lipoprotein receptor (LDL-R) was found to be upregulated in an *in vitro* model of the blood-brain barrier using a coculture of brain endothelial cells and astrocytes (Dehouck et al., 1994). This result indicated a complex interaction between different cell types at the blood-brain barrier which is dependent not only on specific cell-cell communications via secreted factors and cell adhesion molecules but also influences by metabolic changes in the brain (neuronal activity) as well as in the periphery (e.g. diet, hormones, cytokines).

The molecular mechanisms of the induction or down-regulation of the unique characteristics described above are unknown although the brain environment appears to induce the expression of blood-brain barrier characteristics in developing vessels.

Using chicken-quail transplantation experiments, Stewart and Wiley (1981) were first to demonstrate that blood-brain barrier characteristics can be induced in vessels which had invaded avascular brain transplants from 3 day old quail embryos placed in the coelomic cavity of chicken embryos. Using the chicken-specific monoclonal antibody HT7 we observed the upregulation of the blood-brain barrier-specific antigen in chicken endothelial cells which had invaded embryonic mouse brain in a similar setting (Risau et al., 1986b). These findings in combination with the observation that all the endothelial cells that vascularize the brain tissue originate from precursors from outside the neuroectoderm support the hypothesis that central nervous system tissue microenvironment determines the endothelial phenotype of the blood-brain barrier.

The question remains which cell or combination of cells within the neuroectoderm induces blood-brain barrier characteristics in endothelium. The first direct evidence has been provided by Janzer and Raff (1987) who showed that purified astrocytes induce blood-brain barrier-like permeability changes in invading endothelial cell *in vivo*. Experimental evidence from various studies (Arthur et al., 1987; Dehouck 1990b; Meyer et al., 1990; Rubin et al., 1991a and 1991b; Reuterdahl et al., 1991; Tao-Cheng and Brightman, 1988; Tontsch and Bauer, 1991) suggest that the differentation of type I astrocytes in the central nervous system may be critically involved in the induction and maintenance of barrier features. On the other hand, no blood-brain barrier inducing soluble factors have been identified from astrocytes to date. Holash et al. (1993) reported their inability to reproduce the Janzer and Raff data. In addition, the observation that some blood-brain barrier characteristics in brain endothelial cells appear very early and before astrocyte differentiation (Risau et al., 1986b; Bauer at al., 1993; Schinkel et al., 1994; Qin and Sato, 1994, Achen et al., 1995) suggest that other cells and mechanisms are involved possibly in addition to astrocytes.

## MAINTENANCE OF THE BLOOD-BRAIN BARRIER

Pathological conditions within the central nervous system like ischemia, inflammation or tumor growth lead to blood-brain barrier dysfunction, emphasizing that the permeability barrier is not simply "switched on" during embryonic development, but that a continuing regulation of its maintenance, probably provided by the tissue microenvironment, is necessary for a proper barrier function in any circumstance. For example, in many brain tumors morphological irregularities of the perivascular ensheathment such as enlarged perivascular space, gaps in the basal lamina or deficient glial investment correlate with a shift of endothelial balance leading to breakdown of the blood-brain barrier. This supports the concept that endothelial differentiation to form a blood-brain barrier can be modulated by the tissue microenvironment (reviewed by Plate et al., 1994).

During inflammation the vasculature throughout the whole central nervous system develops an inflamed phenotype even it there is only a local inflammation (Sasseville, et al., 1992; Engelhardt et al., 1994; Steffen et al., 1994). The "leakyness" of the blood-brain barrier does not seem to simply correlate with the activation of brain endothelium, but rather with the localization of infiltrating leukocytes around a certain vessel. The local re-expression of the Meca-32 antigen on vessels surrounded by inflammatory cells suggests that an "embryonic" undifferentiated brain endothelial phenotype is regained where inflammatory cells disturb the cellular ensheathment of blood-brain barrier endothelium (Engelhardt et al., 1994).

Thus, development and maintenance of the blood-brain barrier are tightly regulated by the permanent interaction of endothelial cells with the neuroectoderm. In our "New Concept of a Blood-Brain Barrier" we would like to propose that during brain angiogenesis there are at least two phases of endothelial-neuroectodermal interaction leading to blood-brain barrier differentiation (Figure 5): In the first phase endothelial-neuroectodermal interactions induce a blood-brain barrier "commitment" in the proliferating and migrating endothelial cells. Endothelial "commitment" is reflected by the specific expression of lyn and the multidrug resistance gene in brain endothelium at this timepoint. In a subsequent phase, secondary interaction of "committed" brain endothelium with astrocytes and/or pericytes may induce further endothelial differentiation leading to a functional blood-brain barrier. The downregulation of membrane tyrosine kinase receptors on brain endothelium in parallel to the upregulation of a panel of marker molecules listed in Table I and the development of high electrical resistance over brain endothelium are characteristic for this second phase. Whereas this second phase is reversible the commitment of brain endothelial cells for the blood-brain barrier lineage remains stable. This is emphasized by tissue culture experiments, where brain-derived endothelial cells readily loose their blood-brain barrier characteristics such as P face association of tight junctions, electrical resistance, expression of specific proteins and, on the other hand, upregulate molecules not expressed in differentiated blood-brain barrier endothelium. However, the maintenance of complex tight junctions, and of lyn-expression in these primary cultures suggest that brain-derived endothelial cells do not completely "forget" their commitment to form a blood-brain barrier and therefore may be able to reform an intact blood-brain barrier quite rapidly after short periods of inflammation and injury. This new concept implies that we have to search for the true blood-brain barrier - inducing molecules in early neuroectoderm and for their receptors in embryonic endothelium. In addition to brain endothelium, this concept may also be applicable to other organ-specific differentiation processes of endothelium.

## CONCLUSION

The vascular system of the brain develops in three phases (i) vasculogenesis, (ii) angiogenesis and (iii) "barriergenesis" (Farrell and Risau, 1994). The molecular mechanisms involved in "barriergenesis" are the least understood. The specific microenvironment provided by the central nervous system tissue during embryonic development and in the adult

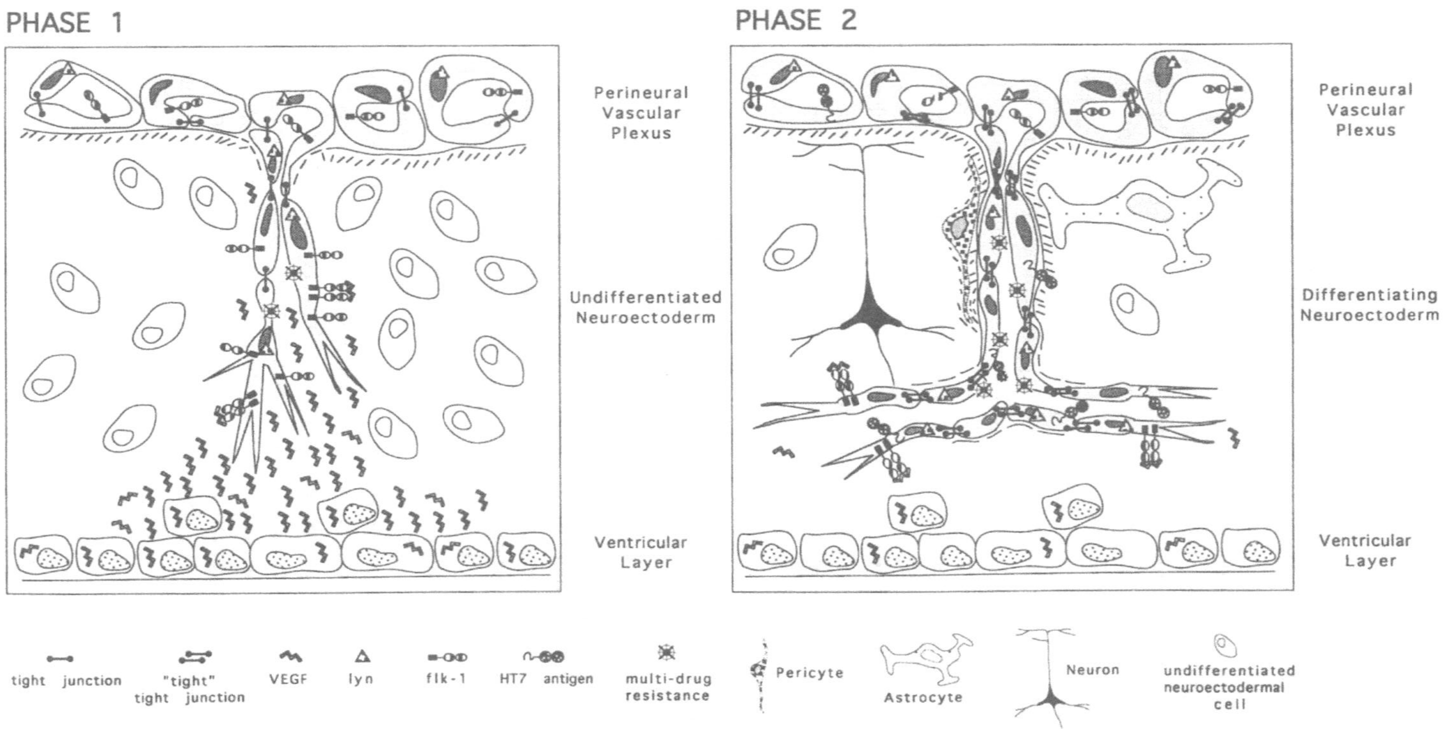


**Figure 5: "Two-Phase Model" of the differentiation of the blood-brain barrier induced by endothelial - neuroectodermal interaction.** Phase 1: Vascular sprouts radially invade the neuroectoderm following a concentration gradient provided by VEGF. VEGF is produced by neuroctodermal cells in the ventricular layer and binds to flk-1 expressed by endothelial cells. Interaction of the undifferentiated neuroectoderm with migrating and proliferating endothelial cells induces endothelial blood-brain barrier "committment" already at this early timepoint. This is suggested by the specific expression of lyn and multidrug resistance in brain endothelium during early development. Phase 2: At a later timepoint when angiogenesis via VEGF-flk-1 interaction is still proceeding, the interaction of differentiating neuroectoderm with "committed" endothelial cells induces the development of the impermeable endothelial barrier as marked by the development of complex tight junctions and the upregulation of surface markers like the HT7 antigen.

animal seems to be critical for development and maintenace of the highly specialized blood-brain barrier endothelium. In our hypothetical "New Concepts of a Blood-Brain Barrier" we suggest a "Two-Phase Model" where in a first step of endothelial-neuroectodermal interaction endothelial cells become committed to a blood-brain barrier lineage, which in a second subsequent step of endothelial-neuroectodermal interaction may lead to a fully differentiated blood-brain barrier phenotype in endothelial cells. A suitable *in vitro* model system is needed to allow the identification of the individual cells and factors involved in the induction and maintenance of the blood-brain barrier.

## ACKNOWLEDGEMENTS

We thank Dr. U. Deutsch for his critical discussion of the manuscript and his patient help with the figures. We are greatful to Dr. G. Breier for providing Figure 2 and Dr. H. Wolburg for the freeze fracture micrographs shown in Figure 3.

## REFERENCES

Achen, M.G., Clauss, M., Schnürch, H. and Risau, W. 1995. The non-receptor tyrosine kinase Lyn is localised in the developing blood-brain barrier. *Differentiation* (In Press).

Albrecht, U., Seulberger, H., Schwarz, H., and Risau, W. 1990a. Correlation of blood-brain-barrier function and HT7 protein distribution in chick brain circumventricular organs. *BrainResearch* 535:49-61.

Anderson, J. M., Stevenson, B. R., Jesaitis, L. A., Goodenough, D. A., and Mooseker, M. S. 1988. Characterization of ZO-1, a protein component of the tight junction from mouse liver and Madin-Darby canine kidney cells. *J. Cell Biol.* 106:1141-1149.

Arthur, F. E., Shivers, R. R. and Bowman, P. D. 1987. Astrocyte-mediated induction of tight junctions in brain capillary endothelium: an efficient in vitro model. *Dev. Brain Res.* 36:155-159.

Bär, T. 1980. The vascular system of the cerebral cortex. *Adv. Anat. Embryol. Cell Biol.* 59:1-62.

Bauer, H. C., Bauer, A., Lametschwandtner, A., Amberger, A., Ruiz, P. and Steiner, M. 1993. Neovascularization and the appearance of morphological characteristics of the blood-brain barrier in the embryonic mouse central nervous system. Brain Res. Dev. Brain Res. 75: 269-278.

Betz, A. L. 1986. Transport of ions across the blood-brain barrier. *FederationProc.* 45:2050-2054.

Betz, A. L. and Goldstein, G. W. 1978. Polarity of the blood-brain barrier: neutral amino acid transport into isolated brain capillaries. *Science* 202:225-227.

Betz, A. L. and Goldstein, G. W. 1986. Specialized properties and solute transport in brain capillaries. *Ann. Rev. Physiol.* 48:241-250.

Bouchaud, C. and Bosler, O. 1986. The circumventricular organs of the mammalian brain with special reference to monoaminergic innervation. *Int. Rev. Cytol.* 105: 283-327.

Bradbury, M. W. B. 1979. The concept of a blood-brain barrier. J. Wiley & Sons.

Bradbury, M.W. B. 1984. The structure and function of the blood-brain barrier. *Fed.Proc.* 43: 186-191.

Breier, G., Albrecht, U., Sterrer, S. and Risau, W. 1992. Expression of vascular endothelial growth-factor during embryonic angiogenesis and endothelial-cell differentiation. *Development* 114:521-532.

Brightman, M. W. and Reese, T. S. 1969. Junctions between intimately apposed cell membranes in the vertebrate brain. *J. Cell Biol.* 40:648-677.

Butt, A. M., Jones, H. C. and Abbott, N. J. 1990. Electrical resistance across the blood-brain barrier in anaesthetized rats: A developmental study. *J. Physiol.* 429: 47-62.

Cereijido, M., González-Mariscal, L. and Contreras, B. 1989. Tight junction: barrier between higher organisms and environment. *NIPS* 4:72-75.

Citi, S., Sabanay, H., Jakes, R., Geioger, B. and Kendrick-Jones, J. 1988. Cingulin, a new peripherlal component of tight junctions. *Nature* 333: 272-275.

Citi, S. 1993. The molecular-organization of tight junctions. *J. Cell Biol.* 121:485-489.

Claude, P. and Goodenough, D. A. 1973. Fracture faces of zonulae occludentes from "tight" and "leaky" epithelia. *J. Cell Biol.* 58:390-400.
Conn, G., Bayne, M. L. , Soderman, D. D., Kwok, P. W., Sullivan, K. A., Palisi, T. M., Hope, D. A. and Thomas K. A. 1990a. Amino-acid and cDNA sequences of a vascular endothelial-cell mitogen that is homologous to platelet-derived growth-factor. *Proc. Natl. Acad. Sc* i. *USA* 87:2628-2632.
Conn, G., Soderman, D. D., Schaeffer, M. T., Wile, M., Hatcher, V. B. , and Thomas, K. A. 1990b. Purification of a glycoprotein vascular endothelial-cell mitogen from a rat glioma-derived cell-line. *Proc. Natl. Acad. Sci. USA* 87:1323-1327.
Coomber, B. L., Stewart, P. A., Hayakawa, K., Farrell, C. L., and Del Maestro, R. F. 1987. Quantitative morphology of human glioblastoma multiforme microvessels: structural basis of blood-brain barrier defect. *J. Neuro-Oncol.* 5:299-307.
Cordon-Cardo, C., O'Brien, J. P., Casals, D., Rittman-Grauer, L., Biedler, J. L., Melamed M. R. and Bertino, J. R. 1989. Multidrug-resistance gene (P-glycoprotein) is expressed by endothelial cells at blood-brain barrier sites. *Proc. Natl. Acad. Sci. USA* 86:695-698.
Crone, C. and Olesen, S. -P. 1982. Electrical resistance of brain microvascular endothelium. *Brain Res.* 241:49-55.
De Vries C., Escobedo, J. A., Ueno, H., Houck, K., Ferrara, N., and Williams, L.T. 1992. The fms-like tyrosine kinase, a receptor for vascular endothelial growth factor. *Science* 255: 989-991.
Dehouck, B., Dehouck, M.P., Fruchart, J-C- and Cecchelli, R. 1994. Upregulation of the low density lipoprotein receptor at the blood-brain barrier: intercommunications between brain capillary endothelial cells and astrocytes. *J. Cell. Biol.* 126: 465-473.
Dehouck, M. P., Meresse, S., Delorme, P., Fruchart, J. C., and Cecchelli, R. 1990b. An easier, reproducible, and mass-production method to study the blood-brain-barrier in vitro. *J. Neurochem.* 54:1798-1801.
Dermietzel, R. and Krause, D. 1991. Molecular anatomy of the blood-brain-barrier as defined by immunocytochemistry. *Int. Review Cytol.* 127:57-109.
Drenckhahn, D. and Dermietzel, R. 1988. Organization of the actin filament cytoskeleton in the intestinal brush border: a quantitative and qualitative immunoelectron microscope study. *J. Cell Biol.* 107:1037-1048.
Duffy, K. R. and Pardridge, W. M. 1987. Blood-brain barrier transcytosis of insulin in developing rabbits. *Brain Res.* 420:32-38.
Dwyer, K. J. and Pardridge, W. M. 1993. Developmental modulation of blood-brain-barrier and choroid-plexus glut1 glucose transporter messenger-ribonucleic-acid and immunoreactive protein in rabbits. *Endocrinology* 132:558-565.
Ehrlich, P. 1904. Über die Beziehung chemischer Constitution, Vertheilung, und pharmakologischer Wirkung. *Gesammelte Arbeiten zur Immunitätsforschung, Berlin.*
Emoto, N., Gonzalez, A. , Walicke, P. A., Wada, E., D. Simmons, M., Shimasaki, S., and Baird, A. 1989. Basic fibroblast growth factor (FGF) in the central nervous system: Identification of specific loci of basic FGF expression in the rat brain. *Growthfactors* 2:21-29.
Engelhardt, B., Conley, F.K., Butcher, E.C. 1994. Cell adhesion molecules on vessels during inflammation in the mouse central nervous system (CNS). *J. Neuroimmunol.* 51:199-208.
Evans, H. M. 1909. On the development of the aortae, cardinal and umbilical veins, and the other blood vessels of vertebrate embryos from capillaries. *Anat. Rec.* 3:498-519.
Ferrara, N. and Henzel, W. J. 1989. Pituitary follicular cells secrete a novel heparin-binding growth-factor specific for vascular endothelial-cells. *Biochem. Biophys. Res.* Comm. 161:851-858.
Fleming, T. P., Hay, M., Javed, Q., and Citi, S. 1993. Localization of tight junction protein cingulin is temporally and spatially regulated during early mouse development. *Development* 117:1135-1144.
Folkman, J. and Klagsbrun, M. 1987. Angiogenic factors. *Science* 235:442-447.
Furuse, M., Hirase, T., Itoh, M., Nagafuchi, A., Yonemura, S., Tsukita, S. and Tsukita, S. 1993. Occludin: A novel integral membrane protein localizing at tight junctions. *J. Cell. Biol.* 123: 1777-1788.
Gaffney, J., West, D., Arnold, F., Sattar, A., and Kumar, S. 1985. Differences in the uptake of modified low density lipopoteins by tissue cultured endothelial cells. *J. Cell Sci.* 79:317-325.

Goldmann, E. E. 1913. Vitalfärbung am Zentralnervensystem. *Abh. Preuss. Akad. Wissensch. Phys. - Math.* 1:1-60.

Goldstein, G. W. and A. L. Betz. 1986. The blood-brain barrier. *Scient. American* 255:70-79.

Gospodarowicz, D. and Lau, K. 1989. Pituitary follicular cells secrete both vascular endothelial growth-factor and follistatin. *Biochem. Biophys. Res. Comm.* 165:292-298.

Hallman, R. Mayer, D.N., Berg, E.L., Broermann, R. and Butcher, E.C. 1995. Novel mouse endothelial cell surface marker is suppressed during differentiation of the blood-brain barrier. *Developmental Dynamics* 202: 325-332.

Harahap, A. R. and Goding, J. W. 1988. Distribution of the murine plasma cell antigen PC-1 in non-lymphoid tissues. *J. Immunol.* 141:2317-2320.

Heuer, J. G., Vonbartheld, C. S., Kinoshita, Y., Evers, P. C., and Bothwell, M. 1990. Alternating phases of FGF-receptor and NGF-receptor expression in the developing chicken nervous-system. *Neuron* 5:283-296.

Holash H. A., Noden, D. M. and Stewart, P. A. 1993. Re-evaluating the role of astrocytes in blood-brain barrier induction. *Dev. Dyn.* 197: 14-25

Ishii, H., Salem, H. H., Bell, C. E., Laposata, E. A., and Majerus, P. W. 1986. Thrombomodulin, an endothelial anticoagulant protein, is absent from the human brain. *Blood* 67:362-365.

Janzer, R. C. and Raff, M. C. 1987. Astrocytes induce blood-brain barrier properties in endothelial cells. *Nature* 325:253-257.

Jefferies, W. A., Brandon, M. R., Hunt, S. V., Williams, A. F., Gatter, K. C., and Mason, D. Y. 1984. Transferrin receptor on endothelium of brain capillaries. *Nature* 312:162-163.

Kalaria, R. N., Mitchell, M. J., and Harik, S. I. 1987. Correlation of 1-methyl-4-phenyl-1,2,3,6-tetrahydropyridine neurotoxicity with blood-brain barrier monoamine oxidase activity. *Proc. Natl. Acad. Sci. USA* 84:3521-3525.

Keck, P. J., Hauser, S. D., Krivi, G., Sanzo, K., Warren, T., Feder, J., and Connolly, D. T. 1989. Vascular-permeability factor, an endothelial-cell mitogen related to PDGF. *Science* 246:1309-1312.

Korhonen, J., Partanen, J., Armstrong, E., Vaahtokari, A., Elenius, K., Jalkanen, M. and Alitalo, K. 1992. Enhanced expression of the tie receptor tyrosine kinase in endothelial-cells during neovascularization. *Blood* 80:2548-2555.

Krause, D., Mischeck, U., Galla, H. J., and Dermietzel, R. 1991. Correlation of zonula-occludens zo-1 antigen expression and transendothelial resistance in porcine and rat cultured cerebral endothelial-cells. *Neurosci. Letters* 128:301-304.

Kuban, K. C. and Gilles, F. H. 1985. Human telencephalic angiogenesis. *Ann. Neurol.* 17: 539 458.

Kunz, J., Krause, D., Kremer, M., and Dermietzel, R. 1994. The 140-kda protein of blood-brain barrier-associated pericytes is identical to aminopeptidase-n. *J. Neurochem.* 62:2375-2386.

Leonhardt, H..1980. In Handbuch der Mikroskopischen Anatomie des Menschen (Oschke, A. Ed.) pp. 177-666, Springer-Verlag.

Leung, D. W., Cachianes, G., Kuang, W. J., Goeddel, D. V., and Ferrara, N. 1989. Vascular endothelial growth-factor is a secreted angiogenic mitogen. *Science* 246:1306-1309.

Lossinsky, A. S., Vorbrodt, A. W., and Wisniewski, H. M. 1986. Chracterization of endothelial cell transport in the developing mouse blood-brain barrier. *Dev. Neurosci.* 8:61-75.

Macdonald, P. N., Bok, D., and Ong, D. E. 1990. Localization of cellular retinol-binding protein and retinol-binding protein in cells comprising the blood-brain-barrier of rat and human. *Proc. Natl. Acad. Sci. USA* 87:4265-4269.

Madara, J. L. 1987. Intestinal absorptive cell tight junctions are linked to cytoskeleton. *Am. J. Physiol.* 253:C171-C175.

Madara, J. L. and Pappenheimer, J. R. 1987. Structural basis for physiological regulation of paracellular pathways in intestinal epithelia. *J. Membrane Biol.* 100:149-164.

Maher, P. A. and Pasquale, E. B. 1988. Tyrosine phosphorylated proteins in different tissues during chick embryo development. *J. Cell. Biol.* 106: 1747-1755.

Maisonpierre, P. C., Goldfarb, M., Yancopoulos, G. D. and Gao, G. X. 1993. Distinct rat genes with related profiles of expression define a tie receptor tyrosine kinase family. *Oncogene* 8:1631-1637.

Matthews, W., Jordan, C. T., Wiegand, G. W., Pardoll, D. and Lemischka I. R. 1991. A receptor tyrosine kinase specific to hematopoietic stem and progenitor cell-enriched populations. *Cell* 65:1143-1152.

Méresse, S., Delbart, C., Fruchart J.C., and Cecchelli, R. 1989. Low-density lipoprotein receptor on endothelium of brain capillaries. *J. Neurochem.* 53:340-345.

Mergia, A., Tischer, E., Graves, D., Tumolo, A., Miller, J., Gospodarowicz, D., Abraham, J. A., Shipley, G. D., and Fiddes, J. C. 1989. Structural-analysis of the gene for human acidic fibroblast growth-factor. *Biochem. Biophys. Res. Comm.* 164:1121-1129.

Meyer, J., Mischeck, U., Veyhl, M., Henzel, K., and Galla, H. J. 1990. Blood-brain-barrier characteristic enzymatic-properties in cultured brain capillary endothelial-cells. *BrainResearch* 514:305-309.

Millauer, B., Wizigmann-Voos, S., Schnürch, H., Martinez, R., Møller, N. P. H., Risau, W., and Ullrich, A. 1993. High affinity VEGF binding and developmental expression suggest Flk-1 as a major regulator of vasculogenesis and angiogenesis. *Cell* 72:835-846.

Mori, S. and Nagano, M. 1985. Electron-microscopic cytochemistry of alkaline-phosphatase activity in endothelium, pericytes and oligodendrocytes in the rat brain. *Histochemistry* 82:225-231.

Mühleisen, H., Wolburg, H., and Betz, E. 1989. Freeze-fracture analysis of endothelial cell membranes in rabbit carotid arteries subjected to short-term atherogenic stimuli. *Virch. Arch. B Cell Pathol.* 56:413-417.

Nagy, Z., Peters, H., and Hüttner, I. 1984. Fracture faces of cell junctions in cerebral endothelium during normal and hyperosmotic conditions. *Lab. Invest.* 50:313-322.

Nico, B., Cantino, D., Bertossi, M., Ribatti, D., Sassoe, M., and Roncali, L. 1992. Tight endothelial junctions in the developing microvasculature - a thin-section and freeze-fracture study in the chick-embryo optic tectum. *J. Submicr. Cytol. Pathol.* 24:85-95.

Noden, D.M. 1991. Development of craniofacial blood vessels. In The development of the vascular system. R.N. Feinberg, G.K. Sherer, and R. Auerbach, editors. Karger, Basel. 1-24.

Orlowski, M., Sessa, G., and Green, J. P. 1974. Gamma-glutamyl transpeptidase in brain capillaries: possible site of a blood-brain barrier for amino acids. *Science* 184:66-68.

Partanen, J., Armstrong, E., Makela, T. P., Korhonen, J., Sandberg, M., Renkonen, R., Knuutila, S., Huebner, K. and Alitalo K. 1992. A novel endothelial-cell surface-receptor tyrosine kinase with extracellular epidermal growth-factor homology domains. *Mol. Cell. Biol.* 12:1698-1707.

Peters, K. G., Werner, S., Chen, G., and Williams, L. T. 1992. 2 FGF receptor genes are differentially expressed in epithelial and mesenchymal tissues during limb formation and organogenesis in the mouse. *Development* 114:233-243.

Phillips, H. S., Hains, J., Leung, D. W. and Ferrara, N. 1990. Vascular endothelial growth-factor is expressed in rat corpus luteum. *Endocrinology* 127: 965-967.

Plate, K. H., Breier, G., Weich, H. A. and Risau, W. 1992. Vascular endothelial growth-factor is a potential tumor angiogenesis factor in human gliomas in vivo. *Nature* 359:845-848.

Plate, K.H., Breier, G. and Risau, W. 1994. Molecular mechanisms of developmental and tumor angiogenesis. *Brain Pathol.* 4: 207-218.

Plouet, J., Schilling, J. and Gospodarowicz, D. 1989. Isolation and characterization of a newly identified endothelial-cell mitogen produced by ATT-20 cells. *EMBO Journal* 8:3801-3806.

Puumala, M., Andersson, R.E. and Meyer, F.B. 1990. Intraventricular infusion of HBGF-2 promotes cerebral angiogenesis in wistar rat. *Brain Res.* 534:283-286.

Qin, Y. and Sato, T. N. 1995. Mouse multidrug resistance 1a/3 gene is the earliest known endothelial cell differentiation marker during blood-brain barrier development. *Develop. Dynamics* 202: 172-180.

Reese, T. S. and Karnovsky, M. J. 1967. Fine structural localization of a blood-brain barrier to exogenous peroxidase. *J. Cell Biol.* 34:207-217.

Risau, W. 1986. Developing brain produces an angiogenesis factor. *Proc. Natl. Acad. Sci. USA* 83:3855-3859.

Risau, W., Hallmann, R. and Albrecht, U. 1986a. Differentiation-Dependent Expression of Protein in Brain Endothelium during Development of the Blood-Brain Barrier. *Dev. Biol.* 117:537-545.

Risau, W., Hallmann, R., Albrecht, U. and Henke-Fahle, S. 1986b. Brain induces the expression of an early cell surface marker for blood-brain barrier specific endothelium. *EMBO J.* 5:3179-3183.

Risau, W. and Lemmon, V. 1988. Changes in the vascular extracellular matrix during embryonic vasculogenesis and angiogenesis. *Dev. Biol.* 125:441-450.

Risau, W., Sariola, H., Zerwes, H. -G., Sasse, J., Ekblom, P, Kemler, R., and Doetschman, T. 1988a. Vasculogenesis and angiogenesis in embryonic-stem-cell-derived embryoid bodies. *Development* 102:471-478.

Risau, W., Gautschi-Sova, P. and Böhlen, P. 1988b. Endothelial cell growth factors in embryonic and adult chick brain are related to human acidic fibroblast growth factors. *EMBO J.* 7:959-962.

Risau, W. 1990. Angiogenic growth factors. *Progr. Growth Factor Res.* 2:71-79.

Risau, W. and Wolburg, H. 1990. Development of the blood-brain barrier. *Trends Neurosci.* 13:174-178.

Risau, W., Engelhardt, B. and Wekerle, H. 1990. Immune function of the blood-brain barrier: Incomplete presentation of protein (auto-)antigens by rat brain microvascular endothelium in vitro. *J. Cell Biol.* 110:1757-1766.

Risau, W. 1991a. Induction of blood-brain-barrier endothelial-cell differentiation. *Ann. NY Acad. Sci.* 633:405-419.

Risau, W. 1991b. Vasculogenesis, angiogenesis and endothelial cell differentiation during embryonic development. In The development of the vascular system. R.N. Feinberg, G.K. Sherer, and R. Auerbach, editors. Karger, Basel. 58-68.

Risau, W. 1991c. Embryonic angiogenesis factors. *Pharm. Ther.* 51:371-376.

Risau, W. and Wolburg, H. 1991. The importance of the blood-brain-barrier in fetuses and embryos - reply. *Trends Neurosci* 14:15.

Risau, W., Dingler, A., Albrecht, U., Öcalan, M., Dehouck, M.-P. and Cecchelli, R. 1992. Blood-brain barrier pericytes are the main source of gamma-glutamyltranspeptidase activity in brain capillaries. *J. Neurochem.* 58: 667-672.

Robertson, P. L., Du Bois, M., Bowman, P. D. and Goldstein, G. W. 1985. Angiogenesis in developing rat brain: an in vivo and in vitro study. *Dev. Brain Res.* 23:219-223.

Romanoff, A. L. 1960. The avian embryo. Macmillan Company, New York.

Roncali , L., Nico, B., Ribatti, D., Bertossi, M. and Mancini, L. 1986. Microscopical and ultrastructural investigations on teh development of the blood-brain barrier in the chick embryo optic tectum. Acta Neuropathol. 70: 193-201.

Rosenstein, J. M., Krum, J. M., Sternberger, L. A., Pulley, M. T. and Sternberger, N. H. 1992. Immunocytochemical expression of the endothelial barrier antigen (eba) during brain angiogenesis. *Dev. Brain Res.* 66:47-54.

Rubin, L. L., Hall, D. E., Porter, S. , Barbu, K., Cannon C., Horner, H. C., Janatpour, M. , Liaw, C. W., Manning, K. , Morales, J., Tanner, L. I., Tomaselli, K. J., et al. 1991a. A cell-culture model of the blood-brain-barrier. *J. Cell Biol.* 115:1725-1735.

Rubin, L. L., Barbu, K., Bard, F., Cannon, C., Hall, D. E., Horner, H., Janatpour, M., Liaw, C., Manning, K., Morales, J., Porter, S., Tanner, L. et al. 1991b. Differentiation of brain endothelial-cells in cell-culture. *Ann. NY Acad. Sci.* 633:420-425.

Saunders, N. R. and Dziegielewska, K. M. 1991. The importance of the blood-brain barrier in fetuses and embryos- letter. *Trends Neurosci.* 14: 14.

Sasseville, V. G., Newman, W. A., Lackner, A. A., Smith, M. O., Lausen, N. C. G., Beall, D. and Ringler, D. J. 1992. Elevated vascular cell adhesion molecule-1 in AIDS encephalitis induced by Simian immunodeficiency virus. *Am J. Pathol.* 141: 1021-1030.

Sato, T. N., Qin, Y., Kozak, C. A. and Audus, K.L. 1993. tie-1 and tie-2 define another class of putative receptor tyrosine kinase genes expressed in early embryonic vascular system. *Proc. Natl. Acad. Sci. USA* 90: 9355-9358.

Schielke, G. P., Moises, H. C. and Betz, A. L. 1990. Potassium activation of the Na,K-pump in isolated brain microvessels and synaptosomes. *BrainResearch* 524:291-296.
Schinkel., A. H., Smit, J.J.M., van Tellingen, O., Beijnen, J. H., Wagenaar, E., van Deemter, L., Mol, C. A. A. M., van der Valk, M. A., Robanus-Maandag, E. C., de Riele, H.P. J., Berns, A. J. M. and Borst, P. 1994. Disruption of the mouse mdr1a P-glycoprotein gene leads to a dificiency in the blood-brain barrier and to increased sensitivity to drugs. *Cell* 77: 491-502.
Schnürch, H. and Risau, W. 1991. Differentiating and mature neurons express the acidic fibroblast growth-factor gene during chick neural development. *Development* 111:1143-1154.
Schnürch, H. and Risau, W. 1993. Expression of tie-2, a member of a novel family of receptor tyrosine kinases, in the endothelial cell lineage. *Development* 119: 957-968.
Schulze, C. and Firth, J. A. 1992. Interendothelial junctions during blood-brain barrier development in the rat: morphological changes at the level of individual tight junctional contacts. *Dev. Brain Res.* 69: 85-95.
Schulze, C. and Firth, J. A. 1993. Immunohistochemical localization of adherens junction components in blood-brain barrier microvessels of the rat. *J. Cell. Sci.* 104: 773-782.
Senger, D. R., Galli, S. J., Dvorak, A. M., Perruzzi, C. A., Harvey, V. S. and Dvorak, H. F. 1983. Tumor cells secrete a vascular permeability factor that promotes accumulation of ascites fluid. *Science* 219: 983-985.
Seulberger, H., Lottspeich, F., and Risau, W. 1990. The inducible blood-brain barrier specific molecule HT7 is a novel immunoglobulin-like cell surface glycoprotein. *EMBO J.* 9:2151-2158.
Seulberger, H., Unger, C., Albrecht, U., and Risau, W. 1991. The inducible blood-brain-barrier endothelium-specific antigen HT7, a novel immunoglobulin-like membrane glycoprotein. *Ann. NY Acad. Sci.* 633:611-614.
Seulberger, H., Unger, C. M. and Risau, W. 1992. HT7, neurothelin, basigin, gp42 and OX-47 - many names for one developmentally regulated immunoglobulin-like surface glycoprotein on blood-brain barrier endothelium, epithelial tissue barriers and neurons. *Neurosci. Letters* 140: 93-97.
Shibuya, M., Yamaguchi, S., Yamane, A., Ikeda, T., Tojo, A., Matsushime, H. and Sato, T. 1990. Nucleotide sequence and expression of a novel human receptor-type tyrosine kinase gene (flt) closely related to the fms family. *Oncogene* 5: 519-524.
Simionescu, M., Ghinea, N., Fixman, A., Lasser, M., Kukes, L., Simionescu, N., and Palade, G.E. 1988. The cerebral microvasculature of the rat: structure and luminal surface properties during early development. *J. Submicrosc. Cytol.* 20:243-261.
Sims, D. E. 1986. The Pericyte - A review. *Tissue and Cell* 18:153-174.
Steffen, B.J., Butcher, E.C., Engelhardt, B. 1994. Evidence for involvement of ICAM-1 and VCAM-1 in lymphocyte interaction with endothelium in experimental autoimmune encephalomyelitis (EAE) in the central nervous system (CNS) in the SJL/J mouse. *Am. J. Pathol.* 145: 189-201.
Stevenson, B. R., Anderson, J. M., Godenough, D. A. and Mooseker, M. S. 1988. Tight junction structure and ZO-1 content are identical in two strains of Madin-Darby canine kidney cells which differ in transepithelial resistance. *J. Cell Biol.* 107:2401-2408.
Stewart, P. A. and Hayakawa, E. M. 1987. Interendothelial junctional changes underlie the developmental 'tightening' of the blood-brain barrier. *Dev. Brain Res.* 32:271-281.
Stewart, P. A. and Wiley, M. J. 1981. Developing nervous tissue induces formation of blood-brain barrier characteristics in invading endothelial cells: A study using quail-chick transplantation chimeras. *Dev. Biol.* 84:183-192.
Stewart, P.A., Magliocco, M., Hayakawa, K., Farrell, C.L., Del Maestro, R. F., Girvin,J., Kaufmann, J.C.E., Vinters, H.V. and Gilbert, J. 1987. A quantitative analysis of blood-brain barrier ultrastructure in the aging human. *Microvasc. Res.* 33:270-282.
Tao-Cheng, J. -H. and Brightman, M. W. 1988. Development of membrane interactions between brain endothelial cells and astrocytes in vitro. *Int. J. Dev. Neurosci.* 6:25-37.

Tischer, E., Gospodarowicz, D., Mitchell, R., Silva, M., Schilling, J., Lau, K., Crisp, T., Fiddes, J. C., and Abraham, J. A. 1989. Vascular endothelial growth-factor - a new member of the platelet-derived growth-factor gene family. *Biochem. Biophys. Res. Comm.* 165:1198-1206.

Tontsch, U. and Bauer, H. C. 1991. Glial-cells and neurons induce blood-brain-barrier related enzymes in cultured cerebral endothelial-cells. *Brain Research* 539:247-253.

Trancard, J., Delamanche , I. S., Ruiz, G., and Bouchaud, C. 1989. Interspecific variations of cerebral endothelial cholinesterases in rodents and carnivores. *Brain Res.* 476:213-219.

Tsukita, S., Oshi, K., Akiyama, T., Yamanashi, Y., Yamamoto, T. and Tsukita, S. 1991. Specific proto-oncogenic tyrosine kinases of src family are enriched in cell-to-cell adherens junctions where the level of tyrosine phosphorylation is elevated. *J. Cell Biol.* 113: 867-879

Ullrich, A. and Schlessinger, J. 1990. Signal transduction by receptors with tyrosine kinase activity. *Cell* 61:203-212.

Umemori, H., Wanaka, A., Kato, H., Takeuchi, M., Tohyama, M., and Yamamoto, T. 1992. Specific expressions of fyn and lyn, lymphocyte antigen receptor- associated tyrosine kinases, in the central-nervous- system. *Mol. Brain Res.* 16:303-310.

Vlodavsky, I., Folkman, J., Sullivan, R., Fridman, R., Ishai-Michaeli, R., Sasse, J., and Klagsbrun, M. 1987. Endothelial cell-derived basic fibroblast growth factor: Synthesis and deposition into subendothelial extracellular matrix. *Proc. Natl. Acad. Sci. USA* 84:2292-2296.

Vorbrodt, A. W. 1986. Changes in the distribution of endothelial surface glyoconjugates associated with altered permeability of brain micro-blood vessels. *Acta Neuropathol. (Berl.)* 70:103-111.

Vorbrodt, A. W., Lossinsky, A. S., and Wisniewski, H. M. 1986c. Localization of alkaline phosphatase activity in endothelia of developing and mature mouse blood-brain barrier. *Dev. Neurosci.* 8:1-13.

Vorbrodt, A. W., Lossinsky, A. S., Dobrogowska, D. H., and Wisniewski, H. M. 1986b. Distribution of anionic sites and glycoconjugates on the endothelial surfaces of the developing blood-brain barrier. *Dev. Brain Res.* 29:69-79.

Vorbrodt, A. W., Dobrogowska, D. H., Lossinsky, A. S. and Wisniewski, H. M. 1986a. Ultrastructural localization of lectin receptors on the luminal and abluminal aspects of brain micro-blood vessels. *J. Histochem. Cytochem.* 34:251-261.

Wakai, S. and Hirokawa, N. 1978. Development of the blood-brain barrier to horseradish peroxidase in the chick embryo. *Cell Tissue Res* 195:195-203.

Wanaka, A., Milbrandt, J. and Johnson, E. M. 1991. Expression of FGF receptor gene in rat development. *Development* 111:455-468.

Weilerguttler, H., Sommerfeldt, M., Papandrikopoulou, A., Mischek, U., Bonitz, D., Frey, A., Grupe, M., Scheerer, J., and Gassen, H. G. 1990. Synthesis of apolipoprotein-a-1 in pig brain microvascular endothelial-cells. *J. Neurochem.* 54:444-450.

Wolburg, H., Neuhaus, J., Kniesel, U., Krauß, B., Schmid, E.-M., Öcalan, M., Farrell C. and Risau, W. 1994. Modulation of tight junction structure in blood-brain barrier endothelial cells. *J. Cell. Sci.* 107: 1347-1357.

Yamanashi, Y., Okada, M., Semba, T., Yamori, T., Umemori, H., Tsunasawa, S., Toyoshima, K., Kitamura, D., Watanabe, T. and Yamamoto, T. 1993. Identification of HS1 protein as a major substrate of protein-tyrosine kinase(s) upon B-cell antigen receptor-mediated signalling. *Proc. Natl. Acad. Sci. USA* 90: 3631-3635.

Yoshida, Y., Yamada, M., Wakabayashi, K. and Ikuta, F. 1988. Endothelial fenestrae in the rat fetal cerebrum. Dev. Brain Res. 44: 211-219.

# ELECTRICAL RESISTANCE MEASUREMENTS OF BLOOD-BRAIN BARRIER PERMEABILITY

Arthur M. Butt

Division of Physiology
U.M.D.S.
St. Thomas' Hospital
Lambeth Palace Road
London SE1 7EH

## INTRODUCTION

In adult mammals, homeostasis of the brain interstitial fluid is provided by the blood-brain barrier, protecting neurones from disturbances in blood composition.[1] The exceptionally low permeability of the blood-brain barrier is determined by tight intercellular junctions between the endothelial cells of the cerebral vasculature which greatly restrict the movement of substances between blood and brain.[2]

### Mediators of Blood-brain Barrier Opening

An opening of the blood-brain barrier can occur under various pathological conditions, such as trauma, ischemia, and allergic or inflammatory diseases, and this can lead to vasogenic brain oedema and the development of secondary brain damage. There is increasing evidence of a role for chemical mediators released locally or at the inflammatory site in these events.[3,4] Increased brain levels of serotonin and bradykinin are found following brain injuries, stroke and ischemia,[5-7] and brain oedema formation can be reduced by antagonists to histamines and bradykinin.[6-8] In the brain, histamine and serotonin may be released from perivascular mast cells, nerves which innervate the cerebral vasculature, the cerebral endothelium itself, or by circulating platelets, whilst bradykinin can be formed via the kallikrein-kinin system in the blood or brain. Additionally, the degranulation of mast cells or activation of the kallikrein-kinin system is implicated in the development of experimental allergic encephalomyelitis, an autoimmune inflammatory disease which is used as an animal model for multiple sclerosis, suggesting histamine-, serotonin-, or bradykinin-mediated opening of the blood-brain barrier may be important in multiple sclerosis.[9,10] Elevated levels of histamine, serotonin and bradykinin have been found in the brain and CSF of humans under numerous conditions, including hepatic encephalopathy,[11] and multiple

*New Concepts of a Blood–Brain Barrier*
Edited by J. Greenwood *et al.*, Plenum Press, New York, 1995

sclerosis.[12] However, there have been few direct studies of inflammatory agents on the cerebral endothelium of the blood-brain barrier.[13-18]

### Electrical Resistance of Brain Surface Vessels

This technique was introduced by Crone and Olesen[19] in the frog and enables the cerebral endothelium to be studied directly, *in situ*. It has been shown to be an accurate and sensitive measurement of small and rapid changes in blood-brain barrier ion permeability, both in the frog[15-17] and the rat.[13,14,20] This preparation has a number of advantages for the investigation of blood-brain barrier function: (1) brain-surface microvessels are an easily accessible part of the cerebral vasculature; (2) they have typical blood-brain barrier characteristics;[13-17,20] (3) the cerebral endothelium can be studied directly because brain-surface microvessels are not ensheathed by glial cells or by smooth muscle;[20,21] (4) properties of both luminal (blood-facing) and abluminal (brain-facing) faces of the blood-brain barrier can be investigated, by introducing test substances into the blood or onto the brain surface, respectively;[13-17] and (5) the electrophysiological measurement of transendothelial resistance provides a particularly sensitive method for investigating endothelial cell physiology *in situ* and for measuring small and rapid changes in permeability of the blood-brain barrier.[13-17] In the present study, electrical resistance was used to measure the effects of chemical mediators on blood-brain barrier permeability in rat pial vessels.

## METHODS

Wistar rats, 30-35 days old, were anaesthetized with pentobarbitone i.p. (60 mg/kg) and maintained at 35°C on a warm water-filled perspex bath. Brain-surface microvessels were exposed as described previously.[20] In brief, a piece of parietal bone was removed with a scalpel following initial thinning using a dental drill, and a window was carefully cut in the dura and arachnoid maters to expose the pial surface of the cerebrum. Measurements were made in brain surface microvessels 15-30 $\mu$m diameter. Throughout the operation and experiments the brain was superfused with filter-sterilized artificial CSF.

Blood vessels were impaled with two 3 M-KCl-filled microelectrodes and the preparation was earthed via a 3 M-KCl-agar bridge. Constant current pulses (200 ms duration pulse of 50-100 nA passed at 2.5 Hz) injected through one microelectrode inside a blood vessel caused a voltage displacement, $V(x)$, which was measured by the second microelectrode (Figure 1A). The voltage decayed exponentially with distance from the current source and was defined by the length constant of the vessel, $\lambda$ (Ref. 18), which was the distance at which the voltage at zero distance, $V(0)$, fell to $1/e$ (Figure 1B, $\lambda_C$). The length constant was the experimentally determined variable from which transendothelial electrical resistance, $R_m$, was calculated, according to the equation:[15]

$$R_m = 2\rho_i\lambda^2/a \qquad (1)$$

where $\rho_i$ is the resistivity of the blood (90 $\Omega$ cm; Ref. 20), and $a$ is the vessel radius, which was measured using an eye-piece graticule with 18 $\mu$m divisions at x50 magnification.

### Administration of Chemicals

The effects of histamine, bradykinin and serotonin (5-hydroxytryptamine, 5-HT) were tested by adding them directly to the artificial CSF superfusing the brain surface. The effects of antagonists to $H_1$ and $H_2$ receptors were tested by pretreatment of rats with promethazine or cimetidine, respectively.[13,14] In each rat, resistance was first measured in five or more vessels in normal superfusate to provide values for control resistance, and then in another five or more vessels during exposure to one of the agents. In addition, the voltage deflection was measured in single vessels during the turnover from CSF to the test agent (Figure 1A). Since the input voltage, *V(0)*, was constant throughout the experiment, then a change in the voltage deflection was directly related to a change in length constant (Figure 1B, $\lambda_B$). Thus, the change in voltage deflection (Figure 2A) provided continuous measurements of electrical resistance during application of test agents (Figure 2B).

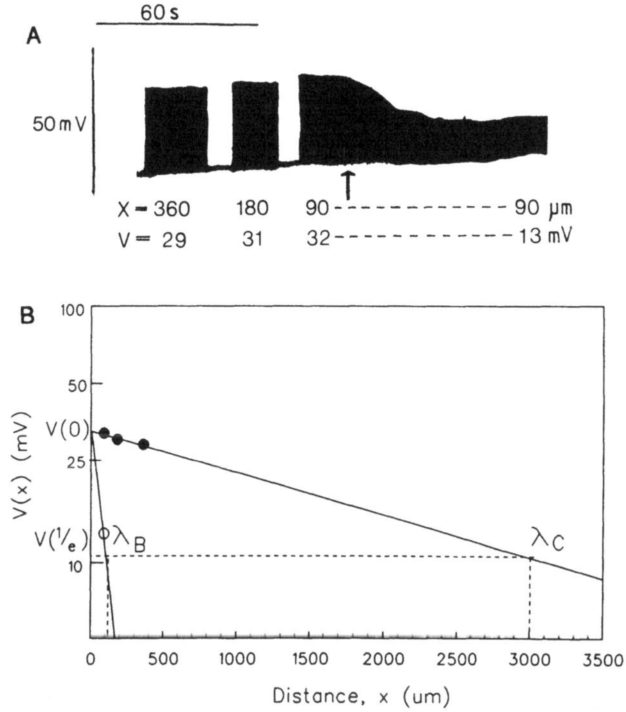


**Figure 1.** Measurement of microvessel electrical resistance. Microvessels were impaled with two microelectrodes and current injected through one electrode into the vessel lumen caused a voltage deflection which was measured with the second electrode (A). The voltage deflection was measured at three decreasing distances and the length constant for the vessel in CSF ($\lambda_C$) was estimated from a plot of the log voltage against distance, as the distance at which *V(0)* fell to $1/e$ (B). At the third distance, the voltage deflection was measured during the turnover to test agent (in this case bradykinin, arrow), to provide the new length constant ($\lambda_B$). Since the input voltage, *V(0)*, was constant throughout the experiment, then a change in the voltage deflection was directly related to a change in length constant, and provided a continuous measurement of electrical resistance (see Figure 2). Electrical resistance, $R_m$, was calculated according to equation 1.

## RESULTS

In normal CSF, mean electrical resistance (± SEM) of pial microvessels (n=49) was 1488 ± 213 $\Omega.cm^2$ (Figure 3). In continuous measurements from single vessels, resistance was observed to fall immediately on superfusion with $10^{-4}$M histamine or bradykinin, but not $10^{-3}$ M serotonin (Figure 2). These effects were completely reversible on return to normal CSF. In histamine and bradykinin, mean resistance was 367 ± 78 $\Omega.cm^2$ (n=28) and 384 ± 77 $\Omega.cm^2$ (n=27), respectively, significantly lower than in CSF ($p<0.001$, unrelated t-test), whereas mean resistance was 1380 ± 532 $\Omega.cm^2$ (n=16) in serotonin, not significantly different than in CSF (Fig. 3). The histamine-mediated decrease in resistance was completely blocked by pretreatment with cimetidine but was unaffected by promethazine.[13]

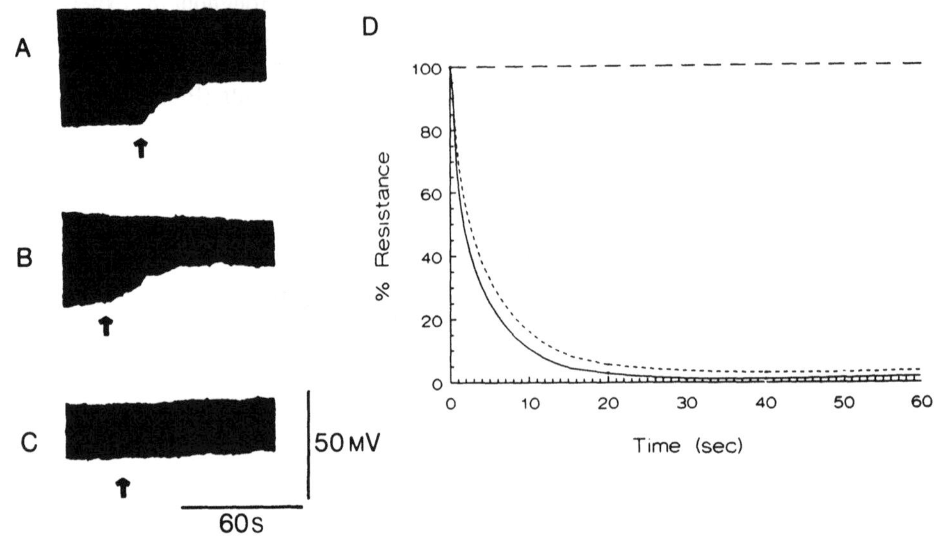


**Figure 2.** Permeability changes in response to inflammatory mediators. The traces illustrate typical records from three rats showing the changes in voltage displacement during exposure to CSF containing (A) $10^{-4}$ M histamine, (B) $10^{-4}$ M bradykinin or (C) $10^{-3}$ M serotonin (5-HT). The change in voltage deflection provided continuous measurement of the change in electrical resistance from single vessels (D). Histamine (dotted line) and bradykinin (filled line), but not serotonin (broken line), caused a large and rapid fall in resistance (D).

## DISCUSSION

Transendothelial electrical resistance in pial vessels was 1500 $\Omega.cm^2$, similar to values from the frog blood-brain barrier[15-17,19] and indicates a tight blood-brain barrier with low ion permeability.[13-17,19,20] Application to the brain surface of histamine or bradykinin, but not serotonin, caused a rapid opening of the blood-brain barrier, which was completely reversible over the short-term[14]. The histamine-induced decrease in resistance was blocked by cimetidine but not by promethazine, indicating the histamine response was mediated via $H_2$ receptors.[13] It is concluded that histamine and bradykinin acted directly on receptors at the basal membranes (brain-facing) of the cerebral endothelium to mediate changes in blood-brain barrier permeability, although it is not excluded that other mechanisms may also have been activated[14].

### Mechanism of Blood-Brain Barrier Opening

It is well known in several vascular beds that chemical mediators increase vascular permeability by an opening of intercellular junctions, resulting from endothelial contraction which is triggered by a receptor-mediated increase in intracellular calcium.[16] The mechanisms by which this may bring about increased blood-brain barrier permeability is of particular interest to brain oedema formation and to the role of the cerebral endothelium in immune diseases of the central nervous system. Direct evidence for an agonist-activated increase in intracellular calcium has come from cultured cerebral endothelium,[18] and *in situ* evidence that these cause an increase in blood-brain barrier permeability comes from electrical resistance measurements in frog brain-surface venules.[15-17] Furthermore,

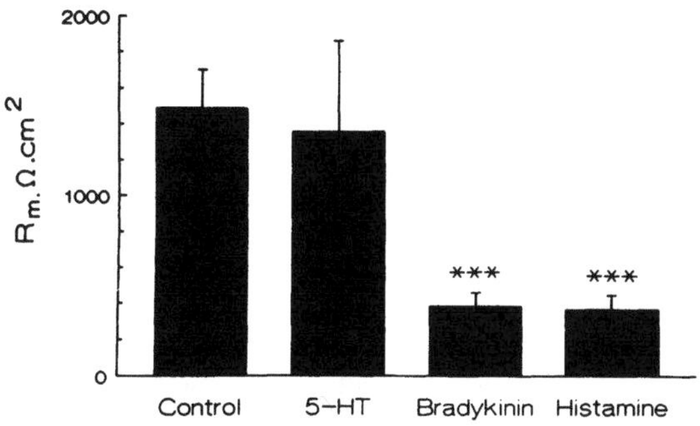


**Figure 3.** Effect of agents on electrical resistance. Mean electrical resistance was 1488 $\Omega.cm^2$ in CSF, indicating a tight blood-brain barrier with low ion permeability. In histamine and bradykinin, mean resistance was 367 and 384 $\pm$ 77 $\Omega.cm^2$, respectively, significantly lower than in CSF (***=$p<0.001$, unrelated t-test), indicating an opening of the blood-brain barrier and increase in ion permeability. Mean resistance was 1380 $\Omega.cm^2$ in serotonin (5-HT), not significantly different than in CSF.

contractile elements are present in rat cerebral endothelium.[22] It is proposed, therefore, that the decrease in electrical resistance indicates an increase in tight junction ion permeability due to receptor-mediated changes in intracellular calcium.

## SUMMARY AND CONCLUSIONS

In summary, this study shows that the measurement of electrical resistance in rat pial vessels is a sensitive technique for determining small and rapid changes in the ion permeability of the blood-brain barrier and is a useful means of studying the cerebral endothelium *in situ.* Pial vessels are an easily accessible part of the cerebral endothelium and, unlike parenchymal vessels, are not enwrapped by astrocytic end-feet, so that the cerebral endothelium can be studied directly. Furthermore, the properties of both apical and basal membranes of the cerebral endothelium can be studied, by superfusing over the brain surface or injecting into the cerebral circulation. These experiments have shown that histamine and bradykinin, but not serotonin, caused a rapid decrease in transendothelial electrical resistance, supporting a role for these two agents in mediating blood-brain barrier opening and the formation of brain oedema.

## REFERENCES

1. M.W.B. Bradbury. "The Concept of a Blood-Brain Barrier", Wiley, Chichester (1979).
2. M.W. Brightman and T.S. Reese. Junctions between intimately apposed cell membranes in the vertebrate brain, *J. Cell Biol.* 40:648 (1969)
3. M. Wahl, A. Unterberg, A. Baethmann, and L. Schilling, Mediators of blood-brain barrier dysfunction and formation of vasogenic brain edema, *J. Cereb. Blood Flow Met.* 8:621 (1988).
4. N.J. Abbott and P.A. Revest, Control of brain endothelial permeability. *Cerebrovasc. Brain Met. Rev.* 3:39 (1991).
5. J.L. Costa, U. Ito, M. Spatz, I. Klatzo, and C. Demirjian, 5-Hydroxytryptamine accumulation in cerebrovascular injury, *Nature* 248:135 (1970).
6. A. Unterberg and A.J. Baethmann, The kallikrein-kinin system as mediator in vasogenic brain edema. Part 1: Cerebral exposure to bradykinin and plasma, *J. Neurosurg.* 6:87 (1984).

7. K. Maier-Hauff, M. Lange, L. Schürer and A. Unterberg, The kallikrein-kinin system as mediator in vasogenic brain edema. Part 2: Studies on kinin formation in focal and perifocal brain tissue, *J. Neurosurg.* 61:97 (1984).
8. E. Dux, P. Temesvari, P. Szerdahelyi, A. Nagy, J. Kovacs and F. Joo, Protective effect of antihistamines on cerebral oedema induced by experimental pneumothorax in newborn piglets, *Neurosci.* 24:317 (1987).
9. L. Germain, J. Barabe, and C. Galeano,Blood levels of kinins in experimental allergic encephalomyelitis, *J. Neuroimmunol.* 13:135 (1986).
10. G.N. Dietsch and D.J. Hinrichs, The role of mast cells in the elicitation of experimental allergic encephalomyelitis, *J. Immunol.* 142:1476 (1989).
11. J. Borg, J.M. Warter, J.L. Sclienger, M. Imler, C. Marescaux, and G. Mack, Neurotransmitter modifications in human cerebrospinal fluid and serum during hepatic encephalopathy, *J. Neurol. Sci.* 57:343 (1982).
12. L. Tuomisto, H. Kilpelainen, and P. Riekkinen, Histamine and histamine-N-methyltransferase in the CSF of patients with multiple sclerosis, *Agents and Actions* 13:255 (1983).
13. A.M. Butt and H.C. Jones, Effect of histamine and antagonists on electrical resistance across the blood-brain barrier in rat brain-surface microvessels, *Brain Res.* 569:100 (1992).
14. A.M. Butt, Effect of inflammatory agents on electrical resistance across the blood-brain barrier in pial microvessels of anaesthetized rats, *Brain Res.* in press (1995).
15. S.-P. Olesen, A calcium-dependent reversible permeability increase in microvessels in frog brain, induced by serotonin, *J. Physiol. (London)* 361:103 (1985).
16. S.-P. Olesen and C. Crone, Substances that rapidly augment ionic conductance of endothelium in cerebral venules, *Acta Physiol. scand.* 127:233 (1986).
17. S.-P. Olesen, Regulation of ion permeability in frog brain venules, significance of calcium, cyclic nucleotides and protein kinase C, *J. Physiol. (Lond.)* 387:59 (1987).
18. P.A. Revest, N.J. Abbott, and J.I. Gillespie, Receptor-mediated changes in intracellular [$Ca^{2+}$] in cultured rat brain capillary endothelial cells, *Brain Res.* 549:159 (1992).
19. C. Crone and S.-P. Olesen, Electrical resistance of brain microvascular endothelium, *Brain Res.* 241:49 (1982).
20. A.M. Butt, H.C. Jones, and N.J. Abbott, Electrical resistance across the blood-brain barrier in anaesthetized rats: a developmental study, *J. Physiol. (Lond.)* 429:47 (1990).
21. M. Bundgaard, Ultrastructure of frog cerebral and pial microvessels and their impermeability to lanthanum ions, *Brain Res.* 241:57 (1982).
22. Y.J. Le Beaux and J. Willemot, Actin-like filaments in the endothelial cells of adult rat brain capillaries, *Exp. Neurol.* 58:446 (1978).

# THE APPLICATION OF QUANTITATIVE IMMUNOCYTOCHEMISTRY FOR THE EVALUATION OF BLOOD-BRAIN BARRIER (BBB) TO ENDOGENOUS ALBUMIN

Andrzej W. Vorbrodt

Laboratory of Cytochemistry
New York State Institute for Basic Research
in Developmental Disabilities
Staten Island
New York 10314
USA

## INTRODUCTION

In morphological studies of the permeability of blood microvasculature, various protein tracers have been used. Among them, horseradish peroxidase (HRP) appeared to be most useful thanks to its absence in animal tissues, ease of handling, and excellent visibility both at light microscopy (LM) and electron microscopy (EM) levels. Since the pioneer work of Reese and Karnovsky (1967), this tracer has also been widely used in ultrastructural studies of the blood-brain barrier (BBB) in normal, pathological, and experimental conditions.

In spite of its wide utilization, HRP does not appear to be the best choice of a macromolecular tracer for animal experiments, because as a plant protein, it is a foreign, heterologous substance with a phylogenetically distant species origin, and it can be transported by the endothelial cells from various types of vasculature mainly by the unspecific, fluid-phase mechanism. It can also induce some undesirable reactions affecting the functions of the endothelial cells. It was shown that HRP molecules are characterized by higher osmotic activity than their molecular weight would suggest (Møllgård and Saunders, 1986), that they induce the release of some stress hormones that can affect the function of the endothelium and also that they increase vascular permeability (Cotran and Karnovsky, 1967; Straus, 1979). In addition, HRP gives all pitfalls related to the formation of the reaction product, such as relocation, secondary adsorption, slightly blurred or fuzzy outlines, and lack of quantitative correlation, with the density of antigen. Due to its high cost, HRP can be used in larger animals (dogs, apes, primates) only on a very limited scale. It is obvious that this tracer cannot be utilized for evaluation of the BBB function in human brain.

All these limitations and disadvantages of the HRP technique prompted scientists to look for more reliable methods that could be applied to this type of studies. Among various macromolecular solutes that eventually could be taken into consideration, the endogenous proteins appear to be the most adequate tracers. Albumin appears to be a prime choice among them because this protein not only is one of the normal components of blood plasma that does not affect the brain function but also normally serves as a carrier for many important and metabolically active substances (Ghitescu et al., 1986; Milici et al., 1987). Moreover, this protein is not transported by normal brain microvasculature, and consequently, it is perfectly suitable for studies of modified or disturbed BBB (Vorbrodt and Lossinsky, 1986; Vorbrodt et al., 1987, 1993).

The utilization of endogenous albumin rules out the possibility of eventual chemical modification of its molecules, but concomitantly, the introduction of any foreign or artificial macromolecular tracers into the blood circulation can be avoided. The previously reported findings from LM (Klatzo et al., 1980; Liu et al., 1989; Salahuddin et al., 1988; Broadwell and Sofroniew, 1993) and EM studies (Maeda et al., 1992; Guy et al., 1994) convincingly indicate that endogenous albumin can be utilized successfully as a natural macromolecular tracer in studies of normal and disturbed BBB.

The purpose of our studies was to adopt a new quantitative immuno-gold procedure and to determine its usefulness in the examination of the functional efficiency of the BBB to endogenous albumin in various physiological, experimental, and pathological conditions.

## METHODOLOGICAL PRINCIPLES

This new procedure was originally used by Ghitescu & Bendayan (1992) for comparative studies of transendothelial transport of endogenous and dinitrophenylated (DPN-tagged) albumin in mouse diaphragm capillaries. Because the endogenous albumin epitopes are sensitive to chemical fixation and particularly to osmium, the specimens should be fixed for approximately 1-1.5 h in a freshly prepared mixture of 2-2.5% formaldehyde (made from paraformaldehyde) and 1-1.5% glutaraldehyde in 0.1M phosphate buffer, pH 7.4. In situ fixation at room temperature (22°C) followed by immersion-fixation in refrigerator (4°C) was used to avoid washing out blood plasma. Small tissue blocks (1-2 $mm^3$) are carefully washed and dehydrated in ethanols with concomitant lowering of the temperature to -35°C. They were finally embedded in hydrophilic resin Lowicryl K4M and polymerized under UV lamp in the low-temperature embedding apparatus TTP 010 (Balzers).

Ultrathin sections attached to Formvar-carbon-coated nickel grids are placed in a moist chamber on a drop of anti-albumin antiserum in PBS (1:200) containing 0.5% ovalbumin and incubated in a refrigerator for 16 h. After careful washing with PBS, the sections are exposed to protein A-gold (10 nm) in PBS for 1h at room temperature. After washing, they are stained with 4% uranyl acetate (3 min) and Reynold's lead citrate (1 min). Controls for the specificity of the immunostaining consisted of sections incubated with protein A-gold solution only or the anti-albumin antiserum preabsorbed with 0.1% albumin.

The density of labelling over four compartments--the vascular lumen; endothelium; subendothelial (perivascular) space including basement membrane (BM); and adjacent tissue (neuropil)--is expressed as the number of gold particles (GP) per unit surface ($\mu m^2$). Morphometry of these compartments was performed with Sigma Scan Software Packages on a PC 486 DX equipped with Jandel JS-2 tablet. Only electron micrographs at the final magnification of 45,000 X were used. Mean values, standard deviation (SD), and standard error (SE) were calculated. The CSS program and non-parametric Mann and Whitney U-test were used for statistical evaluation of the results.

The results obtained by Ghitescu and Bendayan (1992) clearly indicated that the immunolabelling density (immunosignal) accurately represents the tracer concentration in particular compartments. Although these authors claimed relatively low immunodetection efficiency for endogenous albumin, this procedure appeared to be sufficiently sensitive for evaluating the actual functional state of the BBB. Consequently, we decided to use it in some physiological and experimental conditions presented below.

## EXAMPLES OF THE APPLICATION OF QUANTITATIVE IMMUNOCYTOCHEMICAL PROCEDURE FOR EVALUATION OF BBB FUNCTION

### BBB to Endogenous Albumin in Adult, Newborn, and Aged Mice

These studies were performed on Balb/c mice of both sexes divided into three groups: newborn; adult (6-month-old); and aged (2.5-year-old). The results of these studies are presented in detail in a separate, recently published paper (Vorbrodt and Dobrogowska, 1994).

In the cerebral cortex of adult mouse, the distribution of immunosignals for albumin appears to be typical, that is, the highest density is associated with blood plasma located inside the vessel lumen (Fig. 1 A). The mean value (153 GP/$\mu m^2$) is considered to represent 100% concentration of the circulating albumin. The labelling density of the endothelial cell (EC) profiles is much lower, representing only 4.7% of the labelling

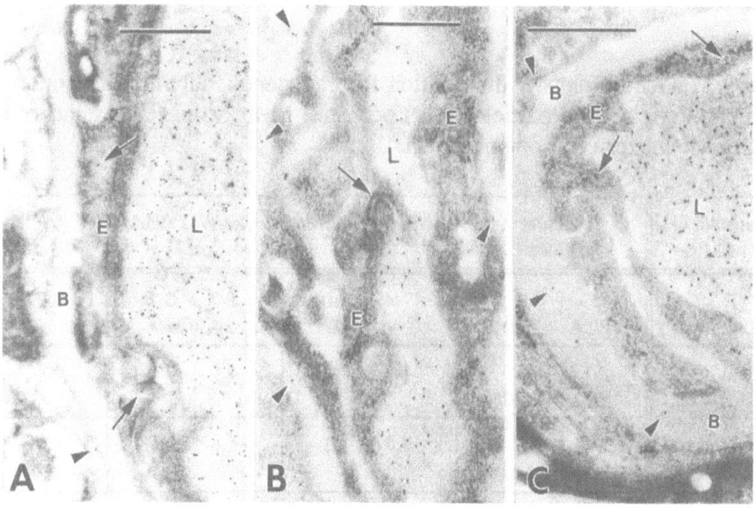


**Figure 1.** Immunolabelling of endogenous albumin in mouse brain capillaries. **A)** Adult mouse cerebral cortex. The highest density of labelling is associated with blood plasma located inside the vessel lumen (L), whereas only a few gold particles (arrows) appear inside the endothelial cell (E). Subendothelial space (B) is not labelled, and in the perivascular neuropil, only solitary gold particles are present (arrowhead). X 37,500. **B)** Newborn mouse brain. Immunosignals are present only inside the vessel lumen (L) and in the narrow perivascular space (arrowheads), where the basement membrane is not yet formed. The inter-endothelial junction (arrow) remains unlabelled. X 36,000. **C)** Aged mouse brain. The density of labelling is highest in the vascular lumen (L), although several gold particles also appear inside the endothelial cell (E) cytoplasm (arrows) and in the thickened (arrowheads) basement membrane (B). X 45,000. Bars = 0.5$\mu$m.

density of circulating albumin (Table 1). Even lower labelling density can be observed in subendothelial space, which is occupied mainly by distinct basement membrane (BM).

In many capillary profiles, the BM remains unlabelled (Fig. 1 A), whereas in others, only a few gold particles occasionally are present, yielding a mean value that corresponds with 1.6% of that found in the vessel lumen. In the adjacent neuropil, the labelling density is even lower, representing merely 0.5% of that found in the blood plasma. These data can be considered convincing evidence of the existence of a barrier to endogenous albumin in brain capillaries of adult mice.

The main features of the cerebral cortex capillaries of newborn mice are evidently the lower density of immunosignals for circulating blood plasma albumin (51 GP/$\mu m^2$) and the absence of distinguishable BM in the narrow perivascular space (Fig. 1 B). In spite of the low density of circulating albumin, the mean number of immunosignals in the perivascular space is higher than in adult brain capillaries. In the adjacent neuropil, however, the labelling density is evidently lower and only insignificantly different from that found in adult animals (Table 1).

Although the density of immunolabelling in aged mouse brain capillaries is essentially similar to that observed in adult animals, some characteristic deviations can be noted both morphologically and quantitatively. In aged mice, the most striking changes relate to the BM, which is evidently thickened in a large proportion of brain capillaries (Fig. 1 C). The labelling density of the endothelial profiles and of the BM in these vessels is also increased. These ultrastructural observations are supported by quantitative analysis of the distribution of immunolabelling density in all compartments examined (Table 1).

Our results indicate that the barrier function of brain microvasculature to endogenous albumin in newborn and aged mice is not significantly different from that in adult animals. Quantitation of these results, however, makes it possible to detect even subtle deviations from normal values in both extreme-age (newborn, aged) groups.

**Table 1.** Labelling density distribution for endogenous albumin in mouse cerebral cortex capillaries. Gold particles (GP) per $\mu m^2$. Mean values and standard errors (SE) are shown for each compartment.

| Experimental Group of Mice | Vascular Lumen | Endothelial Cells | Perivascular Space | Adjacent Neuropil |
|---|---|---|---|---|
| Adult<br>n = 20 | 153 ± 8.9<br>PCA = 100% | 7.4 ± 1.0<br>4.7% | 1.7 ± 0.2<br>1.6% | 0.7 ± 0.17<br>0.5% |
| Newborn<br>n = 12 | 51 ± 4.1<br>PCA = 100% | 1.3 ± 0.09<br>2.5% | 4.6 ± 0.6<br>8.9% | 0.4 ± 0.03<br>0.7% |
| Aged<br>n = 22 | 142 ± 9.2<br>PCA = 100% | 8.7 ± 1.1<br>6.1% | 4.6 ± 0.6<br>3.2% | 1.4 ± 0.20<br>1.0% |

PCA - percentage of circulating albumin; the label of the vessel lumen was considered to represent 100% of the circulating albumin. n = number of samples examined.

## Osmotic Opening of the BBB in Rats

These experiments were performed on Wistar rats (6 months old) of both sexes divided into two groups: one group (experimental) received intracarotid infusion of 1.8M L(+)arabinose in Ringer's solution (30 sec infusion at a rate of 0.12 ml/sec), and the

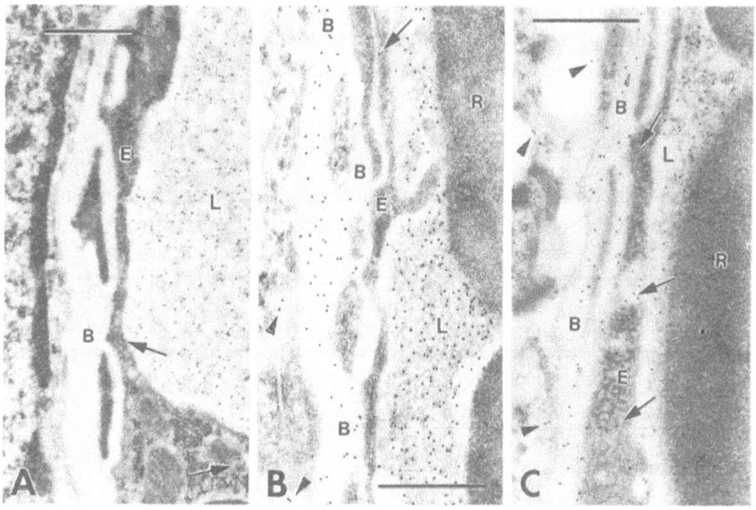


**Figure 2.** Immunocytochemical detection of endogenous albumin distribution in rat cerebral cortex capillaries. **A)** Control rat. Immunosignals are highly concentrated inside the vascular lumen (L), whereas only solitary gold particles appear inside the endothelial cell (E) cytoplasm (arrows). The basement membrane (B) and adjacent neuropil remain unlabelled. X 37,500. **B)** Five minutes after intracarotid infusion of hyperosmolar L(+)arabinose, the immunosignals appear in interendothelial clefts (arrow), shrunken endothelial cells (E), swollen basement membrane (B) and adjacent brain parenchyma (arrowheads). R, red blood cells. X 45,000. **C)** In this capillary, examined 30 min after intracarotid infusion of L(+)arabinose, numerous gold particles are scattered throughout the cytoplasm (arrows) of the endothelial cell (E) and also appear in subendothelial space (B) and adjacent neuropil (arrowheads). X 45,000. Bars = 0.5 $\mu$m.

control group received a similar injection of Ringer's solution only. Rats were sacrificed by decapitation at 5 and 30 min after infusion was completed.

In control rats, the distribution of immunosignals for endogenous albumin is essentially similar to that observed in adult mice, that is, the highest density is associated with blood plasma located inside the vessel lumen (Fig. 2 A).

Quantitative analysis indicates that the labelling density of the perivascular space, essentially the BM, represents barely 0.4% of that found for circulating blood plasma albumin. In the neuropil, this value is even lower, amounting to 0.3%. These values indicate normal function of the BBB to endogenous albumin.

In contrast, as soon as 5 min after infusion of hyperosmolar L(+)arabinose solution, drastic changes in the distribution of immunosignals for endogenous albumin occur in capillaries located in the ipsilateral hemisphere. These changes consist of shrinkage and attenuation of the endothelial lining, the continuity of which is disrupted in many vessels. In addition, in some capillaries, deep invaginations and dilatation (widening) of interendothelial clefts appear. Because numerous gold particles appear in these clefts and cytoplasmic invaginations, one can conclude that they represent main routes for escaping albumin, which ultimately appears in the slightly swollen BM (Fig. 2 B).

At 30 min after L(+)arabinose infusion, the appearance of many ECs is almost normal, suggesting the repair process of the injured endothelial lining. Nevertheless, in some vessels, an uncontrolled passage of albumin manifested by "inundation" of the cytoplasm of injured ECs with gold particles can be observed (Fig. 2 C). At this time, more numerous gold particles also appear in the BM and in the adjacent neuropil. These morphological observations are supplemented with quantitative data shown in Table 2.

A slow increase of the labelling density in the adjacent neuropil, occurring at 30 min after L(+)arabinose infusion, suggests that the BM represents an obstacle for escaping albumin, which apparently is trapped in this structure. These results also indicate that the applied procedure, although based on morphometric analysis of static pictures, can be used for studying dynamic processes, such as transvascular passage of endogenous albumin after disruption of the BBB.

**Table 2.** The effect of intracarotid infusion of hyperosmolar L(+)arabinose on the labelling density distribution for endogenous albumin in rat brain capillaries.

| Experimental Group of Rats | Vascular Lumen | Endothelial Cells | Perivascular Space | Adjacent Neuropil |
|---|---|---|---|---|
| Control<br>n = 16 | 148 ± 6.90<br>PCA = 100% | 4.8 ± 0.40<br>3.2% | 0.6 ± 0.15<br>0.4% | 0.4 ± 0.05<br>0.3% |
| 1.8M L(+)arabinose:<br>5 min.<br>n = 20 | 216 ± 12.40<br>PCA = 100% | 13.7 ± 1.10<br>6.3% | 55.2 ± 9.10<br>25.5% | 1.6 ± 0.20<br>0.7% |
| 30 min.<br>n = 20 | 169 ± 7.13<br>PCA = 100% | 16.4 ± 1.46<br>9.7% | 95.1 ± 4.42<br>56.2% | 7.4 ± 0.69<br>4.3% |

## Human Brain Samples

**Autopsy Specimens.** These specimens were obtained from four cases of accidents or heart failure. All samples of the cerebral cortex, although obtained as early as 6 h after death, show rather similar changes such as coagulation or scarcity of blood plasma inside the vessel lumen, blurred outlines of the ECs, and swelling of the BM and adjacent perivascular astrocytic endfeet.

The density of immunolabelling of the vessel lumen is generally low and irregular, whereas numerous immunosignals appearing inside the EC cytoplasm suggest uncontrolled passage of albumin to the perivascular space, where it presumably gets stuck to the slightly swollen BM. Perivascular structures, mainly swollen astrocytic endfeet, are also labelled with numerous gold particles (Fig. 3 A). These observations indicate that autopsy specimens are inappropriate for evaluating the BBB because it is impossible to distinguish post-mortem changes from changes resulting from previous illness of the living organism.

**Biopsy Specimens.** Small portions of specimens of cerebral cortex obtained during surgical procedures on three patients with brain tumors used previously for our cytochemical studies (Wisniewski et al., 1991; Vorbrodt et al., 1992) were utilized for the present immunocytochemical trials. This material appears to be as suitable for immunocytochemical examination as that obtained from laboratory animals. After incubation of Lowicryl K4M-embedded sections with anti-human albumin antiserum, the distribution of immunosignals is essentially similar to that observed in control mice and rats (see above). Some capillaries show the typical distribution of immunosignals, that is the highest density over the vessel lumen, few immunosignals inside the EC cytoplasm, and essentially no labelling of the BM and adjacent neuropil (Fig. 3 B).

Such a distribution pattern of immunolabelling suggests the normal barrier function of the vessel.

In close vicinity to the tumor (meningioma), however, numerous capillaries show signs of BBB disruption, as evidenced by an increased number of gold particles inside the cytoplasm and cytoplasmic vesicles of the ECs and in the BM (Fig. 3 C). Numerous gold

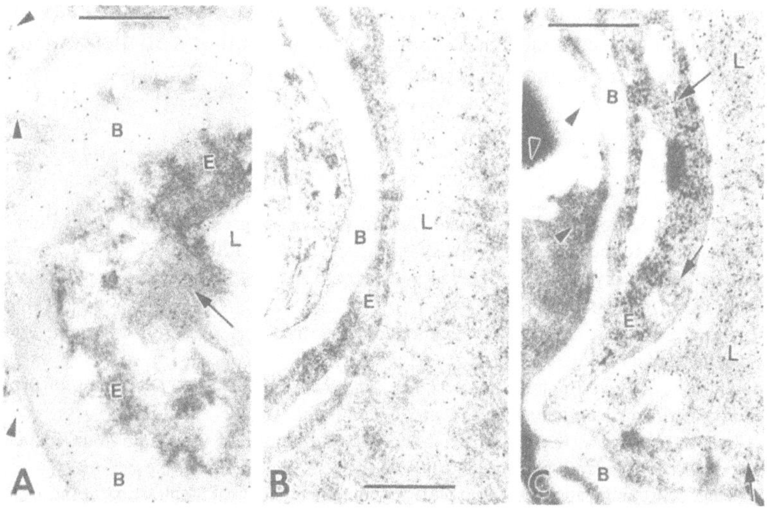


**Figure 3.** Immunodetection of endogenous albumin in human cerebral cortex. **A)** Autopsy specimen. In the vessel lumen (L), the distribution of labelling is uneven, frequently associated with coagulated blood plasma (arrow). The outlines of the endothelial cell (E) are blurred, and their cytoplasm is irregularly labelled with a few gold particles. Swollen basement membrane (B) is labelled with numerous gold particles, which also appear in the adjacent neuropil (arrowheads). X 37,500. **B)** Biopsy specimen. In the capillary, the labelling is restricted solely to the vessel lumen (L). X 37,500. **C)** Biopsy specimen showing the capillary located in close vicinity to the neoplasm (meningioma). Immunolabelling for albumin appears not only in the vessel lumen (L) but also inside the endothelial cell (E) cytoplasm (arrows), in the basement membrane (B), and in the adjacent parenchyma (arrowheads). X 37,500. Bars = 0.5 $\mu$m.

particles also appear in swollen perivascular structures of the neuropil. These observations indicate that biopsy specimens can be used for immunocytochemical evaluation of the BBB to endogenous albumin in human brain.

## CONCLUDING REMARKS

This quantitative immunocytochemical procedure utilizes endogenous blood plasma albumin as a natural protein tracer. Consequently, it appears to be a valuable tool that can help us study the functional state of the BBB at the ultrastructural level without introducing any foreign or artificial macromolecular substances into the blood circulation. The application of this postembedding technique also makes possible for the first time evaluation by the morphological method of the functional state of the BBB in human brain.

Although the method of fixation and embedding, with the omission of osmium tetroxide, is not optimal for demonstrating of the cell membranous structures that are associated with endocytosis and transcytosis, it does make it possible to observe various routes of transendothelial or transvascular passage of endogenous blood plasma albumin after the BBB disruption. The significant advantages of the immunogold procedure used in our study are its relatively high efficiency, the excellent visibility it provides of immunosignals, and its usefulness for quantitative evaluation and statistical analysis.

**Acknowledgments.** The valuable contribution of Ms. Danuta H. Dobrogowska in immunocytochemical and electron microscopical examination of brain samples and in preparation of the final electron micrographs is gratefully acknowledged. The author is deeply indepted to Ms. Maureen Stoddard Marlow for editorial revisions and to Ms. Janis Kay for skillful secretarial assistance. Support for the author's studies was provided by the National Institute on Aging/NIH grant RO1 - AG 10279-03.

## REFERENCES


Broadwell, R.D., and Sofroniew, M.V., 1993, Serum proteins bypass the blood-brain fluid barriers for extracellular entry to the central nervous system, *Exp. Neurol.* 120:245.

Cotran, R.S., and Karnovsky, M.J., 1967, Vascular leakage induced by horseradish peroxidase in the rat, *Proc. Soc. Exp. Biol. Med.* 126:557.

Ghitescu, L., and Bendayan, M., 1992, Transendothelial transport of serum albumin: A quantitative immunocytochemical study, *J. Cell Biol.* 117:745.

Guy, J., Mc Gorray, S., Fitzsimmons, J., Beck, B., and Rao, N.A., 1994, Disruption of the blood-brain barrier in experimental optic neuritis, *Invest. Ophthalmol. Vis. Sci.* 35:1114.

Klatzo, I., Chui, E., Fujiwara, K., and Spatz, M., 1980, Resolution of vasogenic brain edema, *in*: "Brain Edema", J. Cervós-Navarro and R. Ferszt, eds., p. 359, Raven Press, New York.

Liu, H.M., Atack, J.R., and Rapoport, S.I., 1989, Immunohistochemical localization of intracellular plasma proteins in the human central nervous system, *Acta Neuropathol.* 78:16.

Maeda, M., Akai, F., Nishida, S., and Yanagihara, T., 1992, Intracerebral distribution of albumin after transient cerebral ischemia: light and electron microscopic immunocytochemical investigation. *Acta Neuropathol.* 84:59.

Milici, A.J., Watrous, N.E., Stukenbrok, H., and Palade, G.E., 1987, Transcytosis of albumin in capillary endothelium, *J. Cell Biol.* 105:2603.

Møllgård, K., and Saunders, N.R., 1986, The development of the human blood-brain and blood-CSF barriers, *Neuropathol. Appl. Neurobiol.* 12:337.

Reese, T.S., and Karnovsky, M.J., 1967, Fine structural localization of a blood-brain barrier to exogenous peroxidase, *J. Cell Biol.* 34:207.

Salahuddin, T.S., Kalimo, H., Johansson, B.B., and Olsson, Y.,1988, Observations on exudation of fibronectin, fibrinogen and albumin in the brain after carotid infusion of hyperosmolar solutions, *Acta Neuropathol.* 76:1.

Straus, W., 1979, Peroxidase procedure: technical problems encountered during their application, *J. Histochem. Cytochem.* 27:1349.

Wisniewski, H.M., Vorbrodt, A.W., and Epstein, M.H., 1991, Nucleoside diphosphatase (NDPase) activity associated with human β-protein amyloid fibers, *Acta Neuropathol.* 81:366.

Vorbrodt, A.W., and Lossinsky, A.S., 1986, Transport of homologous albumin through various capillary endothelia in mice, *J. Cell Biol.* 103:193a.

Vorbrodt, A.W., Lossinsky, A.S., and Wisniewski, H.M., 1987, Ultrastructural studies of the uptake and transport of homologous albumin by brain endothelia, *Canad. J. Neurol. Sci.* 14:342.

Vorbrodt, A.W., Wegiel, J., and Wisniewski, H.M., 1992, Cytochemical evidence of blood-brain barrier involvement in aging, scrapie and Alzheimer disease, *Neurobiol. Aging* 13, Suppl 1:S86.

Vorbrodt, A.W., Lossinsky, A.S., Dobrogowska, D.H., and Wisniewski H.M., 1993, Cellular mechanisms of the blood-brain barrier (BBB) opening to albumin-gold complex, *Histol. Histopathol.* 8:51.

Vorbrodt, A.W., and Dobrogowska, D.H., 1994, Immunocytochemical evaluation of blood-brain barrier to endogenous albumin in adult, newborn and aged mice, *Folia Histochem. Cytobiol.* 32:63.

# ENDOTHELIN AS A MEDIATOR OF BLOOD–BRAIN BARRIER FUNCTION

Maria Spatz, Danica Stanimirovic, and Richard M. McCarron

Stroke Branch
National Institute of Neurological Disorders and Stroke
National Institutes of Health
Bethesda, MD 20892 USA

## INTRODUCTION

Brain capillary endothelial cells (BCEC), the principal components of the blood–brain barrier (BBB) are interconnected by tight junctions, possess polar membranes and a paucity of pinocytotic vesicles in contrast to capillaries in peripheral organs.[1] BCEC's unique features embrace specialized transport and carrier systems as well as enzymes associated with various metabolic pathways that are essential to sustain the dynamic homeostasis of the brain. Many of these processes are thought to be neuronally regulated. There are a number of excellent reviews regarding the specific function of the BBB and regulation of the cerebral circulation by endothelium.[2-4] It is now widely accepted that BCEC not only constitute a permeability barrier to ions and organic molecules (e.g., water, electrolytes, proteins, neurotransmitters), but are also an important secretory organ. BCEC produce agents and factors that may be involved in autocrine and paracrine regulation of the microvascular function of the brain. In general, the substances [e.g., prostacyclin, nitric oxide (NO), and adenosine] produced by these cells are considered to be cytoprotective. However, several other agents which are formed in BCEC and when released in excess may impair perfusion, alter BBB permeability and/or mediate cellular injury (i.e., endothelin I, angiotensin II, thromboxane, leukotrienes, platelet–activating factor, and superoxide radicals). The secretory function of BCEC can also be modulated by substances released from adjacent cellular elements (other vascular cells, blood and/or brain). There is ample evidence that many of these events play a role in regulating vascular tone, permeability, hemostasis and blood pressure under normal conditions. However, in some pathologic conditions an "oversupply" of some of these agents may adversely affect the BBB permeability and/or surface expression of adhesion molecules (i.e., ICAM–1, VCAM–1, P–selectin) which facilitate binding and ultimately extravasation of circulating leukocytes.

This report will be limited to endothelin 1 (ET–1) as a mediator of BBB function. The summarized review focuses on previously published[5-12] and unpublished studies concerned with ET−1 and human brain endothelial cell (HBEC) responses, namely: (1) secretion; (2) receptors; (3) signal transduction mechanisms; (4) $Ca^{2+}$ uptake; (5) ATPase system; (6) cellular permeability; and (7) expression of surface adhesion molecules and monocytes binding.

## OVERVIEW

### Endothelins

ET−1 is a member of the mammalian endothelin family (ET−1, ET−2, and ET−3), and consists of 21 amino acids with disulphide bonds. The endothelins are produced from prepeptides (200 amino acids) and proendothelin (38 or 39 amino acids) that are cleaved by endothelin−converting enzyme.[13−15] ET−1 formation is regulated by mRNA transcriptase which can be induced by a variety of substances [e.g., hormones, vasoactive agents, cytokines, $Ca^{2+}$, activator of protein kinase C (phorbol esters), thrombin]. ET−1, which originally was isolated and purified from the medium of cultured porcine aortic endothelium, is known as the most powerful vasoconstrictive peptide. This peptide can be secreted by BCEC as well as other cells (e.g., vascular smooth muscle, glia, leukocytes, macrophages).[5,16−19] ET−1 has also been demonstrated in neurons by immunocytochemistry. Due to its unique properties, ET−1 has been considered to be a strong candidate to participate in the pathogenesis of cerebrovascular disorders.

The findings described here were mainly obtained from studies performed on cultured endothelium dissociated from isolated capillaries and/or microvessels of human brain. The purity of the endothelial cultures was >98%, as assessed by positive immunostaining for von Willebrand (Factor VIII−related) antigen and incorporation of acetylated low−density lipoprotein (Dil−Ac LDL) and negative staining for glial fibrillary acidic protein (astrocytes) and ED2 (microglia). Morphologic and biochemical [absent uptake of horseradish peroxidase, presence of alkaline phosphatase, monoamine oxidase, angiotensin−converting enzyme, expression of dopaminergic, cholinergic, adrenergic, and endothelin receptors ($ET_A$), and secretion of ET−1 and prostanoids (ref. 6 and unpublished observations)] characteristics of these cells were preserved in propagated cell cultures for >14 passages.

In addition to HBEC, endothelium derived from human umbilical vein (HUVEC) was used sometimes for comparative studies described in this chapter.

Methods used for culturing BCEC, determining ET−1, prostanoids, phosphoinositol metabolism, arachidonic acid (AA) release, $Ca^{2+}$ uptake, $Na^+$, $K^+$−ATPase activity, cAMP, cGMP, cellular "permeability" (release of $^{51}Cr$ and lactic dehydrogenase activity), expression of cellular adhesion molecules by fluorescence microscopy, FACS analysis, and ELISA are described elsewhere.[5−12]

## RESULTS AND DISCUSSION

### Peptidergic Induction of ET−1, Prostanoid Secretion, and Cellular Signal Transduction Mechanisms

ET−1 is constitutively produced by HBEC. The secretion of ET−1 by HBEC can also be stimulated by vasoconstrictive peptides, arginine vasopressin (AVP) or angiotensin II (Ang II) among other substances [e.g., thrombin, cytokines, $Ca^{2+}$, phorbol esters, activator of protein kinase C (PKC)]. Both of these peptides dose−dependently induced ET−1 secretion. AVP was a more potent activator than Ang II (Fig. 1). A concomitant stimulation of prostanoid production was seen in the medium of HBEC (which constitutively secrete prostanoids) exposed to these peptides. The temporal profile of AVP−induced secretion of ET−1 and prostanoids differed from those stimulated by Ang II in these cells. The maximal rate of AVP−induced ET−1 production coincided with the maximum rates of $PGF_{2\alpha}$ and thromboxane $B_2$ [$TxB_2$, the stable metabolite of $TxA_2$)] secretion (over 0−4 hr) and preceded that of $PGD_2$ and $PGE_2$ secretion (during the period of 0−8 hr). On the other hand, the maximal rate of Ang II−induced ET−1 secretion occurred during the same time period as that of $PGD_2$ (0−4 hr) and preceded that of $PGF_{2\alpha}$, $PGE_2$, and 6−keto−$PGF_{1\alpha}$ (the stable metabolite of $PGI_2$ whose production was unaffected by AVP). Both the AVP− and Ang II−induced secretion of ET−1 and prostanoids was inhibited by their respective specific receptor antagonists [1−(β−mercapto−β,β−cyclopententamethylenpropionic acid)−2−)O−methyl tyrosine] $Arg^8$−vasopressin [(POMT) $V_1$−receptor blocker] and Ang II [($Sar^1$, $Ala^8$), inhibitor of Ang II receptors]. Dexamethasone [(Dxm), the inhibitor of phospholipase $A_2$ ($PLA_2$)/cyclooxygenase II] increased the constitutive ET−1 secretion (not shown) and AVP− or Ang II−stimulated production of ET−1 and decreased formation of $PGF_{2\alpha}$ and $PGD_2$ (Fig. 1) as well as other prostanoids (not shown). Indomethacin (Indo) and nordihydroguaiaretic acid (NDGA), the respective inhibitors of cyclooxygenase and lipoxygenase, affected the AVP− and Ang II−induced secretion of ET−1 and prostanoids in a manner similar to that observed with Dxm. Moreover, an increase of ET−1 and reduction of prostanoid production was seen with nitro−L−arginine, the inhibitor of nitric oxide synthase (NOS) (unpublished results).

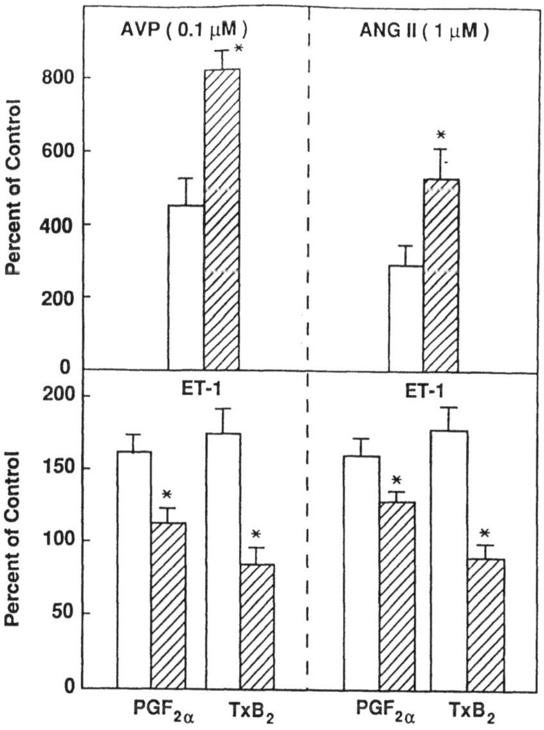


**Figure 1.** Effects of Dxm on AVP− or Ang II−stimulated production of $PGF_{2\alpha}$, $TxB_2$, and ET−1 in HBEC. Results are presented as percentage of control rates. Each bar represents mean ±S.E. of 3−5 independent experiments each performed in triplicate. *Significant difference (ANOVA, $P<0.001$) compared with AVP or Ang alone.

The above findings indicate that HBEC contain $V_1$ and Ang II receptors which mediate the augmented secretion of ET−1 and prostanoids since the AVP and Ang II responses were blocked by their respective specific antagonists. The results also suggest that the ET−1 and vasoconstrictive prostanoids could temporally contribute to the AVP vasopressor activity since the maximal AVP−induced rate of secretion of these substances occurs prior to that of the vasodilatory prostanoids. Modulation of constitutive and peptidergic production of ET−1 and prostanoids by the inhibitors of cyclooxygenase (Dxm and Indo) or lipoxygenase (NDGA) strongly suggest a close interrelationship between ET−1 and prostanoid secretion by HBEC. Furthermore, the observed increased ET−1 production and reduced secretion of prostanoids by inhibiting NOS activity suggest an interaction between NO and ET−1 as well as prostanoids in HBEC.

Both AVP and Ang II also stimulated $IP_3$ formation by HBEC. The maximal effective concentrations ($EC_{50}$) for $IP_3$ production for AVP was lower ($87\pm5$ nM) than that for Ang II ($970\pm17$ nM). AVP and Ang II also increased the release of arachidonic acid [(AA), Fig. 2]. Both the $IP_3$ production and AA release induced by AVP or Ang II were inhibited by their respective specific antagonists (Fig. 2). Pertussis toxin (Ptx) had no effect on either basal or AVP− or Ang II−stimulated production of $IP_3$ and AA release, indicating that Gi protein is not involved in the receptor−mediated signal transduction. Addition of phorbol myristate ester (PMA) to HBEC decreased (about 80%) the formation of $IP_3$ and had little effect of AA release. The ability of PMA to partially inhibit AVP− or Ang II−stimulated $IP_3$ formation without affecting AVP− or Ang II−induced AA release in HBEC indicates that AVP and Ang II receptor linkage to PKC and $PLA_2$ activation might be regulated by independent intracellular processes. Verapamil, the voltage−dependent calcium channel blocker, reduced basal production of $IP_3$ and partially

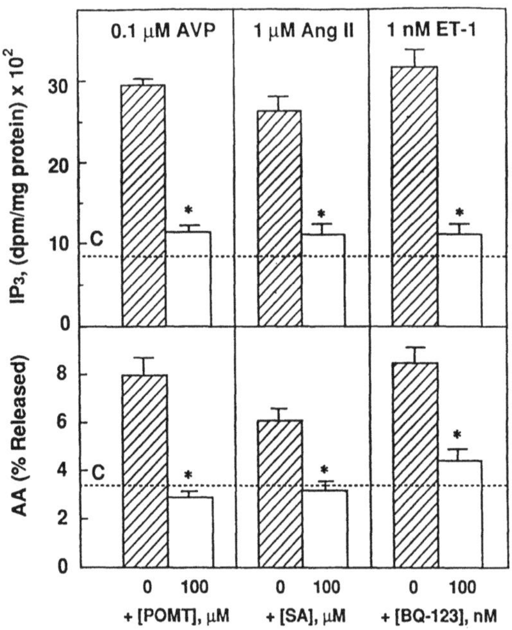


**Figure 2.** Effects of selective $V_1$, Ang II or $ET_A$ receptor antagonists (POMT, SA, and BQ−123, respectively) on AVP−, Ang II−, or ET−1−induced secretion of $IP_3$ formation and AA release in HBEC. Each bar represents mean ±S.E. data obtained from 3 separate dishes from one representative experiment of 3 with similar results. *Significant decrease (ANOVA, $P<0.01$) compared with AVP, Ang II, or ET−1 alone.

inhibited (20%) the Ang II−stimulated but not the AVP−induced formation of $IP_3$. Verapamil also decreased (80%) AA release induced by either of these two peptides. The observed partial decrease of AA release suggests that extracellular $Ca^{2+}$−activated $PLA_2$ participates in prostanoid production.

Thus, the above−described interaction between the AVP− or Ang II−induced ET−1 and prostanoid production activated through the same receptors ($V_1$ and Ang II, respectively) might be modulated by independently regulated signal transduction mechanisms.

### Endothelin−1 Receptor Binding and Cellular Signal Transduction Mechanisms

Saturable binding of [$^{125}$I]−ET−1 to either intact HBEC (Fig. 3, upper panel) or HBEC membranes was observed with increasing concentrations (5−500 pM) of the labeled substance. Scatchard analysis revealed a single class of high−affinity binding sites. The apparent $K_D$ was 68 pM whereas the maximal binding capacity ($B_{max}$) was 124 fmol/mg protein (Fig. 2, lower panel insert). However, the Scatchard plot obtained by transformation of homologous displacement (ET−1/ET−1) data clearly indicated two binding sites [a high−affinity ($K_D$=122 pM and $B_{max}$=112 fmol/mg protein) and a low−affinity ($K_D$=32 nM and $B_{max}$ =909 fmol/mg protein)]. The order of potency for displacement of ET−1 from high−affinity binding sites was

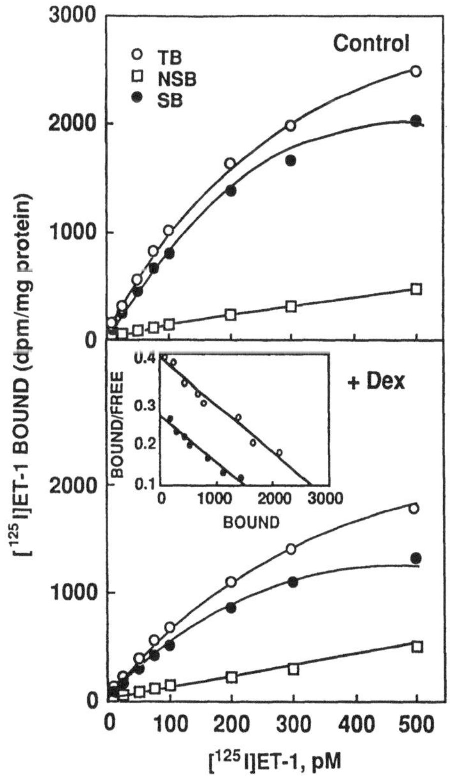


**Figure 3.** [$^{125}$I]ET−1 saturation binding to cultured HBEC. Upper panel: cells were incubated with various concentrations of [$^{125}$I]−ET−1 for 120 min at 37°C. Nonspecific binding was determined in the presence of 1 μM unlabeled ET−1. The data shown represent results obtained from one of three experiments, each performed in triplicate, with similar results. Lower panel: HBEC were incubated with 1 μM Dxm for 24 hr prior to binding experiments, performed as described for upper panel. (Scatchard analysis of saturation binding data of untreated (o) and Dxm−pretreated (•) cells).

ET−1=ET−2>S6b>ET−3. Similar kinetic data were observed with [$^{125}$I]−ET−1 binding to HBEC membranes. The presence of two types of ET receptors was also described on rat BCEC and ET−1 and/or ET−3 receptors on other cells.[7,16–18,20]

Pretreatment of intact HBEC with Dxm decreased the number of ET−1 binding sites without significantly changing its affinity (Fig. 2, lower panel). Competition binding experiments showed a similar displacement profile (ET−1=ET−2>ET−3> S6c) in Dxm−pretreated as in untreated cells.

ET−1−induced dose−dependent intracellular $IP_3$ formation ($EC_{50}$=0.79 ±0.10 nM) and release of AA ($EC_{50}$=59±7 nM) from HBEC. Ptx had no effect on ET−1−stimulated production of $IP_3$ indicating that the inhibitory G protein does not participate in activating the PLC system. The ET−1−stimulated $IP_3$ production and AA release were inhibited by BQ−123, the selective $ET_A$ receptor antagonist (Fig. 3, lower panel). These observations are indicative of $ET_A$ receptor−mediated activation of ET−1−induced $IP_3$ production and AA release. PMA decreased the ET−1−stimulated $IP_3$ production (50%) (without changing basal $IP_3$ levels) and augmented AA release (2.3−fold) from HBEC. Verapamil and neomycin (PLC inhibitor) slightly reduced $IP_3$ production induced by ET−1 in HBEC. Verapamil also slightly decreased (15%) ET−1−stimulated release from these cells.

The $ET_A$−mediated ET−1 stimulation of $IP_3$ formation and increased AA release from HBEC indicate linkage of $ET_A$ receptors to PLC and $PLA_2$ activation. However, ET−1−induced $IP_3$ accumulation and AA release appear to be largely independent of each other since the maximal rate of $IP_3$ formation was detected in the first minute whereas the maximal rate of AA release was seen between 5 and 15 min after addition of ET−1. Moreover, the observed dissociation of ET−1−induced $IP_3$ formation (inhibition) and AA release (stimulation) by PMA, the PKC activator in HBEC, indicates that PLC and $PLA_2$ are activated by different mechanisms. Attenuation of AA release and prostanoid production by verapamil demonstrate that the extracellular component of $Ca^{2+}$ in addition to the $IP_3$−mediated $Ca^{2+}$ mobilization may contribute to $PLA_2$ activation.

The same Dxm pretreatment of HBEC which decreased the number of ET−1 binding sites on these cells also reduced the maximal effect of ET−1−stimulated $IP_3$ formation (40% decrease of $IP_3$ induced by ET−1 at saturating condition of 10 nM). However, the half−maximal effective concentration of ET−1 to induce a response was unchanged by Dxm. The same treatment with 6−hydrocortisone, another glucocorticoid, had a similar effect but decreased to a lesser degree the ET−1−stimulated $IP_3$ production, whereas aldosterone, a mineralocorticoid, Indo and acetylsalicylic acid (ASA) were ineffective. Cortexolone, the inhibitor of intracellular glucocorticoid receptors, reversed the inhibitory effect of Dxm on ET−1−stimulated formation of $IP_3$ when added to the HBEC simultaneously with Dxm. These results indicate that this effect is mediated through a glucocorticoid receptor−specific pathway.

The glucocorticoids (Dxm and 6−hydroxycortisone) reduction of the maximal stimulation of $IP_3$ formation that occurred in parallel with the above−mentioned Dxm induced decrease in the number of ET−1 ($ET_A$) receptors strongly suggest an interaction between endothelin and glucocorticoids that occur *in vivo* at the level of the BBB.

The physiological and pathophysiological significance of the glucocorticoid interaction with ET−1 on HBEC is unknown. Glucocorticoids or Dxm were reported to enhance the properties of an *in vitro* model of BBB, inhibit microvascular morphogenesis, and ET−1−induced release of prostanoids. They were also shown clinically to decrease vasogenic brain edema and ameliorate the development of some inflammatory disorders [i.e., multiple sclerosis and experimental allergic encephalomyelitis (EAE)]. Therefore, it is possible that glucocorticoids may suppress changes in HBEC "permeability", adhesion molecule expression, and adhesiveness induced by ET−1, as described in the subsequent sections, which are implicated in the development of cerebrovascular and/or inflammatory disorders.

**Effects of Endothelin on HBEC Induction of Prostanoids, cAMP, and cGMP**

Exposure of HBEC to ET−1 dose−dependently stimulated production of prostanoids ($TxB_2$, $PGF_{2\alpha}$, 6−keto $PGF_{1\alpha}$, $PGE_2$, and $PGD_2$). The increased secretion of the vasoconstrictive prostanoids $TxA_2$ and $PGF_{2\alpha}$ temporally preceded (at 15 min and 4 hr) that of the vasodilatory prostanoids $PGI_2$, $PGE_2$, and $PGD_2$ (between 4 and 8 hr of incubation). Normalization of both types of prostanoids was observed 24 hr after addition of ET−1. ET−1 also stimulated cAMP and decreased cGMP production in these cells. In contrast, ET−3 had no effect on the HBEC production of prostanoids or cAMP but stimulated cGMP formation (unpublished observation). ET−1 stimulated−prostanoids were inhibited by Dxm and reduced by neomycin (PLC inhibitor) and verapamil but not nifedipine. PMA (activator of PKC) augmented the ET−1−induced prostanoid production whereas it inhibited IP production. These observations are consistent with the above−described findings that ET−1 activates $PLA_2$ and PLC by different intracellular mechanisms. The selective antagonist of $ET_A$ receptors (BQ−123) inhibited both ET−1−induced prostanoid and cAMP formation. Thus, inhibition of the ET−1−induced formation of prostanoids and cAMP by BQ−123 indicates that both events are mediated by $ET_A$ receptors on HBEC.

**Endothelin Effect on $^{45}Ca^{2+}$ Uptake into Rat BCEC**

A specific receptor−mediated rise in intracellular $Ca^{2+}$ in rat BCEC was demonstrated after exposure to either bradykinin, histamine, or ATP, substances which reportedly increase BBB permeability.[21] Since many endothelial responses to endothelins may be activated or modulated by either intracellular $Ca^{2+}$ mobilization and/or calcium influx from extracellular sources, we studied the effect of ET−1 or ET−3 on $^{45}Ca^{2+}$ uptake in BCEC (unpublished observations). Both ET−1 and ET−3 stimulated uptake of $^{45}Ca^{2+}$ into rat BCEC in a concentration−dependent manner. Estimated half−maximal stimulating concentration for ET−1−induced $^{45}Ca^{2+}$ uptake ($EC_{50}$=10 nM) was similar to ET−3 ($EC_{50}$=9 nM). However, a 30−40% increase of $^{45}Ca^{2+}$ uptake was observed in BCEC using S6c, a selective $ET_B$ receptor agonist. The selective antagonist of the $ET_A$ type of endothelin receptor, BQ−123, completely inhibited the ET−1−induced but only partially reduced the ET−3−stimulated $^{45}Ca^{2+}$ uptake into BCEC. Attempts were made to identify the possible pathways responsible for carrying the ET−1 or ET−3−stimulated calcium into BCEC by using various known calcium channel inhibitors. Both ET−1 and ET−3 stimulated BCEC uptake of $^{45}Ca^{2+}$ were insensitive to nifedipine and nitrendipine (highly selective blockers of L−type voltage−dependent calcium channels) whereas they were abolished by the receptor−operated channel blocker SK&F96365. In contrast, only ET−3−induced $^{45}Ca^{2+}$ uptake was inhibited by nickel (a *bone fide* receptor−mediated calcium channel blocker) and suramin (a blocker of purinergic $P_2$−associated channels). Ryanodine, an inhibitor of intracellular $Ca^{2+}$ mobilization, selectively attenuated ET−1−stimulated $^{45}Ca^{2+}$ uptake. This observation together with the above−described inhibition of ET−induced $IP_3$ formation by the selective $ET_A$ antagonist (BQ−123) indicates that the ET−1−induced $^{45}Ca^{2+}$ uptake into BCEC occurs secondary to depletion of intracellular calcium stores. The partial inhibition of ET−3−induced $^{45}Ca^{2+}$ uptake by BQ−123 suggests that the ET−3−stimulated uptake of $^{45}Ca^{2+}$ is mediated by $ET_A$ and/or $ET_B$ receptors. ET−3 activation of the $ET_B$ receptor is supported by the observed moderate induction of $^{45}Ca^{2+}$ uptake by the selective $ET_B$ receptor agonist, S6c , which is not coupled to the PLC system.

## Endothelin Effect on ATPase Activity on BCEC

Capillary and microvascular $Na^+$, $K^+$–ATPase activity have been implicated in maintaining water–electrolyte homeostasis in the brain. Various $Ca^{2+}$ mobilizing peptides and hormones have been shown to activate $Na^+$,$K^+$–ATPase in various tissues. Endothelin activation of $Na^+/H^+$ exchange was demonstrated to be mediated by $ET_B$ receptors via a PKC–independent mechanism in rat BCEC.[22] Our latest studies (unpublished, presented by Kawai et al., 1994) were concerned with elucidating the effects of endothelins (ET–1 or ET–3) on $^{86}Rb^+$ uptake as a measure of $Na^+$, $K^+$–ATPase activity in rat BCEC. Both ET–1 and ET–3 dose–dependently enhanced $^{86}Rb^+$ uptake. The concentration of ET–1 and ET–3 needed for the half–maximal stimulation of $^{86}Rb^+$ uptake ($EC_{50}$) was 0.73±0.17 nM and 12.89±3.69 nM, respectively. The ET–1–induced $^{86}Rb^+$ uptake into BCEC was reduced (35%) in the presence of ouabain (inhibitor of $Na^+$, $K^+$–ATPase), and about 65% with bumetanide (BUMET, inhibitor of $Na^+$–$K^+$–$Cl^-$ cotransport). A complete inhibition of the ET–1–stimulated $^{86}Rb^+$ uptake was seen in the presence of ouabain and BUMET. Both the ouabain–sensitive (OS) and ouabain–insensitive (OI) $Na^+$, $K^+$–ATPase activity induced by ET–1 were inhibited by [–5–(N–ethyl–N–isopropyl)] amiloride (inhibitor of $Na^+/H^+$ antiport exchange system), suggesting a linkage between $Na^+/H^+$ antiport system with $Na^+$, $K^+$–ATPase and $Na^+/K^+/Cl^-$ cotransport systems. PMA, an activator of PKC, induced similar effects to those observed with ET–1 on $^{86}Rb^+$ uptake in BCEC. Staurosporin (PKC inhibitor) also diminished the OS– and OI–stimulated $^{86}Rb^+$ uptake in BCEC indicating a PKC mediation of ET–1–induced $Na^+$, $K^+$–ATPase activity. Similar results were obtained in our studies performed on HBEC. The selective antagonist of $ET_A$ but not $ET_B$ receptors (IRL 1038) inhibited the ET–1 or ET–3–induced OS and OI ATPase activity.

These results indicate that occupancy of $ET_A$ receptors by either endothelin ET–1 or ET–3 activates both OS and OI $Na^+$, $K^+$–ATPase system via a PKC–dependent mechanism. The findings also suggest that overproduction and/or excessive release of ET–1 from the vascular and nonvascular cells may contribute to BBB disturbance of the OS and/or OI ATPase system during pathological conditions. Thus, the changes may be involved in ischemic cerebral edema formation.

## Endothelin Effect on Cellular Permeability

The $^{51}Cr$ loss from prelabeled cells is the most sensitive among various markers (i.e., $^{111}In$, LDH, $^{51}Cr$) for the assessment of peripheral endothelial cell injury. $^{51}Cr$ in the cells is associated with low molecular weight cytosolic components of less than 1,000 daltons. It may be released from the cells without disrupting the cell membrane. Lactic acid dehydrogenase (LDH), a larger molecule (134,000 daltons) than $^{51}Cr$ has been used as a marker for endothelial injury but may be retained in the cell despite appreciable ultrastructural membrane derangement.[23] Moreover, the uptake of Evan's blue–albumin complex has also been used to determine cellular permeability for high molecular weight substances.

Previously we demonstrated that the addition of exogenous AA as well as generators of free radicals ($H_2O_2$, glucose/glucose oxidase) increase the permeability of HBEC for Evan's blue–albumin complex, $^{51}Cr$ and LDH, as well as alter endothelial membrane fluidity.[24, 25] Since ET–1 induced the secretion and release of AA and its various metabolites from HBEC,[5–7] we were interested in determining whether ET–1 can also alter the "permeability" of these cells.

ET–1 which stimulated $IP_3$ formation and AA release, also induced concentration–dependent release of $^{51}Cr$ ($EC_{50}$=7 nM) but not LDH from HBEC. The maximal release of $^{51}Cr$ was observed 15–60 min after ET–1 addition to HBEC. The potency of various endothelins to induce $^{51}$Cr release was the same as that observed for the stimulation of $IP_3$ production, namely ET−1>ET−2>S6b>ET−3>S6c. Interestingly, the known proinflammatory cytokine, TNF$\alpha$, did not affect $^{51}$Cr release from these cells. As shown in Table 1, the ET−1−induced release of $^{51}$Cr and AA from HBEC was inhibited by BQ−123 (antagonist of $ET_A$ receptors), as well as verapamil, and ryanodine, inhibitors of voltage−dependent $Ca^{2+}$ channels and $Ca^{2+}$ mobilization from intracellular stores, respectively. HBEC exposed to the PKC activator, PMA (which decreases the ET−1−induced formation of $IP_3$ and augments ET−1−induced AA release from these cells) facilitated the release of $^{51}$Cr without affecting LDH activity. However, PMA had no additional effect on ET−1−stimulated release of $^{51}$Cr from HBEC. The data clearly indicate that the ET−1−induced change in "permeability" of HBEC is mediated by $ET_A$ receptors since ET−1 stimulated $^{51}$Cr release was abolished by BQ−123. The inhibitors of cyclooxygenase (Dxm and Indo) did not affect ET−1−stimulated release of $^{51}$Cr but reduced ET−1−stimulated AA release.

**Table 1.** Effects of various substances on $^{51}$Cr and AA release from HBEC.

| Stimulus | Percent $^{51}$Cr released | Percent AA released |
|---|---|---|
| M199 | 8.65 ± 0.49 (12) | 3.08 ± 0.07 (9) |
| 10 nM ET-1 | 11.89 ± 0.54 (10) | 4.05 ± 0.09 (6) |
| 100 nM ET-1 | 15.74 ± 0.52 (10) | 5.58 ± 0.05 (6) |
| 100 nM ET-1 + 1 µM BQ-123 | 9.14 ± 0.04 (10)* | 3.87 ± 0.07 (9)* |
| 100 nM ET-1 + 1 µM Dxm | 12.06 ± 0.78 (12) | 3.19 ± 0.13 (9)* |
| 100 nM ET-1 + 10 µM Indo | 15.54 ± 0.78 (9) | 4.12 ± 0.11 (9)* |
| 100 nM ET-1 + 1 µM PMA | 17.18 ± 1.01 (9) | 7.09 ± 0.09 (9) |
| 100 nM ET-1 + 0.2 µM H7 | 8.65 ± 0.44 (12)* | 4.52 ± 0.11 (9) |
| 100 nM ET-1 + 10 µM H8 | 14.98 ± 0.87 (9) | 5.41 ± 0.06 (9) |
| 100 nM ET-1 + 10 µM Verapamil | 9.90 ± 0.63 (12)* | 3.92 ± 0.09 (9)* |
| 100 nM ET-1 + 1 µM Ryonidine | 10.61 ± 0.49 (12)* | 4.01 ± 0.11 (9)* |

Release of $^{51}$Cr and AA were determined 1hour after addition of ET-1 to HBEC, which were preincubated in the presence or absence of indicated inhibitors for 30 min at 37°C. Values are presented as mean ± SEM for the number replicates given in parenthesis.
*-indicates significant difference (ANOVA, p<0.01) from 100 nM ET-1

It is also evident from the above−mentioned findings that ET−1 activates both the PLC and PLA systems. However, based on these results, it is difficult to ascertain which ET−1−activated signal transduction mechanism ($IP_3$−DAG−PKC and/or AA cascade) is principally responsible for the observed ET−1−induced "permeability" change of HBEC. Diacylglycerol (DAG)−mediated activation of PKC is suggested by the observed inhibition of $^{51}$Cr release with H7 (specific PKC inhibitor) but not with H8 (nonspecific PKC inhibitor). This contention is supported by the observation that PMA alone induced changes in "permeability" of HBEC and that the cyclooxygenase inhibitors failed to block the ET−1−stimulated release of $^{51}$Cr but inhibited the ET−1−induced AA release from these cells.

The ET−1−stimulated increase in the release of $^{51}Cr$ but not LDH from HBEC in contrast to the observed increased release of both ($^{51}Cr$ and LDH) from the same cells after exposure to either AA or $Ca^{2+}$ ionophore (data not shown) indicate that a subtle change in cellular permeability can be induced by ET−1 since this event occurs through activation of $ET_A$ receptors mediated by PKC. It is suggestive that other agonists may also similarly affect the BCEC integrity.

It has previously been shown or proposed that agents which activate cAMP, cGMP or PKC alter the BBB permeability.[3] Many of these substances (i.e., forskolin, histamine, serotonin) did not affect the permeability of cultured BCEC (grown on microbeads and/or monolayers) to Evan's blue−albumin complex or $^{51}Cr$ release. As a matter of fact, some of these agents decrease $^{51}Cr$ release from BCEC derived from various species (unpublished observation). It is therefore possible that the contradiction between the *in vitro* and *in vivo* observations concerned with the mechanisms responsible for changes in the BBB may depend upon activation and/or cooperation of more than one mediator and/or intracellular transduction pathway occurring not only in endothelium but also in other cells. Nevertheless, the results obtained from *in vitro* experiments have been helpful in elucidating some mechanism responsible for both passive and/or active (receptor−mediated) changes in BBB permeability.

**Endothelin Effect on Expression of Adhesion Molecules**

It is known that the expression of adhesion molecules by microvascular endothelium plays a role in certain pathological conditions (i.e., multiple sclerosis, EAE, hemorrhagic shock, cerebral ischemia, atherosclerosis, rheumatoid arthritis etc.).[11, 12] These molecules are important in mediating focal adhesion of leukocytes to vascular endothelium, and are considered as a key step in inflammation and vascular disease processes. The intercellular adhesion molecule−1 (ICAM−1) and vascular cellular adhesion molecule (VCAM−1) are both members of the immunoglobulin superfamily. ICAM−1, a ligand for the heterodimeric $\beta 2$ integrins is constitutively expressed on resting peripheral EC and BCEC passaged *in vitro*. E−selectin, an adhesion molecule which binds the carbohydrate ligand sialyated Lewis X, is not constitutively expressed on either cell type.

The investigation of the effect of endothelins on adhesion molecules was prompted by reports that some vasoactive substances and cytokines (which induce ET−1 secretion and are implicated in the pathogenesis of vascular disorders), up−regulate adhesion molecules on endothelial cells.

The examination of HBEC by immunofluorescence microscopy, FACS analysis, or ELISA revealed that these cells constitutively express ICAM−1 and VCAM−1 but not E−selectin. HBEC tested with various concentrations (0.1 − 250 nM) of ET−1, ET−2, ET−3 or cytokines [i.e., TNF$\alpha$ (1−200 U/ml), IFN−$\gamma$ (1−200 U/ml)] or LPS (0.001−10 $\mu$g/ml) up−regulated expression of ICAM−1, VCAM−1, and induced E−selectin. The up−regulation of ICAM−1 expression on HBEC by ET−1 (100 nM) was quantitatively similar to that induced with TNF$\alpha$ (100 U/ml) (Fig. 4). The ET−1 but not the TNF$\alpha$−induced up−regulation of ICAM−1 expression was blocked by the $ET_A$ receptor antagonist (data not shown). These results indicate that only the ET−1−induced up−regulation of ICAM−1 expression is mediated by $ET_A$ receptors even though TNF$\alpha$ is capable of inducing secretion of ET−1 in HBEC. Attempts to further elucidate some of the intracellular events that occur prior to the up−regulation of adhesion molecules induced by different agents such as ET−1 and cytokines (TNF$\alpha$) were made by using inhibitors of intracellular signals thought to be important in expression of adhesion molecules. Signals such as PKC have been implicated in regulating ICAM−1 expression; the PKC inhibitor, H7, dose−dependently blocked ICAM−1 expression induced by ET−1 and TNF$\alpha$, indicating that both substances may utilize similar pathways involving PKC.

The observed partial inhibition (about 25%) of ET−1 and TNFα−induced ICAM−1 expression by verapamil (a voltage−dependent calcium channel blocker) strongly suggest participation of extracellular $Ca^{2+}$ in linking intracellular signals with ICAM−1 expression. The mechanism responsible for ET−3 up−regulation of the adhesion molecules is still unknown and remains to be clarified.

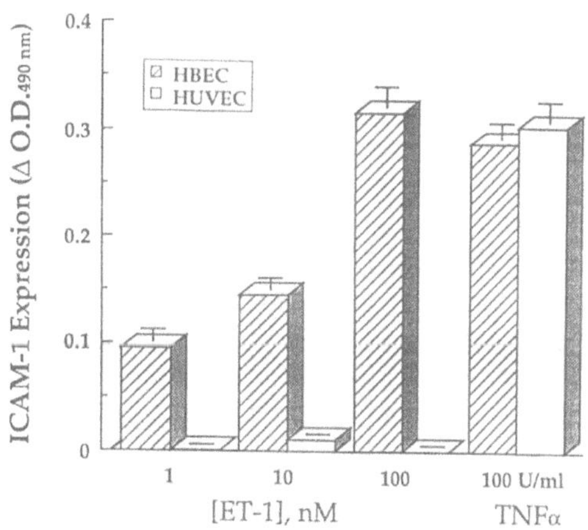


**Figure 4.** Comparative effect of ET−1 and TNFα on ICAM−1 expression of HBEC ▨ and HUVEC □. The cell cultures were incubated for 24 hr in the presence of an indicated concentration of ET−1 or TNFα or in the absence of any additions. The levels of ICAM−1 expression were obtained from ELISA experiments. The data are expressed as Δ O.D. ±S.D. which were obtained by subtracting the mean O.D. of quadriplicate cultures in media (0.327±0.028 and 0.239±0.019 from HBEC and HUVEC, respectively) from the mean O.D. of cultures incubated in the presence of the indicated addition.

Nevertheless, these findings together with the above−mentioned changes in cellular "permeability" demonstrated for the first time that endothelins induce "permeability" changes and expression of E−selectin as well as up−regulation of ICAM−1 and VCAM−1 on HBEC.

**Comparative Studies.** Most of the studies concerned with the induction of adhesion molecules [by cytokines and some vasoactive substances (i.e., substance P, neuropeptide Y)] and their roles in inflammation were studied in peripheral vascular endothelial cells.[11,12] Since many of the microvascular characteristics depend on their anatomic location, we compared the effects of potent vasoactive substances (which have been implicated as playing a role in vascular disorders including inflammation) and cytokines on the HBEC with those on endothelium derived from human umbilical vein (HUVEC).

The levels of ICAM−1 and VCAM−1 constitutively expressed on HUVEC were lower (10−15%) than those on BCEC (25−40%). E−selectin expression was absent in both types of untreated cells. Overall, the same concentration of cytokines (TNFα, IFNγ) or LPS used to induce expression of adhesion molecules on HBEC up−regulated to the same or greater degree the expression of ICAM−1, VCAM−1 and also induced E−selectin on HUVEC. In contrast, none of the molecules of the endothelin family (ET−1, ET−2 or ET−3) which up−regulated ICAM−1 and VCAM−1 and induced E−selectin in HBEC had any effect of HUVEC (Fig. 4). Moreover, these endothelins were ineffective in changing the "permeability" ($^{51}Cr$ release), $IP_3$ formation, and AA release from HUVEC. These findings suggest that HUVEC may not express the same subtypes of endothelin receptors as do HBEC and/or possibly have endothelin receptors which are down-regulated. In support of this contention is the detected expression of $ET_B$ mRNA in HUVEC, and the demonstrated existence of nonselective $ET_B$ receptors in phosphoramidon-treated HUVEC. The HUVEC response to histamine (i.e., increased $IP_3$ formation) was similar to that of HBEC and consistent with the presence of H1 receptors on both cell types (data not shown).

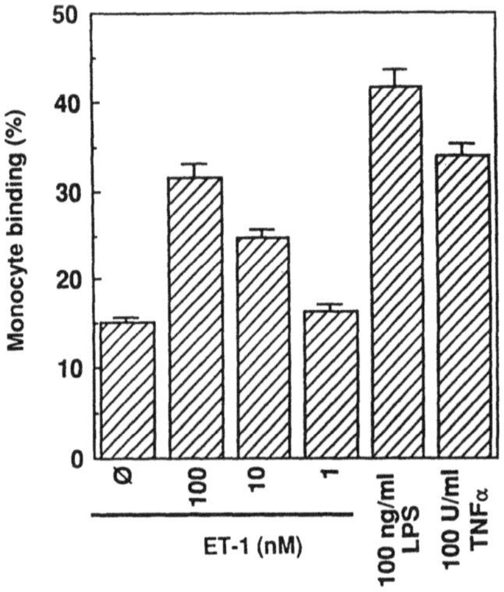


**Figure 5.** Effect of BCEC treatment on monocytes binding. BCEC derived from rat cerebral microvessels were treated for 4 hr with indicated concentrations of ET-1, LPS or TNFα. The data represent the mean ± S.E. of quadriplicate cultures obtained from a single representative experiment of four others with similar results.

These findings indicate that endothelial cells derived from vessels of various sites and/or different anatomic locations respond distinctly to endothelins and implicate variations in the role(s) of endothelial cells derived from dissimilar vessels/and or different anatomic locations in recruitment of blood cells at the site of inflammation.

### Effect of Endothelin on Monocytes Binding to Rat BCEC

The functional significance of the observed induced up-regulation of adhesion molecules on HBEC after exposure to ETs or cytokines has been followed up by studies of their effects on monocyte adhesiveness to BCEC. Due to the limited availability of HBEC cultures, the experiments were performed on rat BCEC (unpublished observations).

The treatment of BCEC with ET-1 dose-dependently increases their adhesiveness to monocytes (Fig. 5). The increase in monocyte adhesion to HBEC treated with 100 nM ET-1 was two-fold over controls, and similar to the level of monocyte adhesiveness of BCEC exposed to 100 $\mu$M of TNFα. However, LPS treatment of BCEC was more effective than either ET-1 or TNFα (Fig. 5). These results correlated well with the capacity of these substances to up-regulate adhesion molecules on BCEC and establish ET-1 as a proinflammatory agent which may contribute to the passage of leukocytes across the BBB under pathological conditions.

## COMMENTS

These studies unequivocally demonstrated that HBEC secretion of ETs and prostanoids can be augmented by vasoconstrictive peptides (AVP or Ang II) through occupancy of their respective receptors and activation of PLC and $PLA_2$ mechanisms. These findings also demonstrate the presence of receptors for ET−1 (high and low affinity) in HBEC. The characteristics of the HBEC high−affinity receptors for ET are consistent with the presence of the $ET_A$ subtype of ET receptors linked to PLC and $PLA_2$ signal transduction mechanisms. The nature and specificity of the low−affinity binding sites for ETs on HBEC as already mentioned is presently unknown and remains to be clarified. ET−1 activation of the $ET_A$ receptors on HBEC stimulates production of prostanoids, cAMP, $Ca^{2+}$ uptake, ATPase activity, and induces changes in cellular "permeability" and expression of adhesion molecules. All these responses are most likely mediated by independent ET−1 activation of PLC and $PLA_2$ through different intracellular mechanisms. Moreover, there is ample evidence of existing cross−talk between the various pathways of intracellular mechanisms which may depend on the cell type.[5,7,26] The presence of $ET_A$ and/or $ET_B$ types of receptors as well as some of the same events induced by ET−1 or ET−3 were described in endothelium and other cells.[7,8,16−18,20] As a matter of fact, a multitude of endothelin studies performed since the discovery of these peptides in 1988, indicate that ET−1 is produced and/or is active in all tissues but most prominently in endothelium.[5,13−19] A wide range of effects (i.e., vasoconstriction, vasodilation, autocrine release, modulation of nervous activity, mitogenesis, ionotropism, and chronotropism) have been documented, depending to some extent, on the "target" cell types or organ(s) studied *in vitro* or *in vivo*. Nevertheless, our observation of ET−1 effects, namely alteration of HBEC "permeability", stimulation of ATPase activity, up−regulation and induction of surface molecules as well as increased leukocyte adhesion to these cells, adds new dimensions to the biological activities of ET−1, especially implicating its role on BBB function under physiological and/or pathological conditions.[4] It is known that peptides operate as chemical signals of intercellular communication.[27] Various peptides including ET have been detected in plasma and extracellular fluid (ECF) in the brain.[28−31] The concentration of ET and prostanoids is low in plasma and they are, therefore, thought to function as local (rather than as circulating) hormones involved in regulating vascular tone.[32] Both substances have also been considered to act as paracrine/autocrine hormones since the "target" cell types which have corresponding receptors are found in the vicinity of the "producer" cell types within the same tissue. Thus, ET−1, whether secreted in cerebral capillary and microvascular endothelium or in the surrounding cells (namely, the glia and/or muscle cells) may be involved in regulating the function of the microvascular bed. The ET−1 signal (provided by endothelium secreted ET), unlike other peptides implicated in BBB transport systems, does not have to cross the barrier for its activities. This notion supports the concept that blood−borne peptides do not have to pass the barrier to reach their target cells.[27] Most importantly, the ET−1 effect on the ATPase system and the previously reported ET−1 action of the $Na^+/H^+$ exchange on BCEC strongly suggest ET−1 participation in regulating water and electrolyte homeostasis of the brain.

As we mentioned before, ET−1 has been thought to participate in the development and/or progression of tissue injury in the brain as well as other organs (i.e., cardiovascular, pulmonary and renal disease). Supportive evidence for this concept has been substantiated by the following observations: (1) ET−1 is the most potent mammalian vasoconstrictor yet discovered, and when superfused onto the middle cerebral artery of anesthetized rats, it produces dose−dependent brain injury;[33] (2) increased levels of ET−1 in plasma and/or cerebrospinal fluid in patients with stroke, hypertension, head trauma, hemorrhage, and vasospasm;[7,29] (3) increased levels of ET−1 in brain tissue, dialysate of

ECF in experimental focal and global ischemia;[31] and (4) neuroprotective effects of $ET_A$ receptor antagonist demonstrated in transient brain ischemia in gerbils.[34] However, despite the known vasoactive properties of ET−1, the exact pathomechanism responsible for the proposed involvement of ET−1 in cerebrovascular disorders is unknown. In general, endothelial activation or dysfunction has been thought to play a major role in "triggering" and/or progression of tissue injury in many organs as well as in the brain. Based on our studies presented here, we would like to propose that ET−1 may contribute to the initiation and/or progression of stroke by its effects on endothelium. This process may involve: (1) the expression or release of various mediators; (2) disturbance of water and electrolyte homeostasis; (3) increases in cellular permeability; and/or (4) up−regulation and induction of adhesion molecules on the endothelial surface as well as increased adhesiveness of leukocytes and other blood cells. These events taken either separately or together may not only alter BBB permeability (both passive diffusion and active transport of water and ions and other low−molecular weight substances) which lead to edema formation. This process may also provide a mechanism for passage of leukocytes across the BBB and could initiate procoagulation and thrombosis. ET−1 activation of cytokines, permeability, procoagulation or thrombosis has been reported in peripheral vascular beds.[12,35,36] Finally, the postulated role of ET−1 involvement in the function of BBB based on *in vitro* studies in BCEC are insufficient without appropriate validation of these phenomena *in vivo*.

In conclusion, this communication describes the HBEC responses and signal transduction mechanism induced by vasoconstrictive peptides particularly by ET−1 which may play an important role as a mediator of physiological and/or pathophysiological function of the BBB.


## Acknowledgements

The authors are grateful to Ms. Joliet Bembry and Nancy Merkel for excellent technical assistance; and to Devera Schoenberg, M.Sc., for outstanding editorial assistance in preparing this manuscript.


## REFERENCES


1. S.I. Rapport. "Blood−Brain Barrier in Physiology and Medicine," Raven Press, New York (1976).
2. A.L. Betz, and G.W. Goldstein. Specialized properties and solute transport in brain capillaries. *Ann. Rev. Physiol.* 48:241 (1986).
3. F. Joo. The blood−brain barrier *in vitro*: the second decade. *Neurochem. Int.* 23:499 (1993).
4. F.M. Faraci. Regulation of the cerebral circulation by endothelium. *Pharmac. Ther.* 56:1 (1992).
5. F. Bacic, S. Uematsu, R.M. McCarron, and M. Spatz. Secretion of immunoreactive endothelin−1 by capillary and microvascular endothelium of human brain. *Neurochem. Res.* 17:699 (1992).
6. M. Spatz, D.B. Stanimirovic, S. Uematsu, and R.M. McCarron. Vasoconstrictive peptides induce endothelin−1 and prostanoids in human cerebromicrovascular endothelium. *Am. J. Physiol.* 266:C654 (1994).
7. D.B. Stanimirovic, T. Yamamoto, S. Uematsu, and M. Spatz. Endothelin−1 receptor binding and cellular signal transduction in cultured human brain endothelial cells. *J. Neurochem.* 62:592 (1993).
8. D.B. Stanimirovic, R.M. McCarron, and M. Spatz. Dexamethasone downregulated endothelin receptors in human cerebromicrovascular endothelial cells. *Neuropeptides* 26:145 (1994).
9. N. Kawai, T. Yamamoto, H. Yamamoto, R.M. McCarron, and M. Spatz. Endothelin stimulates ATPase activity in brain capillary endothelium. *J. Physiol. (Lond).* 480:17P (1994).
10. D.B. Stanimirovic, R.M. McCarron, N. Bertrand, and M. Spatz. Endothelins release $^{51}Cr$ from culture human cerebromicrovascular endothelium. *Biochem. Biophys. Res. Commun.* 26:1 (1993).

11. R.M. McCarron , L. Wang, D.B. Stanimirovic, and M. Spatz. Endothelin induction of adhesion molecule expression on human brain microvascular endothelial cells. *Neurosci, Lett.* 156:31 (1993).
12. R.M. McCarron, L. Wang, D.B. Stanimirovic, and M. Spatz. Differential regulation of adhesion molecule expression by human cerebrovascular and umbilical vein endothelial cells. *Endothelium* (1994) (in press).
13. M. Yanagisawa, H. Kurihara, S. Kimura, et al. Endothelin: a novel potent vasoconstrictor peptide produced by vascular endothelial cells. *Nature* 332:411 (1988).
14. A. Inoue, M. Yanagisawa, S. Kimura, et al. The human endothelin family: three structurally and pharmacologically distinct isopeptides predicted by three separate genes. *Proc. Natl. Acad. Sci. USA* 86:2853 (1989).
15. G.M. Rubany, and L.H. P. Botelho. Endothelins. *FASEB J.* 5:2713 (1991).
16. W.M. MacCumber, A.C. Ross, and H.S. Snyder. Endothelin in the brain: receptors, mitogenesis, and biosynthesis in glial cells. *Proc. Natl. Acad. Sci. USA* 87:2359 (1990).
17. H. Ehrenreich, H.J. Kehrl, W.R. Anderson, et al. A vasoactive peptide, endothelin−3, is produced by and specifically binds to primary astrocytes. *Brain Res.* 538:54 (1991).
18. A. Giaid, J.S. Gibson, T.M. Herrero, et al. Topographical localisation of endothelin mRNA and peptide immunoreactivity in neurones of the human brain. *Histochemistry* 95:303 (1991).
19. W.C. Sessa, S. Kaw, M. Hecker, and J.R. Vane. The biosynthesis of endothelin−1 by human polymorphonuclear leukocytes. *Biochem. Biophys. Res. Commun.* 147:613 (1991).
20. P.O. Couraud, O. Durieu−Trautmann, D.L. Nguyen, P. Marin, F. Gilbert, and A.D. Strosberg. Functional endothelin−1 receptors in rat astrocytoma C6. *Eur. J. Pharmacol.* 206:191 (1991).
21. P.A. Revest, N.J. Abbott, and J.I. Gillespie. Receptor−mediated changes in intracellular [$Ca^{2+}$] in cultured rat brain capillary endothelial cells. *Brain Res.* 549:159 (1991).
22. P. Vigne, A. Ladoux, and C. Frelin. Endothelins activate $Na^+/H^+$ exchange in brain capillary endothelial cells via a high−affinity endothelin−3 receptor that is not coupled to phospholipase C. *J. Biol. Chem.* 266:5925 (1991).
23. J. Chopra, J.H. Joist, R.O. Webster. Loss of $^{51}$chromium, lactate dehydrogenase, and $^{111}$indium as indicators of endothelial cell injury. *Lab. Invest.* 57:578 (1987).
24. A. Villacara, M. Spatz, R.F. Dodson, C. Corn, and J. Bembry. Effect of arachidonic acid on cultured cerebromicrovascular endothelium: Permeability, lipid peroxidation and membrane fluidity". *Acta Neuropathol.* 78:310 (1989).
25. R.M. McCarron, S. Uematsu, S. Merkel, D. Long, J. Bembry, and M. Spatz. The role of arachidonic acid and oxygen radicals on cerebromicrovasculature endothelial permeability. *Acta Neurochem. Suppl.* 51:61 (1990).
26. P. Vigne, L. Lund, and C. Frelin. Cross−talk among cyclic AMP, cyclic GMP, and $Ca^{2+}$−dependent intracellular signalling mechanisms in brain capillary endothelial cells. *J. Neurochem.* 62:2269 (1994).
27. A. Ermisch. Peptide receptors of the blood−brain barrier and substrate transport into the brain. *Prog. Brain Res.* 91:155 (1992).
28. J. Widimsky Jr., K. Horky, and J. Dvorakova. Plasma endothelin−1, −2 levels in mild and severe hypertension. *J. Hypertens.* 9 (Suppl.6):S194 (1991).
29. R. Suzuki, H. Masaoka, Y. Hirata, et al. The role of endothelin−1 in the origin of cerebral vasospasm in patients with aneurysmal subarachnoid hemorrhage. *J. Neurosurg.* 77:96 (1992).
30. A.A. Khraibi, D.M. Heublein, and F.G. Knox. Increased plasma level of endothelin−1 in the Okamoto spontaneously hypertensive rat. *Mayo Clin. Proc.* 68:42 (1993).
31. F.C. Barone, M.Y.−T. Globus, W.J. Price, et al. Endothelin levels increase in rat focal and global ischemia. *J. Cereb. Blood Flow and Metab.* 14:337 (1994).
32. J. Vane, E.E. Anggard, and R.M. Botting. Regulatory functions of the vascular endothelium. *N. Engl. J. Med.* 323:27 (1990).
33. I.M. Macrae, M.I. Robinson, D.I. Graham, J.L. Reid, and J. McCulloch. Endothelin−1 reduction in cerebral blood flow: dose−dependency, time course and neuropathological consequences. *J. Cereb. Blood Flow and Metab.* 13:276 (1993).
34. G.Z. Feuerstein, J.L. Gu, E.H. Olshtein, E.C. Barone, and I.L. Yue. Selective endothelin receptor ($ET_A$) antagonist is neuroprotective in gerbil transient forebrain ischemia. *Stroke* 25:264 (1994).
35. J.G. Filep, M.G. Sirois, E. Foldes−Filip, et al. Enhancement by endothelin−1 of microvascular permeability via the activation of $ET_A$ receptors. *Br. J. Pharmacol.* 109:880 (1993).
36. A. Halim, N. Kanayama, E.E. Maradny, K. Maehara, and T. Terao. Coagulation *in vivo* microcirculation and *in vitro* caused by endothelin−1. *Thromb. Res.* 72:203 (1993).

# THE ROLE OF THE ENDOTHELIAL CELL SURFACE CHARGE FOR BLOOD-BRAIN BARRIER FUNCTION

Barbro B. Johansson

Department of Neurology
Lund University Hospital
S-221 85 Lund, Sweden

## INTRODUCTION

The glycocalyx is a negatively charged layer of fibrous chains of macromolecules, mainly but not exclusively glycoproteins, that covers the luminal surface of endothelial cells (Luft, 1966, 1976; Burry and Wood, 1979; Simionescu and Simionescu, 1986). In addition, anionic sites are present in the plasmalemma and the basement membrane of brain endothelial cells (Nag, 1985; Vorbrodt, 1989). On the luminal surface the main contributors to the anionic sites are sialic acid residues of acidic glycoproteins. In the basal membranes anionic sites are less abundant and represent a mixture of proteoglycan and acid glycopeptides containing hydrophobic amino acids, sialic acid residues and heparan sulphate-bearing glycosaminoglycans.

The permeability of the endothelial cells is influenced by the surface charge. Thus, anionic molecules move less readily than uncharged or cationic molecules of equal size over the capillary walls indicating that endothelial cell surface discriminates among molecules according to their charge. This is well established for other vascular beds (Areekul, 1969; Rennke et al., 1975; Chang, 1975; Haraldsson et al., 1983) and has been shown to be true also in the brain (Houthoff et al., 1984; Triguero et al., 1989, Pardridge et al., 1990; Pardridge and Boado, 1991).

In other vascular beds there is evidence that plasma albumin stabilizes the endothelial-blood interface. Perfusion with albumin-free solutions increases the capillary permeability in the mesenterium (Levick and Michel, 1973; Mason et al., 1977) and in the muscles (Rippe and Folkow, 1977). The mechanism behind this "albumin effect" is not clear (Michel et al., 1985; Rippe and Haraldsson, 1994). It has been suggested that albumin may affect the barrier to macromolecules by occupying space in the glycocalyx. One hypothesis is that serum albumin binds to anionic sites on the glycocalyx and the endothelial cells. In spite of its net negative charge, serum albumin can bind anions, e. g. by positively charged lysine and arginine side chains of the albumin molecule (Kragh-Hansen, 1981). The albumin effect was lost in frog mesenteric microvessels after modification of the positively charged arginine side chains, suggesting that binding of guanidino groups of the arginine residues to glycocalyx and/or anionic sites on the endothelial cells can reduce the permeability in the mesenteric vascular bed (Michel et al., 1985).

The serum α1-acid glycoprotein orosomucoid apparently contributes to the net charge on the microvascular wall in rat skeletal muscles (Haraldsson and Rippe, 1987) and in frog mesenteric capillaries (Curry et al., 1989) thereby reducing its permeability to anionic macromolecules.

There is so far no evidence that serum proteins play an important role for the blood-brain barrier (BBB) function. Greenwood et al. (1985) demonstrated that the integrity of the BBB to the small polar tracer mannitol was maintained for up to 30 minutes during an *in situ* perfusion of the brain with a saline-based solution containing a metabolic inhibitor. In a recent study, the PS product for α-aminoisobutyric acid was higher than that observed *in vivo* during *in situ* perfusion of washed red blood cells. However, it could be reduced to *in vivo* values when neutral amino acids competing for the same carrier mechanism were administered (Ennis et al., 1994).

## EXPERIMENTAL OPENING OF THE BBB WITH POLYCATIONS

### Protamine

Protamines are small polycationic proteins (mol wt 4050-4420) obtained from fish and bird sperm (Scott and Eaglesson, 1988). When infused into a carotid artery, protamine sulphate can increase the permeability of the BBB, an effect that has been attributed to the molecular charge (Nagy et al., 1983; Hardebo and Kåhrström, 1985; Strausbaugh, 1987; Jezova et al., 1989; Westergren and Johansson, 1990). The opening of the BBB is reversable with the duration of the opening depending on the dose used (Strausbaugh, 1987, Westergren and Johansson, 1990). Olesen and Crone (1986) have reported that the protamine-induced BBB alteration is irreversible but their conclusion was based on a study measuring the electrical resistance of pial vessels for less than ten minutes after the protamine infusion had stopped.

Intracarotid infusion of protamine can thus be used to open the BBB unilaterally. One advantage over the more commonly used model of hyperosmolar infusion is that protamine can be administered in smaller amounts of fluid reducing the hemodynamic effects. In pilot studies on how circulating substances can modify the neuroendocrine response when the BBB was disturbed, we found that, in addition to the well known effect on blood pressure (Salahuddin et al., 1988), hyperosmolar solutions induced a massive increase in stress hormones which made it impossible to use them in such studies. In contrast, protamine sulphate in a dose sufficient to open the BBB, did not significantly increase stress hormone levels (Jezova et al., 1989, 1993).

With a low dose of protamine sulphate (5 mg, grade X, Sigma) an opening of the BBB can be obtained with no significant brain oedema whereas a higher dose will significantly decrease brain specific gravity and alter the extracellular fluid environment (Westergren and Johansson, 1991; Westergren et al., 1994). The brain oedema induced by protamine sulphate is similar to that obtained by hyperosmolar solutions and both types can be reduced by blockade of glutamate AMPA receptors (Westergren and Johansson, 1993).

One concern is the possible toxicity of protamine sulphate to the brain. Protamine affects various biological systems and can be toxic in high doses (Horrow, 1985). As reported by Nagy et al. (1983), intracarotid infusion of protamine sulphate grade I (Sigma) results in substantial parenchymatous lesions. However, this can be avoided by using the purest available protamine sulphate (grade X) in a low or moderate dose (Westergren et al., 1993 a).

## Poly-L-lysine and poly-L-arginine

There is evidence that the effect of polycations may not be the same in different vascular beds. Poly-L-lysine, but not protamine, has been reported to cause diffuse epithelial changes and cell damage in the kidney (Seiler et al., 1975). In another study, local arterial infusion of protamine, poly-L-lysine and hexadimetrin in the kidney induced a dose-dependent albuminuria with similar responses for the three polycations (Vehaskari et al., 1982). In contrast, Hardebo and Kåhrström (1985) observed a larger leakage after infusion of protamine sulphate (grade II) than after poly-L-lysine.

The molecular charge of protamine is caused by arginine, which constitutes two thirds or more of the protamine molecule (Alger, 1989). To elucidate whether other factors than endothelial charge are of importance for the opening of the BBB with different cations, we have compared poly-L-lysine and poly-L-arginine of corresponding molecular size (poly-L-lysine hydrobromide, mol wt 10,200 and poly-L-arginine hydrochloride, mol wt 11,600, Sigma). The extravasation of serum albumin into the brain was significantly larger after poly-L-arginine than after poly-L-lysine (Fig. 1). Poly-L-lysine also induced significantly less albumin extravasation than protamine in spite of the larger molecular weight of poly-L-lysine (Westergren & Johansson, 1993 b).

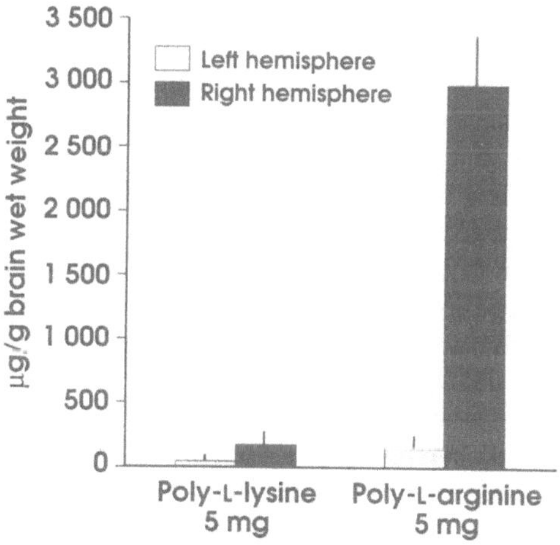


**Figure 1.** Extravasated serum albumin in the left and right hemisphere after intracarotid infusion of poly-L-lysine and poly-L-arginine into the right internal carotid artery.

When the dose of poly-L-arginine is decreased, a threshold effect for the opening of the BBB becomes evident. A similar threshold effect was obtained with a high molecular weight poly-L-lysine (Fig 2). One possible explanation to the threshold effect would be that negatively charged serum constituents competed with the endothelial surface anionic sites. To clarify this issue poly-L-lysine was administered in a different way. Instead of perfusing the substance in 100 µl Krebs solution during 30 seconds, giving the substance time to mix with the blood, the same amount of the substance was given in 1.5 ml Krebs solution infused during the same time. This markedly enhanced the extravasation of serum albumin which is consistent with the hypothesis that negatively charged serum components may compete with the anionic sites in the glycocalyx and the endothelium.

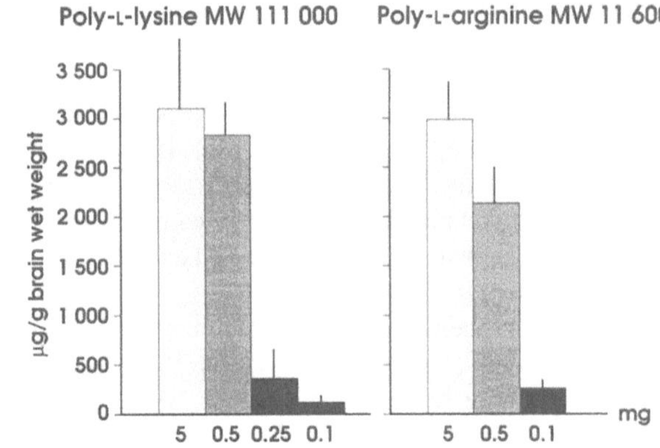


**Figure 2.** A threshold effect is evident for the opening of the BBB with both polycations with a lower threshold for poly-L-arginine.

## WHAT COULD BE A POSSIBLE EXPLANATION FOR THE MARKED EFFECT OF POLY-L-ARGININE ON THE BBB?

The positive charge of lysine (2.6-di-amino-caproic acid) is due to an s-amino group, whereas guanidino groups are responsible for the charge of arginine (2-amino-5-guanidovaleric acid). Aqueous salt solutions of guanidine can denaturate proteins in high concentrations, a process that is dose dependent and reversible (Scott & Eagleson, 1988). This raises the possibility that the guanidino group in poly-L-arginine might enhance the permeability by a direct effect on junctional complexes or endothelial cells in addition to its effect on the endothelial charge.

Another mechanism to consider is the possibility that nitric oxide could be formed from poly-L-arginine. Nitric oxide, which in high concentration can act as a free radical scavenger. is formed from one of the two equivalent terminal guanidino nitrogens of L-arginine (Moncada et al., 1989). If it can be formed from poly-L-arginine is not known. However, polypeptides containing one or more arginine residues have been shown to elicited a strong endothelium-dependent relaxation on aorta *in vitro* and poly-L-lysine caused a much weaker endothelium-dependent relaxation than poly-L-arginine. The authors suggested that arginine-containing peptides may relax vascular smooth muscles by acting directly on the vascular smooth muscles and/or by eliciting release of a relaxing factor from the endothelium (Thomas et al., 1986).

To elucidate whether or not the poly-L-arginine effect on the BBB could be related to nitric oxide formation we have used a nitric oxide inhibitor, N-methyl-L-arginine (NMMA). No significant protective effect on the albumin extravasation was obtained when the inhibitor was administered intravenously or in the carotid artery. However, since NMMA markedly increases the blood pressure the results are not conclusive. Although NMMA did not in itself induce any BBB leakage, we know that the leakage into the brain is highly dependent on the blood pressure once the BBB is open. Thus, we cannot rule out that a moderate reduction in poly-L-arginine-induced leakage could have been concealed by an enhanced extravasation due to the elevated blood pressure. On the other hand, the vasoconstrictor effect of the drug could have an opposite effect. Studies to evaluate the hemodynamic consequences of NMMA with and without concomitant administration of poly-L-arginine are in progress. An independent observation was that the effect of NMMA on blood pressure lasted longer when NMMA was administered via the carotid artery than when given intravenously.

Another way of separating a biological from a chemical effect of the polycations would be to study the stereospecificity. Since poly-D-arginine is not commercially available, poly-L and poly-D-lysine were compared in a pilot study. Although a slight but significant difference was obtained in an early experiment this observation has so far not been confirmed using larger doses of the substances. Considering the discussion on a possible role of nitric oxide, a comparison between poly D-arginine and poly-l-arginine would help to settle the issue since nitric oxid can not be formed from D-arginine.

## ENDOTHELIAL SURFACE CHARGE IN EXPERIMENTALLY INDUCED ALTERATIONS OF THE BBB

A reduction of endothelial surface anionic sites has been observed in experimental altering of the BBB such as acute hypertension (Nag, 1984), cold lesion injury (Vorbrodt et al., 1985), scrapie-infected mice (Vorbrodt et al., 1990) and in hyperosmotic opening of the BBB (Vorbrodt, 1993).

It has been suggested that interaction of aluminum with a negatively charged surface layer of the brain's microvascular endothelial cells is an important mechanism behind aluminum toxicity (Deleers, 1985; Vorbrodt & Trowbridge, 1993). Trivalent aluminum exhibits an amphoteric nature in aqueous solutions, forming several hydrated species that possess a net positive, negative or neutral charge (Deleers, 1985). Anionic groups on the endothelial surface are capable of lateral migration and redistribution under the influence of multivalent ligands. A lateral migration associated with a clustering of anionic sites has been attributed to cross-linking of the cationic groups by negatively charged molecules (Skutelsky and Danon, 1976). Exposure of cultured brain microvascular endothelial cells to aluminum (and also to cadmium) facilitated the lateral migration of surface structures with a net negative charge suggesting an effect on the glycoprotein surface coat (Vorbrodt & Trowbridge, 1993). However, recent *in vivo* studies on the effect of aluminum on the BBB suggest that cytotoxicity of aluminum may be less pronounced in the living organisms (Vorbrodt et al., 1994). Other effects of aluminum on the BBB might be more important (Banks et al., 1988).

## CONCLUSIONS

Interaction of polycations with the negatively charged glycocalyx and endothelium is complex. Poly-L-arginine seems to have an effect on the BBB that not exclusively is related to its negative charge. The interaction between brain endothelial cells, glycocalyx and plasma constituents and the consequences for the BBB under physiological and pathological conditions needs further exploration.

### Acknowledgments

Studies from the author's own laboratory were supported by grants from the Swedish Medical Research Council and King Gustaf V and Queen Victoria's Foundation.

## REFERENCES


Alger, M.S.M., 1989,"Polymer Science Dictionary," pp. 24, 106-107, 399, Elsevier, London and New York.

Areekul, S., 1969, Reflection coefficients of neutral and sulphate-substituted dextran molecules in the isolated perfused rabbit ear, *Acta Soc. Med. Ups.* 74:129-138.

Banks, W.A., Kastin, A.J., Fasold, M.B., 1988, Differential effect of aluminum on the blood-brain barrier transport of peptides, technecium and albumin, *J Tharmacol. Exp. Ther.* 244:579-585.

Burry, R.W., and Wood, J.G., 1979, Contributions of lipids and proteins to the surface charge of membranes, *J. Cell Biol.* 82:726-741.

Chang, R.L.S., Deen, W.M., Robertson, C.R., and Brenner, B.M., 1975, Permselectivity of the glomerular capillary wall. III. Restricted transport of polyanions, *Kidney Int.* 8:212-218.

Curry, P.E., Rutledge, J.C., and Lenz, J.F., 1989, Modulation of microvessel wall charge by plasma glycoprotein orosomucoid, *Am .J. Physiol.* 257:H1354-H1359.

Deleers, M., 1985, Cationic atmosphere and cation competition binding at negatively charged membranes: pathological implications of aluminum, *Res. Commun. Chem. Pathol. Pharmacol.* 49:277-294.

Ennis, S.R., Ren, X.-D., and Betz, A.L., 1994, Transport of α-aminoisobutyric acid across the blood-brain barrier studied with in situ perfusion of rat brain, *Brain Res.* 643:100-107.

Greenwood, J., Luthert, P.J., Pratt, O.E., and Lantos, P.L., 1985, Maintenance of the integrity of the blood-brain barrier in the rat during an in situ saline-based perfusion, *Neurosci. Lett.* 56:223-227.

Haraldsson, B., Ekholm, C., and Rippe, B., 1983, Importance of molecular charge for the passage of endogenous macromolecules across continuous capillary walls, studied by serum clearance of lactate dehydrogenase (LDH) isoenzymes, *Acta Physiol. Scand.* 117:123-130.

Haraldsson, B., and Rippe, B., 1987, Orosomucoid as one of the serum components contributing to normal capillary permselectivity in rat skeletal muscle. *Acta Physiol. Scand.* 129:127-135.

Hardebo, J.E., and Kåhrström, J., 1985, Endothelial negative surface charges areas and blood-brain barrier function, *Acta Physiol .Scand.* 125:495-499.

Horrow, J.C., 1986, Protamine: a review of its toxicity, *Anesth. Analg.* 64:348-361.

Houthoff, H.J., Moretz, R.C., Rennke, H.G., and Wisniewski, H.M., 1984, The role of molecular charge in blood-brain barrier impairment and cerebral edema, *in:* "Brain Edema," K.G. Go, and H. Daetman, eds., Plenum Press, London, pp 67-79.

Kragh-Hansen, U., 1981, Molecular aspects of ligand binding to serum albumin. *Pharmacol Rev* 33:17-33.

Jezova, D., Johansson, B.B., Oprsalova, Z., and Vigas, M., 1989, Changes in blood-brain barrier function modify the neuroendocrine response to circulating substances, *Neuroendocrinology.* 49:428-433.

Jezová, D., Bartanusz, V., Westergren, I., Johansson, B.B., Rivier, J., Vale, W., and Rivier, C., 1992, Rat melanin-concentrating hormone stimulates adrenocorticotropin secretion: evidence for a site of action in brain regions protected by the blood-brain barrier, *Endocrinology* 130:1024-1029.

Levick, J.R., and Michel, C.C., 1973, The effect of bovine albumin on the permeability of frog mesenteric capillaries, *Q .J. Exp. Physiol.* 58:67-85.

Luft, J.H., 1966, Fine structure of capillary and endocapillary layer as revealed by ruthenium red, *Federation Proc.* 25:1773-1783.

Luft, J.H., 1976, The structure and properties of the cell surface coat, *Int. Rev. Cytol.* 45:291-382.

Mason, J.C., Curry, F.E., and Michel, C.C., 1977, The effects of proteins upon the filtration coefficient of individually perfused frog mesenteric capillaries, *Microvasc .Res.* 13:185-202.

Michel, C.C., Phillips, M.E., and Turner, M.R., 1985, The effects of native and modified bovine serum albumin on the permeability of frog mesenteric capillaries, *J. Physiol.* 360:333-346.

Moncada, S., Palmer, R.M.J., and Higgs, E.A., 1989, Biosynthesis of nitric oxide from L-arginine: a pathway for the regulation of cell function and communication, *Biochem. Pharmacol.* 38:1709-1715.

Nag, S., 1984, Cerebral endothelial surface charge in hypertension. *Acta Neuropathol. (Berl.),* 63:276-281.

Nag, S., 1985, Ultrastructural localization of lectin receptors on cerebral endothelium, *Acta Neuropathol. (Berl.)* 66:105-110.

Nagy, Z., Peters, H., and Hüttner, I., 1983, Charge-related alterations of the cerebral endothelium, *Lab Invest.* 49:662-671.

Olesen, S.P., and Crone, C., 1986, Substances that rapidly augment ionic conductance of endothelium in cerebral venules. *Acta Physiol. Scand.* 127:233-241.

Pardridge, W.M., Triguero, D., Buciak, J.L., and Yang, J., 1990, Evaluation of cationized rat albumin as a potential blood-brain barrier drug transport vector, *J. Pharmacol. Exp. Ther.* 255:893-899.

Pardridge, W.M., and Boado, R.J., 1991, Enhanced cellular uptake of biotinylated antisense oligonucleotide or peptide mediated by avidin, a cationic protein, *FEBS* 288:30-32.

Rennke, H.G., Cotran, R.S., and Venkatachalam, M.A., 1975, Role of molecular charge in glomerular permeability. Tracer studies with cationized Ferritins, *J. Cell Biol.* 67:638-646.

Rippe, B., and Folkow, B., 1977, Capillary permeability to albumin in normotensive and spontaneously hypertensive rats, *Acta Physiol. Scand.* 101:72-83.

Rippe, B., and Haraldsson. B., 1994, Transport of macromolecules across microvascular walls: the two pore theory, *Physiol Rev* 74:163-219.

Salahuddin, T.S., Kalimo, H., Johansson, B.B., and Olsson, Y., 1988, Observations on exudation of fibronectin, fibrinogen and albumin in the brain after carotid infusion of hyperosmolar solutions, *Acta Neuropathol.* 76:1-10.

Scott, T., and Eaglesson, M., 1988, "Concise Encyclopedia of Biochemistry," pp. 45, 248, 482-483, W. de Gruyter, Berlin, New York.

Seiler, M.W., Venkatalam, M.A., and Cotran, R.S., 1975, Glomerular epithelium: structural alterations induced by polycations, *Science* 189:390-393.

Simionescu, M., and Simionescu, N., 1986, Functions of the endothelial cell surface, *Ann. Rev. Physiol.* 48:279-293.

Skutelsky, E., and Danon, D., 1976, Redistribution of surface anionic sites on the luminal front of blood vessel endothelium after interaction with polycationic ligand, *J. Cell Biol.* 71:232-241.

Strausbaugh, L.J., 1987, Intracarotid infusions of protamine sulphate disrupt the blood-brain barrier of rabbits, *Brain Res.* 409:221-226.

Thomas, G., Mostaghim, R., and Ramwell, P.W., 1986, Endothelium dependent vascular relaxation by arginine containing polypeptides, *Biochem. Biophys. Res. Commun.* 141:446-451.

Triguero, D., Buciak, J.B., Yang, J., and Pardridge, W.M., 1989, Blood-brain barrier transport of cationized immunoglobulin G: Enhanced delivery compared to native protein, *Proc. Natl. Acad. Sci. USA* 86:4761-4765.

Vehaskari, V.M., Root, E.R., Germuth, F.G., and Robson, A.M., 1982, Glomerular charge and urinary protein excretion: effects of systemic and intrarenal polycation infusion in the rat, *Kidney Int.* 22:127-135.

Vorbrodt, A.W., 1989, Ultracytochemical characterization of anionic sites in the wall of brain capillaries, *J. Neurocytol.* 18:359-368.

Vorbrodt, A.W., 1993, Morphological evidence of the functional polarization of brain microvascular endothelium, *in:* "The Blood-Brain Barrier," W.M. Pardridge, ed.,pp. 137-164, Raven Press, New York.

Vorbrodt, A.W., Dobrogowska, D. H., & Lossinsky, A. S., 1994, Ultracytochemical studies of the effects of aluminum on the blood-brain barrier of mice, *J. Histochem. Cytochem.* 42:203-212.

Vorbrodt, A.W., Dobrogowska, D.H., Lossinsky, A.S., and Wisniewski, H.M., 1990, Changes in the distribution of anionic sites in brain micro-blood vessels with and without amyloid deposits in scrapie-infected mice, *Acta Neuropathol.* 79:355-363.

Vorbrodt, A.W., Losskinsky, A.S., Wisniewski, H.M., et al., 1985, Ultrastructural observations on the transvascular route of protein removal in vasogenic brain edema, *Acta Neuropathol.* 66:265-273.

Vorbrodt, A. W., and Trowbridge, R. S., 1993, Aluminum-induced alteration of surface anionic sites in cultured brain microvascular endothelial cells, *Acta Neuropathol.* 86:371-377.

Westergren, I, and Johansson, B.B., 1990, Albumin content in brain and CSF after intracarotid infusion of protamine sulfate: a longitudinal study, *Exp Neurol.* 107:192-196.

Westergren, I, and Johansson, B.B., 1991, Dixyrazine, a phenothiazine derivate, can prevent brain oedema induced by intracarotid injection of protamine sulphate. *Acta Neurochir (Wien)* 113:171-175.

Westergren, I., and Johansson, B. B., 1993 a, Blockade of AMPA receptors reduces brain edema following opening of the blood-brain barrier. *J. Cereb. Blood Flow Metab.* 13:603-608.

Westergren, I., and Johansson, B. B. ,1993 b, Altering the blood-brain barrier in the rat by intracarotid infusion of polycations: a comparison between protamine, poly-L-lysine and poly-L-arginine. *Acta Physiol. Scand.* 149:99-104.

Westergren, I., Nordborg, C., and Johansson, B.B., 1993, Glutamate enhances brain damage and albumin content in CSF after intracarotid protamine infusion, *Acta Neuropathol.* 85:285-290.

Westergren, I., Nyström, B., Hamberger, A., Nordborg, C., and Johansson, B. B., 1994, Concentrations of amino acids in extracellular fluid after opening of the blood-brain barrier by intracarotid infusion of protamine sulfate, *J. Neurochem.* 62:159-165.

# CAPILLARY PERMEABILITY IN CENTRAL AND PERIPHERAL NERVE TISSUE IN STREPTOZOTOCIN DIABETES IN THE ANAESTHETISED RAT

G. G. Pinter and M. W. B. Bradbury

Physiology Group,
Biomedical Sciences Division,
King's College London,
Strand, London, WC2R 2LS.

In diabetes mellitus of long duration almost all capillary regions are severely affected by microangiopathy including peripheral nerves and the retina of the eye, yet there is little damage to brain capillaries. The experiments we report here were aimed at studying in streptozotocin treated rats potential functional changes in the capillaries of the brain after a long period in untreated diabetes. In co-operation with Professor Sue Lightman of the Institute of Ophthalmology, Moorfields Eye Hospital, University of London, we carried out a study lasting over 14 months after streptozotocin administration.

Since an increase of permeability coincides with or even precedes the appearance of histological changes, we looked for increases of the permeability times surface area (PA) product for sucrose in the brain and in leakage of plasma albumin through the capillary wall in various organs. The measurements were described earlier[1] and were carried out on both diabetic and control rats of the same age, strain and sex. Both experimental and control groups of rats were further subdivided into subgroups one of which received an aldose reductase inhibitor (Ponalrestat, ICI). Measurements of permeability were performed at 3 weeks, 7 months and 13-14 months after induction of diabetes.

At 3 weeks we found increase albumin leakage in the sciatic nerve in the diabetic rats, and at 7 months there was increased albumin leakage also in the skeletal muscle, while in other organs the findings were not different from control[1]. After 13 to 14 months in diabetes, however, there were substantial changes. The capillaries continued to be leaky in skeletal muscle, and became leaky also in the heart and the kidney. At this time we found changes also in the permeability of the capillaries in the nervous system, both in peripheral nerves and in a certain area in the brain. After 13 months duration of the diabetes there was a substantial and statistically highly significant increase in the PA product of sucrose in the retina[2]. At the same time, plasma albumin extravasated excessively in the optic nerve and the sciatic nerve[1,3]. In the sciatic nerve the uptake of sucrose also tended to be larger in diabetic rats than in the controls, however the difference did not reach statistical significance. The use of the aldose reductase inhibitor did not restore to normal the permeability of capillaries in any of these sites[1,2,3].

In several regions of the brain the extravascular plasma albumin space was very small, corresponding approximately to 0.1 ml of plasma in 100 g tissue, and remained at this level in long term diabetes, with the exception of one region: the hypothalamus[3]. Here, after 14 months of untreated diabetes there was a significant increase of the extravascular albumin space compared to the controls of the same age. This observation should be of interest since, among the regions of the brain we tested, it is only the hypothalamus where a tight blood-brain barrier is normally absent. We have speculated that a contributing factor to the microvascular disease in diabetes may be an exposure to high glucose concentration of the capillary wall from the interstitial side. If the capillary wall is tight and prevents free diffusion of glucose, interstitial glucose concentration in such a tissue should remain relatively low even if plasma glucose is elevated. As a consequence, the formation and maintenance of the basement membrane on the interstitial side of the endothelium might be affected to a lesser degree than in other tissues where glucose can diffuse freely through the capillary wall and excess glucose precursor for basement membrane synthesis is available in the interstitium.

An additional interesting finding was that in the brain capillaries at 3 weeks the PA product for sucrose was lower in the diabetic rats than in the controls. We attributed this difference to the condition that in these relatively young rats the blood-brain barrier may be somewhat less tight than in the adult rats, and that the stressful condition created by diabetes might initiate the release of adrenocortical steroids which could bring about a premature tightening of the barrier.

In the course of carrying out these experiments we became intrigued by the finding that the endoneurial fluid of peripheral nerve contains more Na and Cl than the surrounding interstitium, and that in different rat models of diabetes this hyperosmolality would change in different directions. Information in the literature indicated that in the streptozotocin model the hyperosmolality decreases, whereas in the galactose-fed rat model it increases. With the assistance of Mr Criss Cooper, a student in physiology working for his MSc degree, we confirmed that, indeed, in control conditions the endoneurial fluid is hyperosmotic and contains excess Na, and that hyperosmolality is further increased in galactose-fed rats. We attempted to discover how such high ion concentration might be established and maintained in the ocean of isosmotic extracellular fluid. Histological studies by Lundborg[4] depicted adjacent small vessels of the epineurium exhibiting predominantly countercurrent flow. We hypothesized that in analogy with the renal medulla such a countercurrent arrangement might function as an exchanger for water in one direction, and for Na and Cl in the other direction, and thereby contribute to the preservation of the hyperosmolality of the endoneurial fluid, and carried out experiments to test this hypothesis. These experiments totally ruled out any functional role for countercurrent exchange in the perfusion of the endoneurium. Perhaps, we might find comfort in the thought that according to Karl Popper[5], the purpose of experimentation is to falsify hypotheses, i.e. to show that a hypothesis is wrong. In this sense, our experiments were entirely successful and the questions remain: What are the processes that establish and maintain the hyperosmolality of the endoneurial fluid of peripheral nerve, and how are these processes altered in the galactosaemic and other models of diabetes mellitus? In celebrating the retirement of Mike Bradbury, it seems entirely appropriate to pass on this unresolved problem to future investigators with good wishes for framing hypotheses that are not as easily falsified as the countercurrent idea.

In honouring his co-author Professor Bradbury, GGP wishes to express a few personal thoughts. It would be difficult to deny that attitude and a specific value system are important ingredients of being a scientist. In this context, scientists may tend to belong into one of two different categories. One category would include those exemplified by the story of Faust who are easily carried into domains where power and fame are primary rewards. Into the second category would belong the inheritors of the Socratic tradition who love wisdom for wisdom's sake and science for science's sake. The primary reward of these scientists is that they pass on to future generations of investigators a reliable data-base with a treasure-house of thoughtful commentary. Michael Bradbury belonds to the latter category. It is this particular fact that makes it such a great privilege and pleasure for this contributor to participate in this Symposium honouring Michael Bradbury.

## REFERENCES


1. M.W.B. Bradbury, S.L. Lightman, L.Yuen and G.G. Pinter, Permeability of blood-brain and blood-nerve barriers in experimental diabetes mellitus in the anaesthetized rat, *Experimental Physiology* 76:887 (1991).
2. S. Lightman, G. Pinter, L. Yuen and M. Bradbury, Permeability changes at blood-retinal barrier in diabetes and effect of aldose reductase inhibition, *Am. J. Physiol* 259: (Regulatory Integrative Comp. Physiol. 28)R601 (1990).
3. G.G. Pinter, S.L. Lightman, P.D. Wilson, L. Yuen and M.W.B. Bradbury, Microvascular permeability in kidney,heart and skeletal muscle in experimental diabetes in the anaesthetized rat, *Experimental Physiology* 76:899 (1991).
4. G. Lundborg, Ischaemic nerve injury, *Scand.J.Plastic and Reconstr. Surg.* Suppl. No.6 (1970).
5. K. R. Popper,"Conjectures and Refutations: The Growth of Scientific Knowledge," Routledge & Kegan Paul, London (1963)

# SEROTONIN AS A MEDIATOR OF INCREASED MICROVASCULAR PERMEABILITY OF THE BRAIN AND SPINAL CORD

## Experimental Observations in Anaesthetised Rats and Mice

H. S. Sharma[1,2], Y. Olsson[1], and P. K. Dey[2]

[1]Laboratory of Neuropathology, University Hospital,
S-751 85 Uppsala, Sweden
[2]Department of Physiology, Institute of Medical Sciences,
Banaras Hindu University, Varanasi-221 005, India

## INTRODUCTION

Serotonin (5-hydroxytryptamine, 5-HT) is a neurotransmitter in the brain and spinal cord[1]. The amine is present in very high quantity in the brain and spinal cord of various mammalian species[1]. Trauma to the central nervous system results in an increased level of 5-HT in brain, spinal cord and cerebrospinal fluid[2-4]. There are several recent reports indicating that this amine can act as a mediator of increased vascular permeability in different pathological conditions[5-8].

The blood-brain barrier (BBB) regulates the composition of the fluid microenvironment of the CNS within strict limits[9-11]. A breakdown of the BBB occurs in many pathological conditions including trauma to the CNS[1,3,4,6,9] resulting in vasogenic edema and cell reactions[4]. Since 5-HT metabolism is altered in some pathological conditions[1-3] the possibility exists that the amine may influence the function of the BBB.

The hypothesis that 5-HT can serve as a mediator of increased vascular permeability was tested by infusing the amine into the systemic circulation and the internal carotid artery of normal animals. In other experiments 5-HT was applied topically on the exposed pial membrane of the parietal cortex. Vascular permeability was examined using Evans blue, [$^{131}$I]-sodium and lanthanum as tracers. The potential role of 5-$HT_2$ receptors in mediating the increased permeability was examined by pretreating rats and mice with ketanserin, a specific antagonist to this receptor[1].

## MATERIALS AND METHODS

### Animals

Experiments were carried out on young male Wistar rats (100-150 g, age 8-9 weeks) and young male mice (25-35 g) housed at controlled room temperature (21± 1° C) with a 12 h light and 12 h dark schedule. Food and tap water were supplied *ad libitum.*

### Infusion of serotonin

Under urethane anaesthesia (1.5 g/kg, i.p.), a freshly prepared solution of serotonin (5-HT-creatinine sulphate, Sigma Chemical Co., USA) was infused for 10 min either into the right jugular vein of rats or mice (5 or 10 μg/kg/min), or into the internal carotid artery of rats (1 or 2 μg/kg/min) or applied topically in rats on the right parietal cortex (0.05 μg/min). Control groups of rats or mice received equimolar concentration of saline.

### Blood-brain barrier permeability

The BBB permeability in rats was examined using Evans blue (0.3 ml/ 100g body weight of a 2 % solution in physiological saline, pH 7.4) and [$^{131}$I]-sodium (10 μ Ci/100 g)[12]. Both tracers were injected into the right jugular vein 5 min before serotonin application. The intravascular tracer was then washed out by a brief saline rinse through the heart and the brain was removed. Immediately before perfusion, a whole blood sample was withdrawn from the heart to determine blood radioactivity[12]. The brains were counted for radioactivity and the Evans blue dye which had entered the brain was measured[12].

The passage of a tracer across microvessels following 5-HT infusion was examined using lanthanum as an electron dense tracer[4]. For this purpose, immediately after saline rinse, mice were perfused with 50 ml of mixed aldehyde fixative (2 % paraformaldehyde, 1 % glutaraldehyde in 0.1 M phosphate buffer, pH 7.4 containing 2.5 % lanthanum nitrate) at 100 torr perfusion pressure. After perfusion, several tissue pieces from the brain and spinal cord were postfixed in $OsO_4$ and embedded in Epon for transmission electron microscopy[4].

### Influence of drugs

In separate groups of animals, ketanserin (Janssen Pharmaceuticals, Belgium) was administered (10 mg/kg, i.p.) 30 min before serotonin application[1]. The BBB permeability to Evans blue and [$^{131}$I]-sodium was determined in rats and lanthanum permeability was examined in mice.

### Statistical evaluation

Dunnet test for multiple group comparison was used to evaluate the statistical significance of quantitative data obtained. A p-value less than 0.05 was considered to be significant.

## RESULTS

Intravenous infusion of serotonin (10 μg/kg/min) increased the BBB permeability to Evans blue and [$^{131}$I]-sodium in rats (Table 1). The Evans blue staining was seen in cerebral cortex (cingulate, parietal, occipital, temporal and frontal), cerebellum (vermis and lateral hemisphere), dorsal surface of the hippocampus and caudate nucleus. The extravasation of radiotracer was present also in the thalamus, hypothalamus, colliculi and reticular activating system of the brain stem. Intracarotid infusion of 2 μg/kg/min serotonin was sufficient to open the BBB to Evans blue and radioactive iodine (Table 1). Application of 0.5 μg serotonin on the pial membrane resulted in extravasation of Evans blue and radiotracer in the ipsilateral brain parenchyma (Table 1).

**Table 1.** Effect of serotonin on blood-brain barrier permeability after different modes of application in the rat.

| Type of experiment | 5-HT dose/min 10 min | BBB permeability | | | |
|---|---|---|---|---|---|
| | | Evans blue mg % | | [$^{131}$I]-sodium % | |
| | | Right half | Left half | Right half | Left half |
| *Saline infusion* | | | | | |
| Intravenous (6) | - | 0.18±0.02 | 0.20±0.04 | 0.24±0.04 | 0.26±0.03 |
| Internal carotid artery (6) | - | 0.18±0.04[a] | 0.19±0.05 | 0.22±0.06[a] | 0.25±0.04 |
| Topical application (5) | - | 0.22±0.06[a] | 0.18±0.04 | 0.25±0.04[a] | 0.28±0.06 |
| *Serotonin infusion* | | | | | |
| Intravenous (8) | 10 μg/kg | 1.56±0.15** | 1.48±0.16** | 1.88±0.21** | 1.92±0.18** |
| Internal carotid artery (6) | 2 μg/kg | 1.44±0.12**[a] | 0.32±0.12 | 1.64±0.23**[a] | 0.45±0.22 |
| Topical application (7) | 0.05 μg | 0.89±0.14**[a] | 0.26±0.08 | 0.93±0.08**[a] | 0.24±0.06 |

** = P < 0.01, Dunnet test for multiple group comparison (compared from corresponding saline controls). Values are mean±SD. Figures in parentheses indicate number of animals. a = perfused hemisphere.

The increase in the BBB permeability was not present when the tracers were administered 120 min after the serotonin infusion. Treatment with the 5-HT$_2$ receptor antagonist ketanserin 30 min before all the three modes of 5-HT application prevented the extravasation of tracers (Figure 1).

The passage of lanthanum following intravenous infusion of serotonin was examined in mice (Figure 2). Lanthanum had penetrated diffusely into the endothelial cell cytoplasm in many regions of the brain and spinal cord (Figure 2 a). In large vessels, lanthanum was seen in the basal lamina (Figure 2 b). However, the tight junctions remained closed (Figure 2 a, b). In the outer layers of the cerebral cortex (I-III), lanthanum was occasionally present in the brain parenchyma. This extravasation was absent in mice treated with ketanserin.

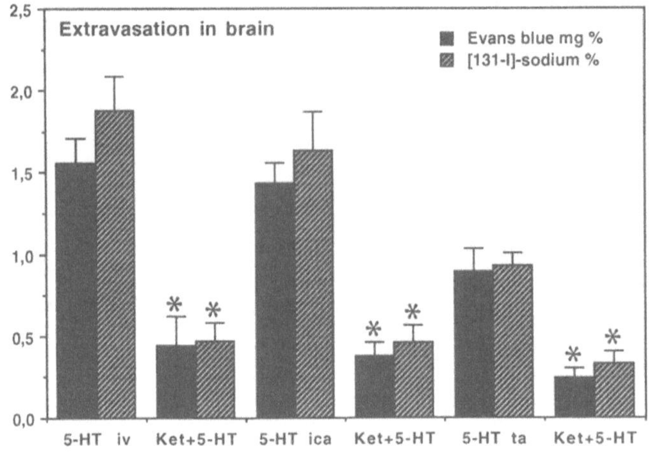


**Figure 1**. Influence of ketanserin (ket) on 5-HT induced increased permeability of the BBB to Evans blue and [$^{131}$I]-sodium. * = $P < 0.01$, Dunnet test for multiple group comparison (iv = intravenous, ica = internal carotid artery, ta = topical application).

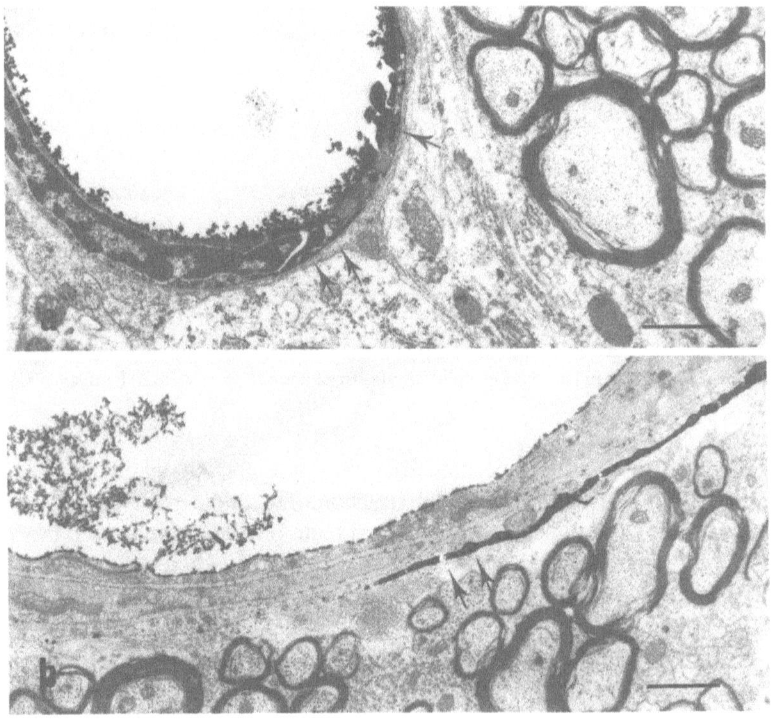

**Figure 2**. Exudation of lanthanum across spinal cord endothelial cells in mice after intravenous 5-HT infusion. Lanthanum has penetrated diffusely into the endothelial cell cytoplasm (a) and is present in the basal lamina . The tight junction is intact (b) (bar = 1 $\mu$m).

## DISCUSSION

Previous studies have demonstrated that serotonin can induce an opening of the BBB to various tracers[5,7,8,12]. Our results show that application of serotonin on the luminal and the abluminal side of CNS microvessels results in a significant increase of vascular permeability to Evans blue and [$^{131}$I]-sodium tracers and indicate that this occurs in a dose dependent manner[12]. The increased permeability was absent 120 min after the serotonin infusion indicating the reversible nature of the BBB opening.

The detailed mechanisms of serotonin induced increased permeability are not clear[6-8,12]. However, our results suggest that 5-$HT_2$ receptors are involved since pretreatment with ketanserin, a specific 5-$HT_2$ receptor antagonist[1] prevented the increased BBB permeability seen after serotonin infusion. It may be that serotonin after binding to its 5-$HT_2$ receptors may induce an increased synthesis of cAMP either directly or indirectly via prostaglandins[12,13]. An increased synthesis of cAMP may be involved in an increased transcellular transport of the tracers[13].

Our results in mice suggest that serotonin has the capacity to induce increased transcellular transfer of lanthanum. Thus, lanthanum had penetrated into the endothelial cell cytoplasm and in the basal lamina without disrupting the tight junctions. However, not all the vessels perfused with serotonin showed an increased permeability to lanthanum. This may be due to regional differences in the sensitivity of microvessels to serotonin or to a difference in the distribution of serotonergic receptor subtypes in various vascular segments.

In conclusion, our studies show that elevation of circulating serotonin can induce a transient breakdown of the BBB. This effect appears to be mediated by 5-$HT_2$ receptors.

## ACKNOWLEDGEMENTS

Supported by grants from the Swedish Medical Research Council project no. 0320, Trygg-Hansa, Göran Gustafsson, Hedlund, Åhlén, Thyring, and O E and Edla Johansson's Foundations; and the University Grants Commission, New Delhi, India.

## REFERENCES

1.W. Essman, "Serotonin in Health and Diseases", vol I-V, Spectrum, New York (1978).

2.P. Kamoun and H. Jerome, Changes in blood serotonin in mental abnormalities, *in:* Serotonin in Mental Abnormality, D. Boullin, ed., Wiley, Chicester (1978).

3.D. Boullin, "Serotonin in Mental Abnormalities", Wiley, Chicester (1978).

4.H. S. Sharma, J. Cervós-Navarro, G. Gosztonyi and P. K. Dey. Role of serotonin in traumatic brain injury. An experimental study in the rat, *in*: The Role of Neurotransmitters in Brain Injury, M. Globus and W. D. Dietrich, eds., Plenum Press, New York, pp. 147 (1992).

5.J. Greenwood, Experimental manipulation of the blood-brain and blood-retinal barriers, *Hand book Exp. Pharmacol.* 103:439 (1992).

6.M. Wahl, A. Unterberg, A. Baethmann and L. Schilling, Mediators of blood-brain barrier dysfunction and formation of vasogenic brain edema. *J. Cereb. Blood Flow Metab.* 8:621 (1988).

7.E. Westergaard, Enhanced vesicular transport of exogenous peroxidase across cerebral vessles, induced by serotonin. *Acta Neuropathol.* 39:27 (1975).

8.S. P. Olesen, An electrophysiological study of microvascular permeability and its modulation by chemical mediators. *Acta Physiol. Scand.* 136 *Suppl.* 579:1 (1988).

9.P. J. Luthert, Opening of the barrier in cerebral pathology, *Hand book Exp. Pharmacol.* 103:439 (1992).

10.S. I. Rapoport, "Blood-Brain Barrier in Physiology and Medicine", Raven Press, New York (1976).

11.M. W. B. Bradbury, "The Concept of a Blood-Brain Barrier", Wiley, Chichester (1979).

12.H. S. Sharma, Y. Olsson and P. K. Dey, Changes in blood-brain barrier and cerebral blood flow following elevation of circulating serotonin levels in anaesthetized rats. *Brain Res.* 517:215 (1990).
13.F. Joò, Z. Rakonczay and M. Wollemann, cAMP-mediated regulation of the permeability in the brain capillaries. *Experientia* 31: 582 (1975).

# MOLECULAR REGULATION OF BLOOD-BRAIN BARRIER GLUT1 GLUCOSE TRANSPORTER

William M. Pardridge

Department of Medicine
Brain Research Institute
UCLA School of Medicine
Los Angeles, CA 90024

## INTRODUCTION

The transport of glucose from the circulation to brain intracellular space involves transport through two membranes in series: the brain capillary endothelial cell, which constitutes the blood-brain barrier (BBB) in vivo, and the brain cell membrane, comprised of neuronal or glial plasma membrane. Owing to the vastly greater surface area of brain cell membranes, relative to the BBB surface area, the rate-limiting transport step of these two membranes in series is at the BBB (Lund-Andersen, 1979). Therefore, modulation of the BBB glucose transporter activity may be an important control point in the overall regulation of cerebral glucose metabolism.

The presence of a saturable carrier-mediated transport system for D-glucose at the BBB has been known for many years (Crone, 1965; Oldendorf, 1971). More recently, several members constituting a sodium-independent glucose transporter gene family have been identified and individual members of this gene superfamily are designated GLUT1-GLUT5 (Table 1). Given the multitude of sodium-independent glucose transporters that are expressed, it is important to identify the transporter isoform responsible for glucose transport through the BBB. This is because the antibodies or cDNAs used to investigate molecular regulation of BBB glucose transporter activity are isoform-specific. The cDNA for the GLUT1 glucose transporter isoform was originally obtained from a human hepatoma library (Mueckler, et al., 1985), and subsequently from rat brain (Birnbaum, et al., 1986). Although earlier studies showed an enrichment of the GLUT1 isoform at the microvasculature of brain (Flier et al., 1987), subsequent studies showed that the gene expression of the GLUT1 isoform in brain is confined, under normal conditions, to the brain microvascular endothelium (Pardridge and Boado, 1993).

*New Concepts of a Blood–Brain Barrier*
Edited by J. Greenwood *et al.*, Plenum Press, New York, 1995

**Table 1.** Sodium-Independent Glucose Transporter (GLUT) Genes

| Gene | Tissue | % Homology to GLUT |
|---|---|---|
| GLUT1 | BBB, human erythrocyte | 100 |
| GLUT2 | liver, $\beta$ cell of pancreas | 68 |
| GLUT3 | neuron | 74 |
| GLUT4 | muscle, fat | 76 |
| GLUT5 | intestinal epithelium | 58 |

From Bell, et al. (1990).

## LOCALIZATION OF GLUT1 GLUCOSE TRANSPORTER ISOFORM TO ENDOTHELIUM IN BRAIN

The initial evidence demonstrating restriction of GLUT1 gene expression in brain to the microvascular endothelium was obtained with Northern blotting analysis coupled with a capillary depletion technique (Boado & Pardridge, 1990). As shown in Figure 1B,

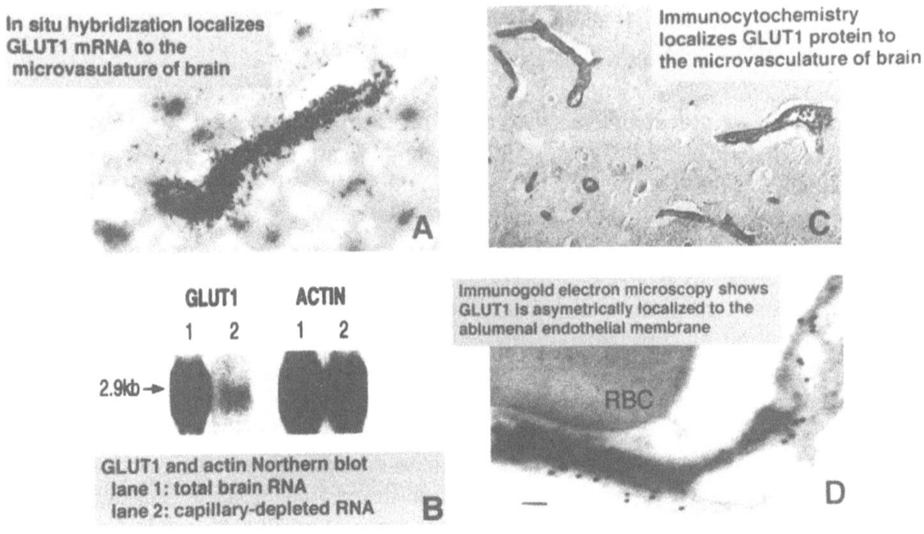


**Figure 1.** Blood-brain barrier glucose transport is mediated by the GLUT1 glucose transporter. (A) In situ hybridization localizes GLUT1 mRNA to the microvasculature of brain. From Pardridge, et al. (1990). (B) GLUT1 and actin Northern blotting using capillary depletion analysis demonstrates that the GLUT1 mRNA is depleted from brain homogenate following removal of microvessels from the homogenate; lane 1 is total rabbit brain RNA and lane 2 is capillary-depleted rabbit brain RNA. The capillary depletion step removes approximately 95% of the GLUT1 mRNA, which parallels the extent of removal of gamma-glutamyl transpeptidase activity from the homogenate using the capillary depletion technique. From Dwyer and Pardridge (1993). (C) Immunocytochemistry using the avidin-biotin peroxidase technique localizes GLUT1 protein to the microvasculature of brain. From Pardridge, et al. (1990). (D) Immunogold electron microscopy shows GLUT1 is asymmetrically localized at the ablumenal endothelial membrane of rat brain. No GLUT1 transporters are detected on the red blood cells (RBC) owing to the paucity of erythrocyte GLUT1 transporters in rodent RBCs. From Farrell and Pardridge (1991).

approximately 95% of the brain GLUT1 mRNA is removed from the brain homogenate by a centrifugation step that quantitatively depletes the homogenate of brain microvascular-derived RNA. The depletion of the GLUT1 mRNA from brain corresponds to the depletion of the enzymatic activity for gamma-glutamyl transpeptidase, an enzyme that is confined to the brain microvasculature.

The Northern blotting study was subsequently confirmed with in situ hybridization (Figure 1A), which showed that riboprobes specific for the GLUT1 isoform hybridized to the microvasculature. No measurable hybridization within brain parenchyma, that was statistically different from the control sense riboprobe, was observed.

In parallel with the GLUT1 mRNA, the immunoreactive GLUT1 protein, as assessed by immunocytochemistry at the light microscopy level or by immunogold electron microscopy, was detectable only over the microvascular endothelium. At the light microscopic level (Figure 1C), only microvessels in brain are illuminated by an antiserum specific to either a synthetic peptide encoding the 13 amino acids at the carboxyl terminus of the GLUT1 isoform or by an antiserum prepared against purified human erythrocyte GLUT1 glucose transporter. No GLUT1 immunoreactivity in brain parenchyma is detectable at the light microscopic level. Although original immunogold silver staining studies detected GLUT1 immunoreactivity in neuropil (Bagley, et al, 1989), subsequent experiments using a light insensitive silver-enhancement stabilizer failed to detect

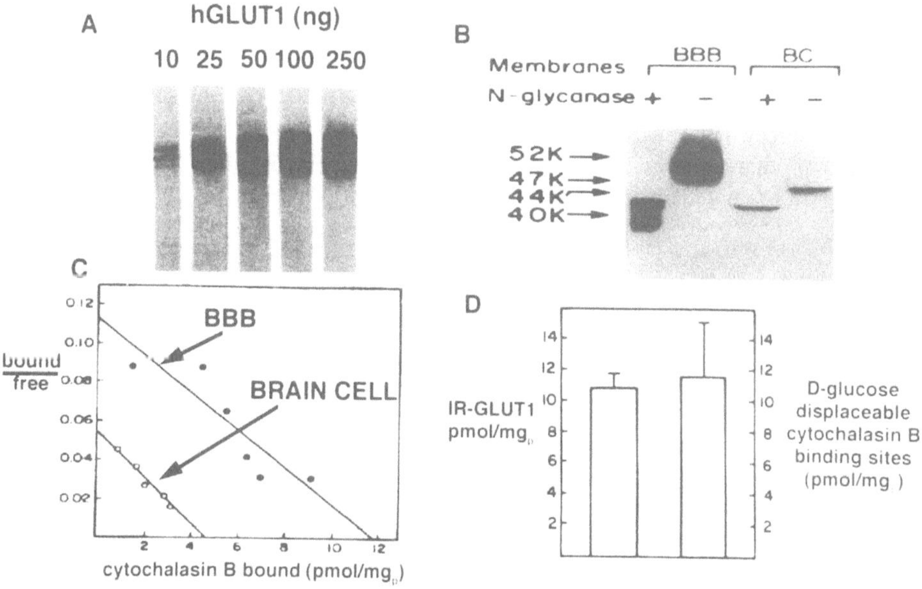


**Figure 2.** Immunoreactive GLUT1 and total D-glucose displaceable cytochalasin B binding sites are equivalent in bovine brain microvessels. Quantitative Western blotting using the human erythrocyte glucose transporter (hGLUT1) as an assay standard (A) allows for quantitation of immunoreactive GLUT1 glucose transporter molecules in bovine or blood-brain barrier (BBB) plasma membranes or in capillary-depleted brain cell (BC) membranes (B). The total number of D-glucose displaceable cytochalasin B binding sites on either bovine brain microvessel or brain cell plasma membrane is measured with $^{3}$H-cytochalasin B radioreceptor assay and Scatchard analysis (C). The concentration of immunoreactive GLUT1 (pmol/mg protein) and D-glucose sensitive cytochalasin B binding sites (pmol/mg protein) is not significantly different in bovine brain capillary plasma membranes (D). From Pardridge, et al. (1990). Similar results have been obtained with intact microvessels isolated from rabbit brain. Dwyer and Pardridge (1993).

immunoreactive GLUT1 within brain neuropil (Farrell, et al., 1992). The use of the immunogold electron microscopic technique shows a decoration of brain capillary endothelial lumenal and ablumenal plasma membranes by the antiserum with approximately a 3:1 asymmetric distribution of the GLUT1 isoform at the ablumenal membranes (Figure 1D). Approximately 40% of the endothelial GLUT1 isoform is localized within the endothelial cytoplasm. The asymmetric localization of the GLUT1 isoform to brain capillary endothelial membranes underscores the importance of using a double membrane model to investigate the physiology of blood-brain barrier glucose transporter activity (Cunningham, et al., 1986).

The asymmetric localization of the GLUT1 isoform at the ablumenal membrane suggests the possibility that glucose transporters other than the GLUT1 isoform are situated at the lumenal membrane. This was evaluated by determining the total number of D-glucose displaceable cytochalasin B binding sites on the blood-brain barrier plasma membranes using $^3$H-cytochalasin B radioreceptor assays (Figure 2C). In parallel, the same bovine brain capillary plasma membranes were evaluated with a quantitative Western blot procedure (Figure 2B), that used purified human erythrocyte GLUT1 glucose transporter as an assay standard (Figure 2A), and the total concentration of immunoreactive GLUT1 molecules in the capillary plasma membranes was measured (Figure 2D). These data show that there is equivalence between the total number of glucose transporter binding sites and the number of immunoreactive GLUT1 glucose transporter molecules in bovine brain capillaries. A similar equivalence has been demonstrated for rabbit brain capillaries (Dwyer and Pardridge, 1993). Given these findings, the GLUT1 glucose transporter isoform would appear to be the major, if not the only isoform, mediating BBB glucose transporter activity in vivo, at least under normal conditions.

Immunoreactive GLUT1 with reduced molecular weight is detectable in capillary-depleted brain cell (BC) membranes by Western blotting (Figure 2B). The function and cell origin of this protein is not known since immunoreactive GLUT1 protein is not measurable in brain cells using morphologic techniques (Figure 1C).

## MODULATION OF BLOOD-BRAIN BARRIER GLUT1 GLUCOSE TRANSPORTER

The BBB GLUT1 glucose transporter isoform is subject to modulation under a variety of pathophysiologic conditions (Table 2). One very pronounced form of modulation occurs in human epileptic foci. Immunogold electron microscopic studies were used to demonstrate an approximately 30-fold increase in the number of GLUT1 glucose transporters in human brain microvascular endothelium perfusing active epileptic foci (Figure 3). Early studies have shown that the rate of glucose utilization in convulsions is increased to levels that cannot be supported by glucose transporter activities found under normal conditions (Chapman, et al., 1977). This led to the hypothesis that the BBB glucose transporter was upregulated in seizures and this hypothesis has recently been verified in the case of human epilepsy (Figure 3). The molecular mechanisms by which the BBB GLUT1 glucose transporter is regulated in epilepsy or the other pathophysiologic conditions listed in Table 2 are at present unclear. Nevertheless, the preliminary data listed in Table 2 show a discordant regulation between the immunoreactive GLUT1 protein and the GLUT1 mRNA found in brain microvascular endothelium. This discordant regulation between the protein and the mRNA suggests there are post-transcriptional mechanisms regulating the activity of the BBB GLUT1 glucose transporter. Post-transcriptional mechanisms may pertain to either the efficiency of translation of the GLUT1 mRNA into the protein or the level of stability of the GLUT1 mRNA within the endothelial cytosol. For example, recent studies have shown that glucose deprivation in cultured brain capillary endothelium results in an increase of the GLUT1 mRNA through a post-transcriptional mechanism related to enhanced GLUT1 mRNA stability (Boado and Pardridge, 1993).

**Table 2.** Modulation of BBB GLUT1 Glucose Transporter

| Condition | GLUT1 mRNA | GLUT1 protein |
|---|---|---|
| experimental diabetes | increased | decreased |
| chronic hypoglycemia | increased | increased |
| development (7 days) | unchanged | decreased |
| (28 days) | unchanged | increased |
| human glioma (grade IV) | increased | decreased |

## POST-TRANSCRIPTIONAL REGULATION OF GLUT1 GENE EXPRESSION

There are two principle levels at which gene expression may be regulated: at the nuclear level and at the cytoplasmic level (Figure 4). Within the nucleus, gene expression may be regulated via transcriptional mechanisms involving interaction of nuclear trans-acting factor (TAF) proteins with specific sequences within the 5'- or 3'-flanking sequences of the gene. Alternatively, in the cytosol, gene expression may be regulated at the post-transcriptional level involving the interaction of cytosolic factors (CF) with specific cis-regulatory sequences within the 5'- or 3'-untranslated region (UTR) of the specific mRNA molecule. The hypothetical interaction between the mRNA and cytosolic CF may be

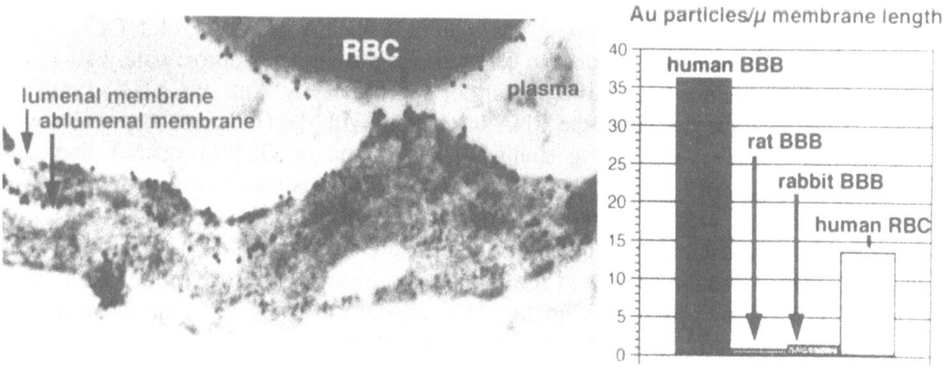


**Figure 3.** Blood-brain barrier GLUT1 glucose transporter is subject to modulation that regulates glucose availability to brain cells in pathophysiologic states such as seizures. Immunogold electron microscopy shows marked upregulation of GLUT1 on brain endothelial lumenal and ablumenal membranes in human epileptic foci. The number of gold particles per micron of membrane length is increased approximately 30-fold in the human BBB within the epileptic foci relative to the rat or rabbit BBB and exceeds the GLUT1 glucose transporter density on human red blood cells (RBC). Electron micrograph from Cornford, et al. (1994).

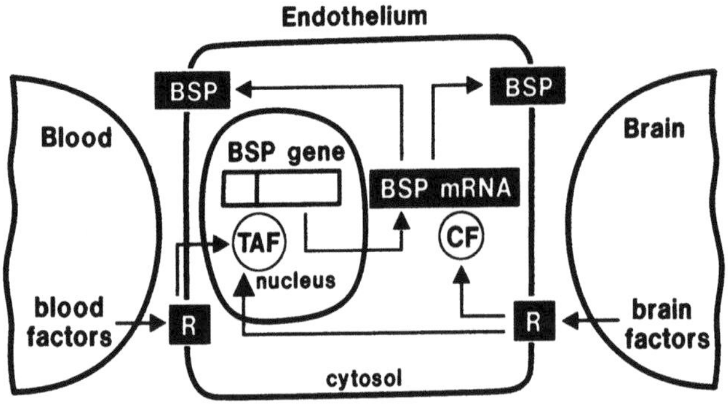


**Figure 4.** Blood-brain barrier tissue-specific gene expression is believed to be regulated by both blood- and brain-derived factors, which trigger specific receptors (R) on the brain capillary endothelial plasma membrane. Subsequent signal transduction cascade may then result in the elaboration of nuclear trans-acting factors (TAF) or cytosolic factors (CF) that interact with specific cis-regulatory sequences within the gene at the nuclear level (i.e., transcriptional regulation) or with the mRNA in the cytosol (i.e., post-transcriptional regulation). These combined factors may then regulate the ultimate translation of the mRNA encoding for a brain capillary-specific protein (BSP), which is subsequently deposited on the lumenal or ablumenal membrane of the brain capillary endothelium. From Pardridge (1991).

characterized using RNA gel-shift mobility assays. In our laboratory, we have prepared synthetic, radiolabeled RNA corresponding to the approximate 1000 bases of the GLUT1 3'-UTR using an in vitro transcription plasmid. This radiolabeled RNA was bound by brain capillary cytosolic proteins and subsequently degraded by RNase T1, which cuts the RNA opposite guanosine or G-residues. Specific sequences within the RNA may be protected from the RNase T1 action if BBB CFs are bound to the RNA. When such assays are performed with the GLUT1 mRNA, cytosolic CFs are found within brain capillary cytosol (Dwyer and Pardridge, 1994). The concentration of these CFs are increased in the rat glioma C6 cell. Subsequent studies used C6 cytosolic proteins to further characterize these cis/trans mechanisms involving the GLUT1 mRNA and cytosolic CFs. The interaction was saturable with respect to C6 cytosolic protein and approximately 10-15 $\mu$g of C6 cytosolic protein saturated the binding of approximately 1 ng of GLUT1 synthetic mRNA. The cis sequence within the RNA interacting with the C6 cytosolic proteins was localized with RNase T1 mapping studies. The region of GLUT1 mRNA that was protected from RNase T1 activity by the C6 cytosolic proteins was eluted from 7% native polyacrylamide gel, extracted, and subsequently electrophoresed on a 7M urea 20% polyacrylamide sequencing gel in parallel with an oligodeoxynucleotide sizing ladder. The RNase T1 mapping studies localized the initial cis-sequence of the GLUT1 mRNA to a specific sequence of nucleotides within the RNA molecule. The cis-regulatory element was further localized using oligodeoxynucleotide competition studies. The C6 cytosolic factor was characterized with respect to molecular weight using ultraviolet (UV) crosslinking of [$^{32}$P]GLUT1 mRNA to the C6 cytosolic protein followed by separation using sodium dodecyl sulfate polyacrylamide gel electrophoresis (SDS-PAGE) and autoradiography. These studies demonstrated that the molecular weight of the cytosolic CF that bound to the cis-regulatory sequence within the GLUT1 3'-UTR was approximately 88 kDa (Dwyer and Pardridge, 1994).

## CONCLUSIONS

In summary, the principle, if not only, glucose transporter mediating glucose flux through the BBB under normal conditions is the GLUT1 isoform. There is minimal, if any, expression of the GLUT1 gene in brain cells, apart from the capillary endothelium under normal conditions. The BBB GLUT1 isoform is subject to modulation at the molecular level under a variety of pathophysiologic conditions and recent studies suggest that a principle mechanism underlying this modulation is at the post-transcriptional level. Gel-shift mobility assays, RNase T1 mapping studies, and UV crosslinking experiments have provided the initial characterization of specific cis/trans mechanisms involving sequences within the 3'-UTR of the GLUT1 mRNA and specific cytosolic factors present in both brain capillary cytosol and rat glioma cells.

## ACKNOWLEDGEMENTS

This work was supported by NIH grant P01-NS-25554.

## REFERENCES


Bagley, P.R., Tucker, S.P., Nolan, C., et al., 1989, Anatomical mapping of glucose transporter protein and pyruvate dehydrogenase in rat brain: an immunogold study, *Brain Res.* 499:214.

Bell, G.I., Kayano, T., Buse, J.B., et al., 1990, Molecular biology of mammalian glucose transporters, *Diabetes Care* 13:198.

Birnbaum, M.J., Haspel, H.C., and Rosen, O.M., 1986, Cloning and characterization of a cDNA encoding the rat brain glucose-transporter protein, *Proc. Natl. Acad. Sci. USA* 83:5784.

Boado, R.J., and Pardridge, W.M., 1990, The brain-type glucose transporter mRNA is specifically expressed at the blood-brain barrier, *Biochem. Biophys. Res. Comm.* 166:174.

Boado, R.J., and Pardridge, W.M., 1993, Glucose deprivation causes post-transcriptional activation of brain capillary endothelial glucose transporter gene expression via GLUT1 mRNA stabilization, *J. Neurochem.* 60:2290.

Chapman, A.G., Meldrum, B.S., and Siesjö, B.K., 1977, Cerebral metabolic changes during prolonged epileptic seizures in rats, *J. Neurochem.* 28:1025.

Cornford, E.M., Hyman, S.H., and Swartz, B.E., 1994, The human brain GLUT1 glucose transporter: ultrastructural localization to the blood-brain barrier endothelia, *J. Cereb. Blood Flow Metab.* 14:106.

Crone, C., 1965, Facilitated transfer of glucose from blood into brain tissue, *J. Physiol.* 181:103.

Cunningham, V.J., Hargreaves, R.J., Pelling, D., and Moorhouse, S.R., 1986, Regional blood-brain glucose transfer in the rat: a novel double-membrane kinetic analysis, *J. Cereb. Blood Flow Metab.* 6:305.

Dwyer, K.J., and Pardridge, W.M., 1993, Developmental modulation of blood-brain barrier and choroid plexus GLUT1 glucose transporter mRNA and immunoreactive protein in rabbits, *Endocrinol.* 132:558.

Dwyer, K.J., and Pardridge, W.M., 1994, Novel cis-element/trans-acting protein interaction with the 3'untranslated region of GLUT1 glucose transporter mRNA, *FASEB J.* 8:A374.

Farrell, C.L., and Pardridge, W.M., 1991, Blood-brain barrier glucose transporter is asymmetrically distributed on brain capillary endothelial lumenal and ablumenal plasma membranes; an electron microscopic immunogold study, *Proc. Natl. Acad. Sci.* 88:5779.

Farrell, C.L., Yang, J., and Pardridge, W.M., 1992, GLUT1 glucose transporter is present within apical and basolateral membranes of brain epithelial interfaces and in microvascular endothelia barriers with and without tight junctions, *J. Histochem. Cytochem.* 40:193.

Flier, J.S., Mueckler, M., McCall, A.L., and Lodish, H.F., Distribution of glucose transporter messenger RNA transcripts in tissues of rat and man, *J. Clin. Invest.* 79:657-661.

Lund-Andersen, H., 1979, Transport of glucose from blood to brain, *Physiol. Rev.* 59:305.

Mueckler, M., Caruso, C., Baldwin, S.A., et al., 1985, Sequence and structure of a human glucose transporter, *Science* 229:941.
Oldendorf, W.H., 1971, Brain uptake of radiolabeled amino acids, amines, and hexoses after arterial injection, *Am. J. Physiol.* 221:1629.
Pardridge, W.M., 1991, "Peptide Drug Delivery to the Brain," Raven Press, N.Y.
Pardridge, W.M., and Boado, R.J., 1993, Molecular cloning and regulation of gene expression of blood-brain barrier glucose transporter, *in*: "The Blood-Brain Barrier: Cellular and Molecular Biology," W.M. Pardridge, ed., Raven Press, N.Y.
Pardridge, W.M., Boado, R.J., and Farrell, C.R., 1990, Brain-type glucose transporter (GLUT-1) is selectively localized to the blood-brain barrier. Studies with quantitative Western blotting and in situ hybridization, *J. Biol. Chem.* 265:18035.

# THE DEPENDENCY OF INFLUX ACROSS THE BLOOD-BRAIN BARRIER ON BLOOD FLOW AND THE APPARENT FLOW-INDEPENDENCE OF GLUCOSE INFLUX DURING STRESS

J. D. Fenstermacher,[1,2] L. Wei,[2] V. Acuff,[2] S.-Z. Lin,[2] J.-L. Chen,[2] D. Bereczki,[2] T. Otsuka,[2] H. Nakata,[2] A. Tajima,[2] F.-J. Hans,[2] J.-F. Ghersi-Egea,[2] W. Finnegan,[2] G. Richardson,[2] H. Haspel,[1] and C. Patlak[3]

[1]Department of Anesthesiology
Henry Ford Hospital
Detroit, MI 48202-2689

[2]Department of Neurological Surgery
[3]Department of Surgery
State University of New York
Stony Brook, NY 11794-8122

## INTRODUCTION

The major functions of the blood circulating through microvascular beds are to: 1) delivery nutrients, blood-borne messengers, and immunoactive materials to the surrounding cells and 2) remove locally produced metabolites, hormones, toxic substances, and heat from the tissue. Useful changes in blood flow through local microvascular networks would be expected to be linked to alterations in one or more of these delivery and/or removal needs (luxury perfusion - often observed following severe ischemia or stroke - is a condition in which blood flow apparently greatly exceeds tissue need and is an example of a seemingly useless or "futile" change in blood flow).

Under conditions that appreciably alter brain activity and metabolism, local cerebral blood flow (LCBF) can change by 50% or more. These changes may be highly localized as occurs when neuroelectrical activity increases within a brain structure or global as happens with hypoxia and hypercapnia. Although many studies have shown variations in LCBF under various conditions, the mechanism or mechanisms of changing blood flow at the cerebral capillary network level has not been firmly established.

In theory, alteration in cerebral blood flow can be effected by either changes in the linear velocity of blood flow through the perfused capillaries or variations in the total number of the perfused capillaries or changes in both velocity and number of perfused capillaries. In the case

*New Concepts of a Blood–Brain Barrier*
Edited by J. Greenwood *et al.*, Plenum Press, New York, 1995

of varying intracapillary flow velocity, the number of perfused capillaries in the individual brain areas and the volume of blood circulating and contained within the parenchyma (often referred to as the tissue distribution space or simply the tissue space) would be essentially constant when flow changes. In the case of the variable number of perfused capillaries, which is generally referred to as the capillary recruitment hypothesis, blood flow changes would cause an increase (recruitment) or decrease (retirement) in the number of perfused capillaries and the blood volume or space of the parenchyma.

The preceding "theory" suggests two ways of examining the mechanism by which blood flow through local cerebral microvascular systems changes. Firstly and more obviously, capillary recruitment could be assessed by measuring the number of capillaries labeled by a circulating intravascular marker such as fluorescent dextran or the volume of blood circulating and contained in the parenchyma. If blood flow increases by capillary recruitment, then the number of labeled capillaries or the volume of radiotagged blood would rise; conversely, these two parameters would become smaller when blood flow decreases and capillaries retire from the business of transporting blood.

Secondly and less obviously, the capillary recruitment hypothesis could be tested by measuring blood flow and influx rate (Ki) of a moderately permeable substance and subsequently calculating the latter's permeability-surface area (PS) product. If blood flow increases by recruitment, then a greater amount of capillary surface area (S) would be perfused and both Ki and PS would be larger than in controls. On the other hand, if blood flow increases by raising flow velocity, then the influx rate would become larger but the PS product would remain unchanged.

To diverge a bit on the coupling between influx rate and flow velocity, when blood flow rises by this means, then the greater velocity of blood convection establishes a higher concentration within the capillaries - usually expressed as the mean intracapillary concentration. The capillary concentration is the driving force for influx, and an elevation in it leads to an increase in solute transfer from blood to brain without altering PS. Mathematically this relationship is described for the Krogh single capillary model by the rearranged Renkin-Crone equation, $PS = -Fv \ln (1 - Ki/Fv)$, where Fv is the rate of volume flow through the capillary of the portion of the perfusing fluid - either blood or plasma - that contains exchangeable material (Chen et al., 1994a, 1994b; Otsuka et al., 1991b).

With respect to the effective flow rate or Fv, plasma water is the source fluid for the exchange of any material from blood to parenchyma. Material that is lost from plasma water can, however, be partially or nearly completely replenished by release from plasma proteins and/or blood cells. Accordingly, Fv is equal to or greater than the flow rate of plasma water. Antipyrine is an example of substance that is rapidly released from both plasma proteins and red cells (Chen et al., 1994a, 1994b).

To return to the capillary recruitment hypothesis, the reports of Pawlik et al. (1981), Rosenblum and Zweifach (1963), Tajima et al. (1992), and Villringer et al. (1991) are consistent with a microvascular system in which: 1) some but not all capillaries are perfused by both plasma and red cells (hematic capillaries); 2) others are perfused only by plasma (plasmatic capillaries); and 3) the remaining capillaries, if there are any, contain fluid that does not flow. About 85% of brain capillaries appear to be hematic in normal conditions (Tajima et al., 1992; Villringer et al., 1991). In view of this, the capillary recruitment hypothesis perhaps should be amended to cover recruiting unperfused capillaries for plasma and red cell flow as well as recruiting plasmatic capillaries for blood cell flow, thus becoming hematic.

Hereafter, both published and unpublished data are presented that challenge the capillary recruitment hypothesis for five conditions (given below) that markedly change local cerebral blood flow. The hypothesis tested consists of four parts: 1) changes in LCBF are mainly accomplished by modifying the velocity of plasma and red cell flows through the microvessels and not by the alterations in the local number of perfused capillaries; 2) red cells and plasma proteins (and by extension plasma *per se*) do not pass through cerebral microvascular beds via identical pathways at identical velocities, and plasma and erythrocyte flow patterns may change in different ways during stress; 3) when blood flow is modified, the influx of highly permeable substances are proportionally changed, that of moderately permeable materials are somewhat less - but still appreciably - altered, and that of slightly to modestly permeable compounds are virtually unchanged; and 4) glucose influx alterations under some conditions of stress are seemingly independent of blood flow changes.

This hypothesis has been tested by measuring LCBF plus labeled red cell and albumin distribution within the parenchyma (done in a way to indicate the volume of radiolabeled blood in microvessel and not all parenchymal vessels) and estimating microvessel hematocrits and mean transit-times of whole blood, red cell, and plasma passage through parenchymal microvascular beds. In addition, blood-to-brain influx rate constants of two solutes, antipyrine (a marker of lipid-mediated transport) and 3-O-methyl-D-glucose (3OMG, a marker of carrier-mediated D-glucose transport) were determined and PS products were estimated from Ki and LCBF values. Tissue radioactivity was assessed by quantitative autoradiography. The five models of blood flow perturbation are hypercapnia, hypoxia, acute nicotine and pentobarbital treatment, and 24 hours of dehydration.

## METHODS

Animal care, surgical preparation and experimental procedures were approved by the Institutional Animal Care and Use Committee of the University at Stony Brook (New York). They were designed in accordance with federal guidelines developed by the American Association for Accreditation of Laboratory Animal Care (AAALAC). The general procedures have been previously described (for example: Otsuka et al., 1991a, 1991b; Tajima et al., 1992; Bereczki et al., 1993a, 1993b; Chen et al., 1994a, 1994b).

Male Sprague-Dawley rats, weighing around 300 to 350 g, were anesthetized with 1.5% halothane and a mixture of 70% nitrous oxide and 30% oxygen. While anesthetized, femoral arteries and veins were catheterized, and the incisions were infiltrated with lidocaine and closed with suture. Subsequently, the rats were immobilized by enclosing the hindquarters and abdomen in a plaster cast. The physiological status of the rats was monitored frequently during the 2-3 hr period of recovery. These measurements included rectal temperature arterial blood pressure, blood gases, plasma osmolality, and plasma glucose concentration.

Cerebral blood flow was altered, either globally or just in some brain areas (local brain activation) by one of the following experimental manipulations: 2) hypercapnia, produced by respiring 8% $CO_2$ (Bereczki et al., 1993b; Chen et al, 1994a); 2) hypoxia, induced by respiring 10% $O_2$ (Bereczki et al., 1993a; Chen et al., 1994b); 3) nicotine, injected into the peritoneal cavity (Hans et al., 1993); 4) pentobarbital, administered via an intraperitoneal catheter (Otsuka et al., 1991a, 1991b; Wei et al., 1993); and 5) dehydration, one day without water (unpublished to date). Local cerebral blood flow, distribution volumes of RBC's and RISA in parenchymal microvascular systems, and local blood-brain transfer constants of radiolabeled antipyrine and 3-O-methyl-D-glucose were measured in the five experimental groups plus controls and shams.

Local cerebral blood flow was measured by $^{14}$C-iodoantipyrine (IAP)-autoradiographic technique of Sakurada et al. (1978) as modified by Otsuka et al. (1991a). A solution of $^{14}$C-IAP was intravenously infused over 30 sec according to a schedule that produces a continuously increasing concentration of $^{14}$C-IAP in arterial blood over time. During this time, arterial blood samples were obtained every 5 sec. The rats were decapitated at the completion of the last sampling period, which came approximately 30 sec after starting the IAP infusion. The brains were rapidly removed and frozen in 2-methylbutane cooled to -45°C. $^{14}$C-radioactivity was evaluated in the blood and plasma samples by liquid scintillation counting and in brain by quantitative autoradiography (QAR).

The distribution of circulating erythrocytes within the parenchyma (reported as a volume or space) was measured with $^{55}$Fe-labeled red cells, which were radiotagged with an in vivo labeling technique (Lin et al., 1990). To begin these experiment, 4 ml of $^{55}$Fe-labeled blood was administered into one femoral vein over 20-25 sec; simultaneously 3 ml of blood were withdrawn from the femoral artery, thereby maintaining whole body blood volume constant and mean arterial pressure steady.

For the measurement of the volume of plasma circulating and contained in the parenchyma (also known as the albumin distribution space), 100 mCi of $^{125}$I-bovine serum albumin in 0.25 ml of saline was infused per animal. For both the RBC and RISA experiments, small samples of blood (50 ul) were obtained at one min intervals. Three minutes after beginning tracer infusion, the rats were decapitated, and the whole head was frozen in 2-methylbutane cooled to -45°C. The brain was removed from the skull in a cold box at -15°C, covered by embedding medium while still frozen, and later sectioned for QAR determination of tissue radioactivity. Whole blood and plasma samples were counted by liquid scintillation ($^{55}$Fe-activity) or gamma ($^{125}$I-activity) counting.

The blood-to-brain transfer constants of radiolabeled antipyrine and 3OMG were measured by intravenously infusing 50 uCi of $^{14}$C-labeled antipyrine or 3OMG in 1 ml saline over 30 sec. The experimental procedure thereafter was the same as previously described for the LCBF experiments.

As indicated above, tissue radioactivity was measured by QAR. To do this, coronal sections of brain, each 20 mm thick, were cut in a cryostat set at -17°C starting at the level of the area postrema, repeated at 400 um intervals thereafter, and ending at the rostral part of the caudate-putamen. Those brain sections designated for autoradiography were placed in cassettes with a set of standards plus a sheet of x-ray film and sealed. A pair of brain sections adjacent to the ones taken for QAR analysis was stained with cresyl violet for histological examination and area identification. After sufficient exposure, the x-ray films were developed and analyzed for radioactivity using an image analysis system (MCID, Imaging Research; St. Catharines, Ontario, Canada). Densitometric readings were taken from 40-50 brain areas using a cursor controlled circular reading frame. Areas of interest were identified using a stereotaxic atlas and the adjacent histological sections.

For the measurement of RISA and RBC radioactivity, optical density was assessed in the areas between the large dark spots and streaks on the autoradiograms using a circular reading frame. As presented by Bereczki et al. (1992), this technique does not measure total blood volume in the brain, but the volume of blood flowing and contained within the smaller parenchymal microvessels (small arterioles, capillaries, and small venules) at the time of decapitation and head freezing.

Calculations of the various parameters and statistical analysis of the data are given in detail in the various publications cited above; they will not be repeated here.

## RESULTS AND DISCUSSION

### Hypercapnia

The rats treated with 8% $CO_2$ were significantly hypercapnic ($PaCO_2$ = 97 mm Hg), acidotic (pH = 7.14), and hyperoxic ($PaO_2$ = 142 mm Hg) (Bereczki et al., 1993b). In the hypercapnic rats, glucose concentration was insignificantly higher (10.1 mM) than in controls (9.2 mM), and mean arterial blood pressure was slightly, but not significantly lower than in controls.

In the control rats, LCBF ranged from around 45 ml/100g/min in various white matter structures to 270 ml/100g/min in the superior olive and temporal cortex (Bereczki et al., 1993b). Inhalation of 8% $CO_2$ resulted in a greater than two-fold rise in LCBF in all areas; in more detail, LCBF was around 2.5 times control in most brain structures, and was greater than 3 times control in a few areas. The degree of LCBF elevation varied significantly among brain areas. This model of hypercapnia clearly induces large elevations in LCBF throughout the brain and might lead to sizable capillary recruitment.

The volume of labeled RBC's in the parenchyma ranged from around 1.2-1.6 ul/g in the white matter areas to 6.3 ul/g in the lateral habenular nucleus. Hypercapnia raised RBC distribution spaces by 8-20% in most brain areas, but did not change these spaces in other parts of the brain (Bereczki et al., 1993b). The effects of hypercapnia on the volume of labeled RBC's in cerebral microvessels were found to be either statistically insignificant (Bereczki et al., 1993b) or marginally significant ($p = 0.052$; Chen et al., 1994a).

As with the parenchymal RBC volumes, the RISA (plasma) volumes of distribution in controls were lowest in white matter and highest in the lateral habenular nucleus. The RISA distribution spaces were 10-30% larger in hypercapnic rats than controls for most brain areas (Bereczki et al. 1993b); the differences in parenchymal plasma volumes between control and hypercapnia were significant ($p < 0.02$).

In nearly all brain areas, the volume of radiolabeled blood in the parenchyma (the sum of the RISA and RBC spaces) was higher in hypercapnic than control rats. The hypercapnia-induced increases in the microvascular blood volumes averaged 23% in hindbrain gray matter, 21% in white matter and 22% in forebrain gray matter. The differences in the blood spaces of the parenchyma between control and hypercapnic groups were significant in gray matter for 15 of 20 forebrain areas and 7 of 14 hindbrain ones and in all 4 forebrain white matter structures (Bereczki et al., 1993b).

The mean transit-times of whole blood passage through parenchymal microvessels (Tt) for controls ranged from 0.3 to 0.7 sec in forebrain and hindbrain gray matter (average = 0.45 sec) and varied from 0.5 to 0.9 sec in white matter. Because blood flow increased much more than microvascular blood volume, the microvascular Tt was much shorter in the hypercapnic group than the control group, becoming as low as 0.15 sec in some areas. Respiring 8% $CO_2$ lowered Tt to less than 45% of control throughout the brain.

Before proceeding with the antipyrine and 3OMG influx data, it is important to emphasize that Ki is the operational expression of transfer from blood to brain (influx). When Ki rises, influx has increased, and this increase is due to a rise in either the rate of blood flow or the PS product or to some combination of changes in these two variables of the local capillary network.

In control rats, antipyrine influx (Ki-An) ranged from around 35 ml/100g/min in the white matter areas to 134 ml/100g/min in the superior olive and inferior colliculus. Hypercapnia clearly raised antipyrine influx; the mean increase was 52% (Chen et al., 1994a). The percentage increase was significant but not uniform among brain areas ($p < 0.001$). The effect of hypercapnia on antipyrine influx was, thus, big and "global" and could have been caused by changes of either LCBF, PS product for antipyrine, or both.

The PS products for antipyrine (PS-An) in controls varied from 59-197 ml/100g/min and were lowest in white matter areas (< 90 ml/100g/min) and highest in the superior colliculus and mammillary nucleus. In most gray matter areas, PS-An was around 120-160 ml/100g/min. Hypercapnia modestly raised PS-An in almost all brain areas, increasing it on the average 24% in hindbrain gray matter, 15% in forebrain gray matter, and 7% in white matter (Chen et al., 1994a). Statistical analysis indicated a significant difference in PS-An between the control and hypercapnic groups ($p < 0.05$). In general, the areas in which hypercapnia increased PS-An the most (14 of 16 hindbrain and 10 of 22 forebrain gray matter areas) were also the ones where the blood space were elevated the most by hypercapnia.

As for 3OMG transfer across the BBB, the influx constants (Ki-3OMG) in controls were < 10 ml plasma/100g/min for all white matter areas and were around 15-25 ml plasma/100g/min for most gray matter areas (Chen et al., 1994a). Influx of 3OMG in controls tended to be higher in hindbrain gray matter than the forebrain gray matter, averaging 22.6 ml plasma/100g/min and 18 ml plasma/100g/min, respectively. The largest control Ki-3OMG values were found in the superior olive and inferior colliculus.

Hypercapnia created by respiring 8% $CO_2$ did not greatly affect the rate of 3OMG influx; modest increases in 3OMG influx were seen only for some areas where local plasma flow was markedly elevated. Overall, the differences in Ki-3OMG between the control and hypercapnic groups were not significant.

The PS products of 3OMG (PS-3OMG) ranged from 8 to 34 ml plasma/100g/min in hypercapnic rats as well as controls and were fairly similar in both groups (Chen et al., 1994a). The permeability-surface area products were lowest in the white matter areas and highest in the superior olive and inferior colliculus. The differences in PS-3OMG values between the control and hypercapnic groups were small and not significant.

Because D-glucose competes with labeled 3OMG for the BBB hexose transporter (GLUT-1 isoform), the concentration of cold glucose must be taken into account when analyzing the rate of influx and PS product for 3OMG. The concentration of cold glucose was slightly but not significantly higher in the hypercapnic 3OMG group (mean $\pm$ SE = 10.6 $\pm$ 0.5 mM, n = 9) than in the control 3OMG group (9.7 $\pm$ 0.5 mM, n = 8). If this difference is real and all other things are equal, then the slightly higher plasma glucose would reduce 3OMG influx by 5% in the hypercapnic group relative to control (Chen et al., 1994a). To look at this in a different way, PS-3OMG becomes about 5% bigger in hypercapnic rats than controls when a glucose concentration correction is applied to the former set of data (Chen et al., 1994a).

To summarize these observations with respect to the capillary recruitment hypothesis, cerebral blood flow was increased 2.0-3.5 times normal by hyperoxic-hypercapnia, yet the volume of radiolabeled parenchymal blood and the PS product of antipyrine only were raised by 7-24% throughout the brain. The blood volume and antipyrine data indicate that functional (perfused) capillary surface area increased modestly (generally < 25%) and suggest that the LCBF elevations produced by this model of hypercapnia are partly due to some capillary recruitment or some dilation of small microvessels or, most probably, some of each (Chen et al., 1994a). The major conclusion is, however, that blood flow was mainly raised by more than doubling the velocity of blood flow through the perfused capillaries and other small microvessels. This is consistent with the greatly diminished mean transit-times in the hypercapnic group.

In view of the appreciable increase in PS-An and the likelihood that this was the result of an increase in capillary surface area or S, the marginal alteration in the PS product of 3OMG is a bit troubling. It is possible that, even though capillary surface area was apparently enlarged, glucose permeability (P) and transporter activity per unit endothelial membrane area was sufficiently decreased by this model of acute hypercapnia to keep 3OMG (and glucose) influx relatively unchanged. The finding by Hertz and Paulson (1982) of a paradoxical increase in glucose transport when blood flow was lowered by hypocapnia lends a degree of plausibility to our suggested action of $CO_2$ on GLUT-1 function.

As for the effects of hypercapnia-induced elevation of blood flow on solute influx across the BBB, LCBF as measured in these studies is actually the influx rate constant of iodoantipyrine (IAP). With the IAP technique, it is tacitly assumed that the influx of IAP is limited by the rate of delivery to the tissue (namely, blood flow), the blood-to-brain extraction (Ei) of iodoantipyrine is nearly complete (Ei $\geq$ 0.95), and Ki-IAP = LCBF. The influx of IAP (LCBF) was increased by 2- to 3.5-fold by hypercapnia; similar increases in influx (or efflux) would be expected under these conditions for other flow-limited or virtually flow-limited substances (0.8 < Ei < 1.0) such as $O_2$, $CO_2$, NO, ethanol, nicotine, and water.

The blood-to-brain (unidirectional) extraction of antipyrine averaged 61% among the brain areas in control animals (Chen et al., 1994a); antipyrine influx is, therefore, significantly dependent on both blood flow and permeability (PS product). The hypercapnia-produced doubling of LCBF caused antipyrine influx to rise on the average by 52%. This rise in influx is driven partly by an increase in PS (capillary recruitment), as indicated by the 15-25% change in PS-An and microvessel blood volume, and, more importantly, by the huge elevation in blood flow. The influx of substances with similar extraction fractions (e.g., caffeine, Ei = 0.7) would also be raised sizably by hypercapnia.

The influx of 3OMG was only slightly increased by hypercapnia. This holds because the rate of 3OMG influx is only slightly dependent on blood flow and is mainly set by its PS product, which was apparently unchanged by hypercapnia. Rather strangely, this constant PS seems to be the result of a modest fall in BBB permeability (P) - perhaps mostly the result of a decrease in GLUT-1 activity and a small increase in the number of perfused capillaries (recruitment) and capillary surface area (S), both changes seemingly <25%. Because of the modest recruitment of capillaries and the strong dependence of their influx on PS (and not blood flow), the blood-to-brain flux of other moderately to slightly permeable solutes (0.4 > Ei > 0.2) would also be only modestly affected by hypercapnia.

In conclusion, hypercapnia raises local cerebral blood flow mainly by increasing the velocity of blood convection through capillaries and to a relatively small extent by perfusing more capillaries or recruitment. This means that the fluxes of highly permeable solutes and water will be greatly elevated, of moderately permeable materials will be appreciably raised, and of slightly permeable to virtually impermeable ones will be increased to the extent of capillary recruitment and surface area enlargement, probably 15-20% at most. With such a response to hypercapnia, low influx of slightly permeable solutes and barrier function of the BBB are retained. Seemingly independent of this, carrier function of the BBB - particularly glucose transfer - may actually be lowered by severe hypercapnia.

## Hypoxia

At the time of experimentation, the rats treated with 10% $O_2$ were strongly hypoxic ($PaO_2$ = 32 mm Hg) and hypocapnic ($PaCO_2$ = 26 mm Hg) and somewhat alkalotic (pH = 7.5) and hypotensive (MABP = 104 mm Hg)(Bereczki et al., 1993a). Glucose concentration was significantly higher in hypoxic rats (13.9 mM) than in controls (8.9 mM)(Chen et al., 1994b).

The control values of the various measured parameters such as LCBF and parenchymal blood volume have been given above in the section on hypercapnia; they will not be repeated hereafter unless necessary. Local cerebral blood flow was significantly elevated by hypoxia throughout the brain; the percentage increases in LCBF among areas ranged from 23 to 92% (Bereczki et al., 1993a). The degree of LCBF elevation varied significantly among brain structures, but the average increases among the major groupings of brain areas - hindbrain gray matter, forebrain gray matter and white matter - were all around 50%. Such global and sizable elevations in LCBF might occur, at least partly, by capillary recruitment. The following data do not, however, support this suggestion.

The distribution volumes of RBC's within the parenchyma were not significantly changed by hypoxia (Bereczki et al., 1993a). The RISA distribution volumes were significantly elevated by hypoxia in some but not all brain areas (Bereczki et al., 1993a; Chen et al., 1994b). The RISA volume differences between the control and hypoxic groups were significant ($p = 0.038$).

In nearly all brain areas, the volume of radiolabeled blood in the parenchyma was higher in hypoxic rats than in controls. The differences were significant and sizable in about 40% of the brain areas (Bereczki et al., 1993a; Chen et al., 1994b). Among the areas where hypoxia sizably raised parenchymal blood volume, the average increase was 17% in hindbrain gray matter, 14% in white matter, and 18% in forebrain gray matter. The differences in blood spaces between control and hypercapnic groups were significant in gray matter for 15 of 20 forebrain areas and 7 of 14 hindbrain ones and in all 4 forebrain white matter structures (Bereczki et al., 1993a).

The control whole blood mean transit-times through parenchymal microvascular systems (Tt) ranged from 0.3 to 0.9 sec; with hypoxia, Tt dropped by about 30% in most brain areas (Bereczki et al., 1993a).

As for blood-brain exchange, hypoxia modestly raised antipyrine influx; the mean increases in Ki-An were 25% in hindbrain gray matter, 14% in forebrain gray matter, and 28% in white matter (Chen et al., 1994a). The percentage increase was significant but not uniform among brain areas ($p < 0.001$). Antipyrine influx in 8 of 22 forebrain gray matter structures was virtually unaffected by hypoxia.

Somewhat surprisingly, the PS products for antipyrine (PS-An) were not appreciably altered by hypoxia (Chen et al., 1994b).

Hypoxia decreased the rate of 3OMG influx and raised the concentration of glucose in plasma from 8.9 to 13.9 mM. The average drop in Ki-3OMG was 11% in hindbrain gray matter, 14% in forebrain gray matter, and 3% in white matter (Chen et al., 1994b). The difference in Ki-3OMG between the control and hypoxic groups was significant ($p < 0.05$). The major cause of the decrease in Ki-3OMG by hypoxia was, undoubtedly, the higher level of plasma glucose, which leads to increased competition of cold glucose with radiolabeled 3OMG for GLUT-1 at the blood-brain barrier.

The PS products of 3OMG (PS-3OMG) were significantly decreased by hypoxia ($p < 0.05$; Chen et al., 1994b); the drops in PS-3OMG averaged 19% in both hindbrain and forebrain gray matter and 10% in white matter. This lowering of PS is likely to be the result of the higher glucose concentration in the plasma of the hypoxic rats, as just discussed for Ki-3OMG and hypercapnia. In theory, the higher plasma glucose in the hypoxic group would reduce 3OMG influx by 22% with respect to control (Chen et al., 1994b). With this correction, the PS products of 3OMG becomes nearly identical in control and hypoxic groups for all brain areas, which fits well with the similarity of control and hypoxia PS products of antipyrine reported above (Chen et al., 1994b).

To summarize the testing of the capillary recruitment hypothesis for the case of hypoxia, local cerebral blood flow was increased 23-92% by hypocapnic-hypoxia, but the volume of radiolabeled blood perfusing and trapped in the parenchyma and the PS product of antipyrine were virtually unchanged in most brain areas and were only slightly elevated in the few remaining ones. When corrected for the differences in cold glucose concentration in plasma, the PS products of 3OMG were identical for the hypoxic and control rats in virtually all brain areas. Collectively these results indicate that few, if any, capillaries were recruited when cerebral blood flow was increased by hypoxia and that LCBF was raised by elevating the velocity of blood flow through already perfused capillaries. Incidentally, some dilation of small parenchymal microvessels including capillaries (Duelli and Kuschinsky, 1993) may be the cause of the small increases in parenchymal blood volumes and PS products in the few brain areas in which these parameters were affected by hypoxia.

The effects of hypoxia-induced elevation of blood flow on solute influx across the BBB can be simply summarized. The influx of highly permeable solutes and water would be enlarged as much as or nearly as much as LCBF increased (around 50% for most areas in the present instance). For moderately permeable compounds, their influx is modeled fairly well by antipyrine and would be raised by approximately 25% throughout the brain by this degree of hypoxia. The influx of modestly permeable to relatively impermeable solutes would be essentially unchanged by acute hypoxia, and the barrier function of the BBB is apparently maintained. In addition to barrier function of the BBB, carrier function - particularly glucose transfer - may also be unchanged by severe hypocapnic hypoxia.

## Pentobarbital Treatment

The administration of 50 mg/kg pentobarbital made the rats hypoxic ($PaO_2$ = 76 mm Hg), hypercapnic ($PaCO_2$ = 48 mm Hg), and hypotensive (MABP = 92 mm Hg)(Otsuka et al., 1991a, 1991b). Plasma glucose concentration was insignificantly higher in treated (9.5 mM) than control (8.2 mM) rats.

Local cerebral blood flow was significantly lowered by pentobarbital in 24 of 25 forebrain gray matter areas (Otsuka et al., 1991a); LCBF in all of the affected areas dropped to around 50-70 ml/100g/min regardless of the normal rate of flow (control range = 90-290 ml/100g/min). Pentobarbital significantly lowered LCBF in only 2 of 18 hindbrain areas but in 3 of 4 white matter structures. The effects of pentobarbital on LCBF obviously varied among brain structures. Evidence for capillary "retirement" might, thus, be found in some structures but not in others.

The volumes of radiolabeled RBC's in the parenchyma were significantly raised ($p < 0.05$) by treatment in most brain areas, including some where LCBF was not decreased by pentobarbital (Wei et al., 1993); these increases in RBC volume tended, however, to be greater in forebrain than in hindbrain gray matter. In contrast to the RBC results, the RISA distribution volumes were significantly lowered ($p < 0.05$) by pentobarbital throughout the brain and the extent of lowering varied among areas. The volume of radiolabeled blood perfused and retained in parenchymal microvessels (the sum of the RISA and RBC spaces) was identical in control and treated groups, which hints that no capillaries were retired (no longer perfused) when blood flow was diminished by pentobarbital (Wei et al., 1993).

Pentobarbital dropped the influx of antipyrine by 25% or more in most forebrain and a few hindbrain areas but did not appreciably alter Ki-An in a number of hindbrain areas (Wei, Otsuka, and Fenstermacher, unpublished observations). With pentobarbital treatment, the extraction fractions of antipyrine (= Ki/LCBF) rose to more than 0.75 in many areas, mostly forebrain ones. In this instance, the estimation of PS-An, which is made from local Ki values from one group of animals and LCBF's from another group, becomes strongly dependent on blood flow; even modestly inappropriate LCBF numbers in this situation yield large errors in PS. The treated/control ratios of antipyrine PS products scattered broadly (range = 0.5-2.0), most probably due to the weakness inherent in estimating PS from group averages; these ratios tended, however, to clump around 0.95, which suggests that perfused capillary surface area was not appreciably decreased and few capillaries were retired when blood flow was dropped by pentobarbital.

The influx of 3OMG was lowered by 20-30% throughout the brain by pentobarbital, and the concentration of glucose in plasma rose slightly but not significantly (Otsuka et al., 1991b). The PS products of 3OMG were also decreased to the same degree in both hindbrain and forebrain areas by pentobarbital. This uniform lowering of PS-3OMG may to some slight extent (< 5%) be the result of the higher glucose concentration in the plasma of the treated rats (discussed above for hypercapnia) but probably is mainly caused by pentobarbital diminishing the ability of GLUT-1 to transport glucose and glucose analogs across the BBB (Otsuka et al. 1991b). Preliminary observations with cultured mammalian cells and human red cells suggest that barbiturates interact directly with the GLUT-1 transporter in the membrane to elicit this change (Honkanen, Fenstermacher, and Haspel, unpublished data).

With respect to the capillary recruitment hypothesis, the volume of radiolabeled blood in the microvessels and the PS product of antipyrine were essentially unchanged following pentobarbital treatment. The number of unperfused capillaries, hence, did not appear to increase even though LCBF dropped dramatically in most parts of the forebrain. The PS products of 3OMG were fairly uniformly reduced in virtually all brain areas by pentobarbital, a change that undoubtedly reflects a decrease in glucose transporter activity and not a drop in the number of perfused capillaries and associated glucose transporters. In conclusion, pentobarbital lowers

LCBF in the affected areas mainly by decreasing the velocity of both red cell and plasma flow within already perfused microvessels.

When blood flow drops because of a decrease in convection through capillaries, then influx of all materials becomes more dependent on flow rate and less on PS product. The changes in Ki-An produced by pentobarbital resemble the changes in LCBF. In other words, the influx rates of moderately permeable compounds ("antipyrine") decrease less than those of highly permeable ones ("iodoantipyrine") and the the Ki's of the more permeable solutes approach those of the less permeable substances. Lowering blood flow makes the influx across the BBB of all substances even more dependent on LCBF rate.

**Nicotine Administration**

The subcutaneous injection of nicotine (1.75 mg/kg) raised arterial $PO_2$ (the result of mild hyperventilation) and mean blood pressure modestly but significantly (Hans et al., 1993). Plasma glucose concentration was similar in nicotine-treated and control rats.

Local cerebral blood flow was significantly raised ($p < 0.01$) by nicotine in 16 of 44 gray matter areas (Hans et al, 1993); LCBF in the affected areas increased by 40-150%. Previous work indicated that local cerebral glucose utilization was either marginally or signifcantly increased in these 16 areas (see Hans et al., 1993) and that these changes in LCBF were driven by elevation of local neuroactivity.

The volume of radiolabeled RBC's in the parenchyma was unaltered by treatment with nicotine (Hans et al., 1993). The RISA and radiolabeled blood distribution volumes were comparable in control and treated rats for the 28 areas of unchanged blood flow and for half (7) of the areas where LCBF increased. For the remaining areas of increased LCBF, the RISA and blood spaces were raised by 20-30%. These observations suggest for just half of the nicotine-activated areas either some capillary recruitment or some microvessel dilation or both changes.

Nicotine raised the influx of antipyrine by 7-40% in the flow-stimulated brain areas but had no effect on Ki-An in the rest of the brain (Chen, Wei, and Fenstermacher, unpublished data); this change in Ki-An was significant ($p < 0.001$). The PS products of antipyrine tended to be higher in the flow-increased areas relative to control (average elevation = 7%), and the percentage rise ranged from 0-20%. As might be expected, the variations in PS-An between control and nicotine rats were not statistically significant.

The 3OMG influx constants generally appeared to be larger in the areas of nicotine-elevated flow relative to controls (increases in Ki-3OMG varied from 0 to 25%) but were identical in the two groups for the rest of the brain (Chen, Wei, and Fenstermacher, unpublished results). The differences between controls and treated rats were not significant when all 44 areas were considered but were almost statistically significant when just the activated areas were included in the testing. Virtually identical with the 3OMG influx pattern, the PS products of 3OMG were generally larger in the flow-increased areas for nicotine-treated rats than for controls, but PS-3OMG was similar in the two groups for each of the other brain areas.

Regarding nicotine stimulation of LCBF and the capillary recruitment hypothesis, the volume of radiolabeled blood in the microvessels and the PS products of antipyrine and 3OMG were slightly higher in about half of the areas of increased flow but were essentially unchanged in the others. These findings indicate a little (<10%) capillary recruitment and/or microvessel dilation in half of the flow-stimulated areas but no recruitment or vasodilation in the remaining half of them. Again, as with global changes in cerebral blood flow, the major mechanism of local, nicotine-driven increased blood flow is elevated velocity of convection.

As presented before for global elevation of cerebral blood flow, the local influx of highly permeable solutes and of water are greatly and similarly increased when and where LCBF is increased. In the case of nicotine stimulation, local fluxes of such materials would rise by approximately 40-150%, depending on the brain area and extent of flow change. For less permeable compounds such as antipyrine, the changes in influx would be smaller, e.g., Ki rises of 7-40% when LCBF increases 40-150%. For molecules and ions of low BBB permeability, their influxes would rise imperceptibly - basically only to the extent of capillary recruitment in the local micrascular network (<10% in the present case of nicotine administration).

## Dehydration

This study of dehydration was begun with the idea that blood flow, tissue activity, and cerebral glucose utilization might be increased in a few areas of the brain, presumably those involved in water balance, and another model of localized blood flow alteration would emerge. The data are still being gathered and are quite different than expected. They will be given briefly below without reference to authors and generally without statistical analysis.

After 24 hours of dehydration, plasma osmolality had risen by 10-12 mOsm ($p < 0.01$), and arterial hematocrit had changed from 46.5% to 50% ($p < 0.01$). Arterial $PO_2$, $PCO_2$, and pH plus plasma glucose concentration and mean arterial blood pressure were unchanged by 24 hours of dehydration.

Local cerebral blood flow was lowered by 10-12% in nearly all brain areas. The volume of RBC's in the parenchyma was about 11% higher in dehydrated rats than in controls in all areas, whereas the plasma distribution volumes were about 10% lower in dehydrated rats than in controls. Overall, 24 hours of dehydration slightly (but apparently not significantly) lowered the volume of radiolabeled blood in the parenchyma (a decrease of about 4%) and raised the hematocrit of the blood in the parenchymal microvessels by several percent.

Following 24 hours of dehydration, the influx of antipyrine was lowered by about 4% in all gray matter areas but was not changed in white matter. This slight drop in Ki-An was caused by the lowered LCBF because the PS products of antipyrine were essentially identical in control and dehydrated rats for each part of the brain examined.

In contrast, the influx of 3OMG in dehydrated animals was increased by 5-15% in most gray matter structures but was unchanged in a few gray matter areas and in all white matter structures. This general increase in Ki coupled with the decreased LCBF yielded PS products that were 15-25% higher in most gray matter areas of the dehydrated rats relative to controls. As pointed out above, the concentration of glucose was similar in control and dehydrated rats, a finding that does not explain the apparent dehydration-driven rise in the influx and PS product of 3OMG.

In summary, the radiolabeled blood volumes and PS products of antipyrine indicate that 24 hours of dehydration did not appreciably alter the number of perfused capillaries in any part of the brain. The increase in Ki and PS product of 3OMG in most gray matter areas must, therefore, involve a specific enlargement in the transport capacity of the glucose carrier system in the local cerebral capillary beds and be independent of both blood flow and the number of perfused capillaries.

## SUMMARY

This work tests the capillary recruitment hypothesis by measuring local cerebral blood flow, the parenchymal distribution volumes of radiolabeled red cells, plasma albumin, and blood, plus the influx rate constants and permeability-surface area products for antipyrine and 3OMG when LCBF is changed in one of five different ways - hypercapnia, hypoxia, pentobarbital administration, nicotine injection, and 24 hours of dehydration. Four conclusions are offered.

Firstly, local cerebral blood flow mainly changes in all five instances by altering flow velocity through parenchymal microvessels including capillaries. Some capillary recruitment appeared to occur when LCBF was increased by hypercapnia (< 25% increase in many brain areas) and nicotine (<10% in a few brain areas); blood flow changes caused by hypoxia, pentobarbital, and dehydration did not seem to alter the number of perfused capillaries.

Secondly, pentobarbital treatment and dehydration lowered blood flow throughout the brain, increased the distribution volumes of red cells in the parenchyma, and decreased those of albumin (plasma). When blood flow changes, the effects on red cells and plasma are, therefore, not always the same.

Thirdly, when blood flow is modified, the influxes of highly permeable substances across the blood-brain barrier are similarly changed, and those of moderately permeable solutes are also altered but less so than those of the more permeable materials. The influxes of slightly to virtually impermeable substances across the BBB are essentially unchanged when blood flow is altered.

Finally, the changes in glucose influx may not resemble or be closely coupled to LCBF. During hypercapnia and dehydration and following pentobarbital administration, the changes in 3OMG influx appeared to be different than the changes in local blood flow.

## REFERENCES


Bereczki, D., Wei, L., Acuff, V., Gruber, K., Tajima, A., Patlak, C., and Fenstermacher, J., 1992, Technique-dependent variations in cerebral microvessel blood volumes and hematocrits in rat brain, *J. Appl. Physiol.* 73:918.

Bereczki, D., Wei, L., Otsuka, T., Acuff, V., Pettigrew, K., Patlak, C., and Fenstermacher, J., 1993a, Hypoxia increases velocity of blood flow through parenchymal microvascular systems in rat brain, *J. Cereb. Blood Flow Metab.* 13:475.

Bereczki, D., Wei, L., Otsuka, T., Hans, F.-J., Acuff, V., Patlak, C., and Fenstermacher, J., 1993b, Hypercapnia slightly raises blood volume and sizably elevates flow velocity in brain microvessels, *Am. J. Physiol.* 264:H1360.

Chen, J.-L., Acuff, V., Bereczki, D., Hans, F.-J., Otsuka, T., Wei, L., Finnegan, W., Patlak, C., and Fenstermacher, J., 1994a, Slightly altered permeability-surface area products imply some cerebral capillary recruitment during hypercapnia, *Microvasc. Res.* (in press).

Chen, J.-L., Wei, L., Bereczki, D., Hans, F.-J., Otsuka, T., Acuff, V., Richardson, G., Patlak, C., and Fenstermacher, J., 1994b, Virtually unaltered permeability-surface area products imply little capillary recruitment in brain with hypoxia, *Microcirculation* 1:35.

Duelli, R., and Kuschinsky, W., 1993, Change in brain capillary diameter during hypocapnia and hypercapnia, *J. Cereb. Blood Flow Metab.* 13:1025.

Hans, F.-J., Wei, L., Bereczki, D., Acuff, V., DeMaro, J., Chen, J.-L., Otsuka, T., Patlak, C., and Fenstermacher, J., 1993, Nicotine increases microvascular blood flow and flow velocity in three groups of brain areas, *Am. J. Physiol.* 265:H2142.

Hertz, M., and Paulson, O., 1982, Transfer across the human blood-brain barrier: evidence for capillary recruitment and for a paradox glucose permeability increase in hypocapnia, *Microvasc. Res.* 24:364.

Lin, S.-Z., Nakata, H., Tajima, A., Gruber , K., Acuff, V., Patlak C., and Fenstermacher, J., 1990, Quantitative autoradiographic assessment of $^{55}$Fe-RBC distribution in rat brain, *J. Appl. Physiol.* 69:1637.

Otsuka, T., Wei, L., Acuff, V., Shimizu, A., Pettigrew, K., Patlak, C., and Fenstermacher, J., 1991a, Variations in local cerebral blood flow response to high-dose pentobarbital sodium in the rat, *Am. J. Physiol.* 261:H110.

Otsuka, T., Wei, L., Bereczki, D., Acuff, V., Patlak, C., and Fenstermacher, J., 1991b, Pentobarbital produces dissimilar changes in glucose influx and utilization in the brain, *Am. J. Physiol.* 261:R265.

Pawlik, G., Rackl, A., and Bing, R.J., 1981, Quantitative capillary topography and blood flow in the cerebral cortex of cats: an in vivo microscopic study, *Brain Res.* 208:35.

Rosenblum, W. I., and Zweifach, B., 1963, Cerebral microcirculation in the mouse brain, *Arch. Neurol.* 9:414.

Sakurada, O., Kennedy, C., Jehle, J., Brown, J.D., Carbin, G.L., and Sokoloff, L., 1978, Measurement of local cerebral blood flow with iodo[14C]antipyrine, *Am. J. Physiol.* 234:H59.

Tajima, A., Nakata, H., Lin, S.-Z., Acuff, V., and Fenstermacher, J., 1992, Differences and similarities in albumin and red cell flows through cerebral microvessels, *Am. J. Physiol.* 262:H1515.

Villringer, A., Dirnagl, U., Gebhardt, R.E., and Einhaupl, K.M., 1991, An in vivo approach to assess the capillary recruitment hypothesis in the brain microcirculation using laser scanning microscopy (Abstract), *J. Cereb. Blood Flow Metab.* 11(Suppl.2): S441.

Wei, L., Otsuka, T., Acuff, V., Bereczki, D., Pettigrew, K., Patlak, C., and Fenstermacher, J., 1993, The velocities of red cell and plasma flows through parenchymal microvessels of brain are decreased by pentobarbital, *J. Cereb. Blood Flow Metab.* 13:487.

## BRAIN-BLOOD BARRIER REMOVAL OF DOPA: ROLE IN REGULATION OF DOPAMINE SYNTHESIS AND TREATMENT OF PARKINSON'S DISEASE

Albert Gjedde, Jakob Reith, Gabriel Léger, Paul Cumming, Yoshifumi Yasuhara, Mark Guttman, and Hiroto Kuwabara

McConnell Brain Imaging Center
Montreal Neurological Institute
Canada
and
Department of Neurology and Neurosurgery
McGill University Faculty of Medicine
Canada

## ABSTRACT

To test the hypothesis that brain-blood export of DOPA impairs dopamine synthesis and hence provides a rationale for upregulation of the activity of DOPA decarboxylase, we first used HPLC to confirm that [$^{14}$C]DOPA, generated endogenously from [$^{14}$C]tyrosine, is subject to removal from rat brain tissue in which DOPA decarboxylation is blocked by NSD-1015. Second, we measured the relative rates of decarboxylation and blood-brain transfer of [$^{18}$F]fluoro-DOPA (FDOPA), a radiolabeled analog of DOPA. In Parkinson's disease, the rate of brain-blood removal of FDOPA from the striatum was normal, compared to the 50% reduction of the rate of decarboxylation, causing additional loss of DOPA and consequent reduction of the rate of dopamine synthesis.

## INTRODUCTION

Facilitated diffusion is a mechanism that mediates the transport into brain of nutrients in the bloodstream. However, for substrates generated in the brain, the question of transport out of the brain also arises. This question has particular relevance to the integrity of monoaminergic systems in the brain for which both precursors, DOPA and hydroxytryptophan,

*New Concepts of a Blood–Brain Barrier*
Edited by J. Greenwood *et al.*, Plenum Press, New York, 1995

are generated in brain tissue and may leave brain tissue by facilitated diffusion through the blood-brain barrier. The common facility for these amino acids is the transporter of large neutral or aromatic amino acids (LNAA). This facility raises the question of the influence of the blood-brain barrier transport of neurotransmitter precursors on the regulation of neurotransmission. Specifically, does the efflux of DOPA from the striatum impair dopamine homeostasis and hence dopaminergic neurotransmission?

The blood-brain barrier transport of large neutral amino acids appears to be entirely passive and under no known regulation. The large neutral amino acids participate in protein synthesis and monoamine transmitter synthesis but play no role in energy metabolism. Tyrosine is the precursor of DOPA which in turn is decarboxylated to dopamine. Dopamine is the precursor of another neurotransmitter, norepinephrine. Tyrptophan is the precursor of 5-hydroxytryptophan which in turn is decarboxylated to 5-hydroxytryptamine, or serotonin, using the same enzyme which decarboxylates DOPA. The rate limiting steps in the synthesis are firmly believed to be the tyrosine and tryptophan hydroxylases which tend to be fully saturated *in vivo*. DOPA, or Aromatic Amino Acid Decarboxylase on the other hand, is believed to follow the lead of the hydroxylases passively because it tends to be so unsaturated that the rate of decarboxylation is proportional to the concentration of the substrate. For this reason, the activity of the enzyme is not believed to influence the synthesis of dopamine or serotonin, except when it is so low that the concentration of the substrates tends to saturate the enzyme.

However, not all DOPA and 5-hydroxytryptophan is decarboxylated. Some DOPA is transported from dopaminergic terminals in forearm tissue (Goldstein et al., 1987) and from dopaminergic terminals in brain tissue, through the blood-brain barrier (Eisenhofer et al., 1989).

In Parkinson's disease, dopaminergic cells and terminals disintegrate and the enzymes responsible for dopamine synthesis disappear in part from the substantia nigra and the striatum. The activity of DOPA decarboxylase declines as the cells disintegrate but not in parallel with the numerical loss of the cells and their terminals (Gjedde et al., 1993). The content of dopamine in the striatum declines as well but much more than expected from the cellular loss alone (Hornykiewicz and Kish, 1986). Since the activities of tyrosine hydroxylase and DOPA decarboxylase both fall, the latter of the two enzymes probably remains unsaturated until late in the disease.

It is not known whether the blood-brain transfer of large neutral amino acids changes in Parkinson's disease. If changes occur, they must affect the concentration of DOPA, depending on the relative magnitudes of decarboxylation and transport. The competition between dopamine synthesis and DOPA egress from striatum must favor DOPA egress if decarboxylation declines more than amino acid transport. Such a relative augmentation of DOPA egress suggests a role for DOPA decarboxylase in the regulation of DOPA export and therfore in dopamine synthesis.

## METHODS

To test the hypothesis that blood-brain transfer of large neutral amino acids affects dopamine synthesis and hence provides a rationale for regulation of the activity of DOPA decarboxylase, we first injected male Wistar rats with NSD-1015 which inhibits DOPA decarboxylase activity and determined the clearance of labeled DOPA generated from labeled tyrosine. The [$^{14}$C]tyrosine was adminstered as an intravenous bolus 30 min after the injection of NSD-1015.

In a subsequent study, we injected normal volunteers and patients with Parkinson's disease with a radiolabeled analog of DOPA, fluorine-18-labeled fluoro-DOPA (FDOPA), and recorded regional brain radioactivity, as a function of time by means of positron emission tomography (Gjedde et al., 1993).

With a particular mathematical model of the relationship between the radioactivity recorded in brain tissue and the circulation, we simultaneously estimated the rates of decarboxylation ($k_3$) and brain-blood export ($k_2$) of DOPA (rats) or FDOPA (humans) (Reith et al., 1990; Gjedde et al., 1991). In the rodent study, DOPA in striatum was separately analzyed by HPLC (Cumming et al., 1994). In humans, total radioactivity, including all FDOPA metabolites, was recorded in striatum by means of positron emission tomography (Kuwabara et al., 1993). During the circulation of FDOPA, 3-O-methyl-FDOPA was formed by the action of catechol-O-methyltransferase (COMT) in the liver (Guttman et al., 1992; 1993). This generation of a labeled metabolite in the periphery necessitated the separation of the concentrations of FDOPA and 3-O-methyl-FDOPA by HPLC of plasma samples (Cumming et al., 1993).

Several mathematical models have been applied to the studies reported in the literature (Hoshi et al., 1993). The present model is one that accurately distinguishes between the processes of blood-brain barrier transport and decarboxylation.

## RESULTS

In normal male Wistar rats injected with NSD-1015 half an hour previously, DOPA but no dopamine was formed in the striatum of the rat brain, as shown in Figure 1 which illustrates the accumulation and subsequent wash-out of [$^{14}$C]tyrosine and the resulting [$^{14}$C]DOPA in these rats. The labeled DOPA escaped from the striatum at the rate listed for the rate constant $k_2$ in Table 1. In the rats, the blood plasma concentration of tyrosine was measured to be 65 μM. This value was used to calculate the flux associated with each rate constant, as listed in Table 1.

The nature of the DOPA 'trapping' was not determined. It did not represent dopamine synthesis because no labeled dopamine was detected and because the trapped DOPA was part of the DOPA compartment separated by HPLC.

In patients suffering from Parkinson's disease, as expected, Gjedde et al., (1993) found the calculated DOPA decarboxylase activity to be reduced by 50% in the putamen, when measured with FDOPA, as listed for the rate constant $k_3$ in Table 2.

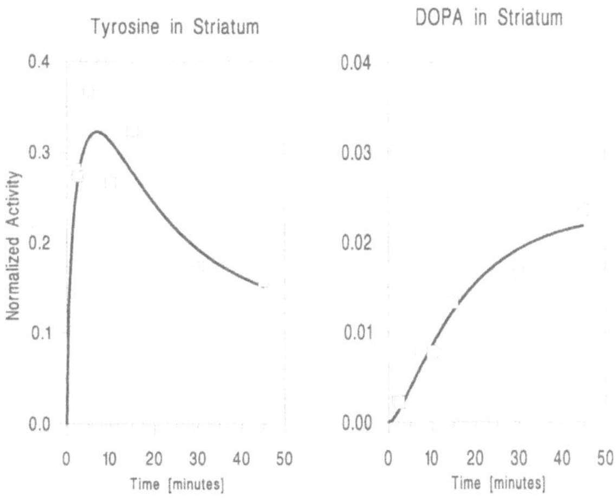


**Figure 1.** Accumulation of labeled tyrosine and labeled DOPA in rat brain following i.v. injection of a bolus of [14C] tyrosine 30 minutes after administration of NSD-1015 to block DOPA decarboxylase. Abscissa: Time (30 minutes). Ordinate: Normalized Activity (ratio). Note 10-fold difference between ordinate scale of left and right panels.

**Table 1.** Striatal Tyrosine and DOPA Metabolism in Rats 30 Min Post-NSD-1015.

| Substrate | Conc [μM] | Step | Rate ($min^{-1}$) | Flux ($nmol\ g^{-1}\ min^{-1}$) |
|---|---|---|---|---|
| p-Tyrosine | 65 | Clearance into Striatum ($K_1$**) | 0.058 ± 0.004 | 3.77 |
| br-Tyrosine | 26* | Protein Synthesis ($k'_3$) | 0.034 ± 0.004 | 0.88 |
| | | DOPA Synthesis ($k''_3$) | 0.019 ± 0.008 | 0.54 |
| br-DOPA | 0.4* | Clearance from Striatum ($k_2$) | 1.2 ± 0.7 | 0.48 |
| | | DOPA 'Trapping' (?) | 0.15 ± 0.1 | 0.06 |
| | | Dopamine Synthesis ($k_3$) | n.d. | - |

p-plasma; br-brain, *calculated [nmol g-1]; **[ml g-1 min]; n.d. not detected

In the patients, the rate of egress of the DOPA analog ($k_2$) was unaffected by the disease (Gjedde et al., 1993). The unchanged egress increased the ratio between export and decarboxylation by an average factor of two and reduced the fraction of DOPA decarboxylated to dopamine by one-third, as listed in Table 3, reducing dopamine synthesis more than predicted by the decline of the rate-limiting enzyme tyrosine hydroxylase. The decarboxylation fraction was calculated as the ratio between $k_3$ and and the sum of $k_2$ and $k_3$.

**Table 2.** FDOPA Decarboxylation rate ($k_3$) in human striatum in vivo (adapted from Gjedde et al., 1993).

| Hemisphere | Normal Volunteers ($min^{-1}$; n = 7) | Parkinson's Disease ($min^{-1}$; n = 7) |
|---|---|---|
| Left and Right | 0.070 ± 0.004 | 0.030 ± 0.004* |

*, p<0.0001, Parkinson's Disease vs. Normal Volunteers. Student's two-tailed *t*-test

The decline of the decarboxylation fraction meant an additional 33% reduction of the rate of decarboxylation, over and above the basic 50% reduction of the activity of the enzyme due to loss of terminals.

**Table 3.** Striatum-to-blood FDOPA removal rate ($k_2$) in Parkinson's disease (adapted from Gjedde et al., 1993).

| | Normal Volunteers | | Parkinson's Disease | |
|---|---|---|---|---|
| Hemisphere | $k_2$ [min$^{-1}$] | Decarboxylation Fraction [ratio] | $k_2$ [min$^{-1}$] | Decarboxylation Fraction [ratio] |
| Left | 0.063 ± 0.002 | 0.52 ± 0.01 | 0.055 ± 0.003 | 0.34 ± 0.02* |
| Right | 0.063 ± 0.003 | 0.52 ± 0.01 | 0.060 ± 0.004 | 0.35 ± 0.02* |

*, $p<0.0001$, Parkinson's Disease vs Normal Volunteers. Student's two-tailed $t$-test with correction for multiple comparisons.

## DISCUSSION

The finding that the brain-blood export of DOPA reduced the fraction of DOPA decarboxylated in Parkinson's disease indirectly provides a rationale for the postulated regulation of DOPA decarboxylase. It provides an explanation for the puzzling discrepancy between the extreme decline of dopamine levels of the striatum and the moderate loss of nigrostriatal neurons in Parkinson's disease. It also suggests that the activity of DOPA decarboxylase may be specifically elevated to counteract the decline of dopamine.

The interpretation of the present data requires proof that the clearance of DOPA is in part by transport from the neurons, and not be other means, also in brains in which DOPA decarboxylation is completely blocked. Goldstein et. al., (1987) showed that the origin of DOPA in the circulation is neuronal, and Eisenhofer et al., (1989) showed that the neuronal origin is in part cerebral. Eisenhofer et al., measured the cerebral arteriovenous deficit for DOPA and found it to average -1.5 nM for the internal jugular vein.

With an arterial DOPA concentration of 8nM, the arteriovenous deficit indicated a net extraction fraction of -20%. With a blood flow of 1.5 l min$^{-1}$, the normal human brain secretes approximately 2-3 nmol DOPA per minute. The caudate nucleus and putamen together weigh about 30g (Gjedde et al., 1991). If the exported DOPA is synthesized entirely in the basal ganglia, the egress from the human brain would be about 0.1 nmol g$^{-1}$ min$^{-1}$. This level of export is about one third of the suspected rate of dopamine decarboxylation of 0.3 nmol g$^{-1}$ min$^{-1}$.

The values of the Michaelis constants $T_{max}$ and $K_t$ for DOPA transport across the blood-brain barrier are known, as are the values of the Michaelis constants $V_{max}$ and $K_m$ for DOPA decarboxylation. The decarboxylation has no native competitors but the endothelial transport is shared by all nine of the aromatic, or large neutral, amino acids. This causes the apparent $K_t$ to be much higher than the uninhibited $K_t$ (Miller et al., 1985). Little is known of the neuronal membrane transporter but if it is similar to to the endothelial transporter, then each of the large neutral amino acids has plasma and brain concentrations which are close to the $K_t$ value of the respective amino acid. The apparent *in vivo* $K_t$ towards DOPA is therefore likely to be approximately 10-fold higher than in the absence of competitors. The affinity of the transporter towards DOPA is also about one-tenth of its affinity towards phenylalanine. For these reasons, the apparent $K_t$ of DOPA is likely to be close to 350 nmol g$^{-1}$while the endothelial $T_{max}$ is close to 70 nmol g$^{-1}$ min$^{-1}$ (Wade and Katzmann, 1975; Gjedde, 1988; Miura et al., 1991). The best estimate of the endothelial $k_2$ of DOPA at low concentrations is therefore 0.1-0.2min$^{-1}$.

The neuronal membrane $T_{max}$ for DOPA is not known but it is likely to be higher than the endothelial $T_{max}$ due to the larger surface area. Based on the combined neuronal membrane and endothelial removal rate constant ($k_2$), the neuronal membrane constant may be as high as 2 $min^{-1}$.

The estimates of $V_{max}$ and $K_m$ of DOPA decarboxylase were summarized by Reith et al., (1990). At low DOPA concentrations, the best estimate of $k_3$ is 0.5$min^{-1}$ which is not directly comparable to the rate constant of DOPA trapping estimated in the rodent study.

In the NSD-1015 blocked rats, the rate of DOPA synthesis was 0.5 nmol $g^{-1}$ $min^{-1}$, in agreement with current estimates of the rate of dopamine synthesis in the species (Reith et al., 1990). The calculated DOPA concentration was 0.4 nmol $g^{-1}$. This value is not different from the value of 0.4 nmol $g^{-1}$ measured by Nissbrandt and Carlsson (1987). At this level, the DOPA export would be 0.48 nmol $g^{-1}$ $min^{-1}$, i.e. 90% of the DOPA synthesized.

The steady-state rate of trapping of DOPA was only 0.06 nmol $g^{-1}$ $min^{-1}$. The rate of trapping is less than measured in previous experiments but it is not clear why. The relationship between the rate of escape and the rate of trapping indicates that the escape of DOPA causes the NSD-1015 blocking experiments to underestimate the rate of DOPA synthesis. The nature of this 'trapping' is not known but it is not in the form of dopamine synthesis because the trapped DOPA was detected with the free DOPA by HPLC. The trappping may be related to the mechanism of DOPA decarboxylation, or may occur in the vicinity of the enzyme from which DOPA may not be able to escape. However, the rate of 'trapping' would have to be very substantially greater in normal rats, i.e., the rate constant of trapping would have to rise to the value of $k_2$ to establish an approximate one-to-one ratio between dopamine synthesis and DOPA efflux. The rate of dopamine synthesis would then be half of the rate of DOPA synthesis, or 0.25 nmol $g^{-1}$ $min^{-1}$.

In the human study, the $k_2$ of FDOPA approximately agreed with the predicted value of 0.1 $min^{-1}$ for the endothelial transport but the $k_3$ was considerably less than the predicted value of 0.5 $min^{-1}$. The most likely explanation is a partial volume effect on the quantitation of positron emission tomography data. Correction for this effect increased the value of $k_3$ by a factor of two or more (Rousset et al., 1993).

## ACKNOWLDGEMENTS

Supported by Grants (SP-5, PG-41, SP-30) from the Medical Research Council of Canada.

## REFERENCES

Cumming, P., Leger, G.C., Kuwabara, H., and Gjedde, A., 1993, Pharmacokinetics of plasma 6-[$^{18}$F] fluoro-L-3,4-dihydroxy-phenylalanine ([$^{18}$F] FDOPA) in humans, *J. Cereb. Blood Flow Metab.* 13:668.

Cumming, P., Kuwabara, H., and Gjedde, A., 1995, Kinetic analysis of 6-[$^{18}$F] fluoro-L-dihydroxyphenylalanine ([$^{18}$F] DOPA) metabolism in the rat, *J. Neurochem.* in press.

Eisenhofer, G., Brush, J.E., Cannon, 3rd R.O., Stull, R., Kopin, I.J., and Goldstein, S., 1989, Plasma dihydroxyphenylalanine and total body and regional noradrenergic activity in humans. *J. Clin. Endocrinol. Metab.* 68:247.

Gjedde, A., 1988, Exchange diffusion of large neutral amino acids between blood and brain, *in*: "Peptide and Amino Acid Transport Mechanisms in the Central Nervous System," L. Rakic, D.J. Begley, H. Davison, and V.V. Zlokovic, eds., Stockton Press, New York, pp. 209.

Gjedde, A. Reith, J., Dyve, S., Diksic, M., Evans, A.C., and Kuwabara, H., 1991, DOPA decarboxylase activity of the living human brain, *Proc. Natl. Acad. Sci. USA* 88: 2721.

Gjedde, A., Léger, G. Cumming, P., Yasuhura, Y., Evans, A.C., Guttman, M., and Kuwabara, H., 1993, Striatal L-DOPA decarboxylase activity in Parkinson's disease in vivo: Implications for the regulation of dopamine synthesis, *J. Neurochem.* 61: 1538.

Goldstein, D.S., Udelsman, R., Eisenhofer, G., Stull, R., Keiser, H.R., Kopin, I.J., and Folio, C.J., 1987, Neuronal source of plasma dihydroxyphenylalanine, *J. Clin. Endocrinol. Metab.* 64: 856.
Guttman, M., Léger, G., Cedarbaum, J.M., Reches, A., Woodward, W., Evans, A.C., Diksic, M., and Gjedde, A., 1992, 3-*O*-Methyldopa administration does not alter fluorodopa transport into the brain, *Ann. Neurol.* 31:638.
Guttman, M., Léger, G., Reches, A., Evans, A., Kuwabara, H., Cedarbaum, J.M., and Gjedde, A., 1993. Administration of the new COMT inhibitor OR-611 increases striatal uptake of fluorodopa, *Movement Disorders* 8:298.
Hornykiewisz, O., Kish, S.J., 1986. Biochemical pathophysiology of Parkinson's disease, *Adv. Neurol.* 45:19.
Hoshi, H., Kuwabara, H., Léger, G., Cumming P, Guttman, M., and Gjedde, A., 1993, 6-[$^{18}$] fluoro-L-DOPA metabolism in living human brain: A comparison of six analytical methods, *J. Cereb. Blood Flow Metab.* 13:57.
Kuwabara, H, Cumming, P., Reith, J., Léger, G., Diksic, M., Evans, A., and Gjedde, A., 1993, Human striatal L-DOPA decarboxylase activity estimated in vivo using 6-[$^{18}$F] fluoro-DOPA and positron emission tomography: Error analysis and application to normal subjects, *J. Cereb. Blood Flow Metab.* 13:43.
Miller, L.P., Pardridge, W.M., Braun, L.D., and Oldendorf, W.H., 1985, Kinetic constraints for blood-brain barrier amino acid transport in conscious rats, *J. Neurochem.* 45:1427.
Miura, S., Murakami, M., Kanno, I., Iida, H., Shishido, F., and Uemura, K., 1991, Transport of phenylalanine in human brain by L-[2-$^{18}$F] fluorophenylalanine and PET, Japan. J. Nucl. Med. 49:1537.
Nissbrandt, H, and Carlsson, A., 1987, Turnover of dopamine and dopamine metabolism in rat brain: Comparison between striatum and substantia nigra, *J. Neurochem.* 49:959.
Reith, J., Dyve, S., Kuwabara, H., Guttman, M., Diksic, M., and Gjedde, A., 1990, Blood-brain transfer and metabolism of 6-[18F] fluoro-L-DOPA in rat, J. Cereb. Blood Flow Metab. 10:707.
Reith, J., Benkelafat, C., Sherwin, A., Yashuhara, Y., Kuwabara, H., Andermann, F., Bachneff, S., Cumming P., Diksic, M., Dyve, S., Etienne, P., Evans, A., Lal, S., Shevell, M., Savard, G., Wong, D., Chouinard, G., and Gjedde, A., 1994. Elevated DOPA decarboxylase activity in living brain of patients with psychosis, *Proc. Nat. Acad. Sci. USA*. 91: 11651.
Rousset, O., Ma., Y., Léger, G., Gjedde, A., and Evans, A.C., 1993, Correction for parital volume effects in PET using MRI-based 3D simulations of individual human brain metabolism, *in*: "Quantification of Brain Funciton, Tracer Kinetics and Image Analysis in Brain PET, "K. Uemura N.A., Lassen, T. Jones, and I. Kanno, Elsevier International Congress Series No. 1030, Amsterdam, pp.113.
Wade, L.A., and Katzmann, R., 1975, Rat brain regional uptake and decarboxylation of L-DOPA following carotid injection, *Am. J. Physiol.* 228-352.

# PEPTIDE TRANSPORT SYSTEM-1

William A. Banks and Abba J. Kastin

Veterans Affairs Medical Center and
Tulane University School of Medicine
New Orleans, LA 70146

## INTRODUCTION

Saturable transport across the blood-brain barrier (BBB) has been described for several peptides (Table 1) and regulatory proteins from numerous laboratories[1-5]. The existence of transporters at the BBB for regulatory molecules suggests that the BBB may be involved in controlling communication between the central nervous system (CNS) and peripheral organs and tissues. Tyr-MIF-1 (Tyr-Pro-Leu-Gly-$NH_2$) was the first peptide to be definitively shown to cross the BBB by a saturable process[6]. To distinguish the transporter for Tyr-MIF-1 from that for other, unrelated peptide transporters, the term peptide transport system (PTS)-1 was coined[2]. We review here some of the salient features of PTS-1 as currently understood.

## GENERAL CHARACTERISTICS

PTS-1 has been described in several strains of rats and mice to transport a limited number of endogenous peptide ligands in the direction of brain to blood[2]. When studied with radioactively iodinated Tyr-MIF-1 (I-Tyr-MIF-1) as the ligand, PTS-1 does not have a blood to brain, or influx, component[7]. Competition studies indicate that Met-enkephalin (Tyr-Gly-Gly-Phe-Met) and Tyr-MIF-1 compete equally well for PTS-1, but that only about half of the Leu-enkephalin transported by saturable pathways from the brain is due to PTS-1[8].

Several investigators may have anticipated the initial description of PTS-1. Studies by Huang and Lathja[9] and Firemark[10] found saturable uptake of enkephalin analogs by the in vitro choroid plexus. However, this uptake could have represented binding to receptors rather than a transport system. Begley and Chain found a rapid disappearance of Leu-enkephalin from the cerebrospinal fluid (CSF) during intraventricular perfusion that could have been due to either an efflux system or to sequestration by brain tissue[11]. In all three of these studies, but not our studies of PTS-1, the amino acid tyrosine inhibited the proposed transport. This inhibition by tyrosine suggests that the ligand being transported could have been the degradation product iodotyrosine being transported by an efflux system

*New Concepts of a Blood–Brain Barrier*
Edited by J. Greenwood *et al.*, Plenum Press, New York, 1995 

for amino acids[12]. However, this is not likely in the study involving intraventricular perfusion because efflux of radioactivity was not inhibited by other amino acids, as would have been excepted, if transport was by the amino acid efflux system. A more likely explanation is that the results demonstrate a tyrosine-sensitive efflux system for Leu-enkephalin that may transport the residual Leu-enkephalin not transported by PTS-1.

**Table 1.** Some peptides# transported across the BBB by saturable systems.

| PEPTIDE | DIRECTION OF SATURABLE TRANSPORT INFLUX | EFFLUX |
|---|---|---|
| TYR-MIF-1 | NO | YES |
| MIF-1 | YES | NO |
| Peptide T analog | YES | NO |
| Arginine vasopressin | YES | YES |
| Delta sleep-inducing peptide | YES* | NT |
| Dynorphin analog E2078 | YES | NT |
| Ebiratide | YES | NT |
| Enkephalin, Met | NT@ | YES |
| Enkephalin, Leu | YES | YES |
| Glutathione | YES | NT |
| Insulin | YES | NT |
| LHRH | YES | YES |
| Oxytocin | NT | YES |
| PACAP 27 | NO | YES |
| PACAP 38 | YES | YES |
| Somatostatin analog RC-160 | NO | YES |
| Tyr-W-MIF-1 | NT | YES |
| Pancreatic polypeptide | YES | NO |

#Regulatory proteins omitted; MIF = melanocyte stimulating hormone release inhibitory factor, LHRH = luteinizing hormone releasing hormone; *in guinea pig but not rat; @uptake demonstrated but saturability not tested; NT = not tested; PACAP = pituitary adenylate cyclase activating polypeptide.

Dual efflux systems for Leu-enkephalin are consistent with a previous review of this topic, which suggests that a family of transport systems may exist in the BBB for enkephalins and related peptides[2]. For example, Tyr-MIF-1[7], unlike Leu-enkephalin[13], is not transported into the brain. A conservative estimate has suggested that at least two other transport systems for enkephalins occur in the BBB in addition to PTS-1[2].

The major ligands transported by PTS-1 (Table 2) are Tyr-MIF-1 and Met-enkephalin[14]. Lesser amounts of Leu-enkephalin[8,15], oxytocin[16], Tyr-W-MIF-1[17], and probably dynorphin$_{1\text{-}8}$[15] and ß-casomorphin[15] are also transported. So far, no peptide smaller than 4 amino acids or larger than 9 has been found to be transported by PTS-1. All peptides transported have an N-terminal L-tyrosine except for oxytocin, which has a tyrosine in its peptide ring. Substitution of the L-tyrosine by D-tyrosine or by phenylalanine, which differs from tyrosine in the absence of a hydroxyl group, will abolish transport by PTS-1. However, iodination of the tyrosine does not interfere with transport since both I-Tyr-MIF-1 and tritiated Tyr-MIF-1 are transported[18]. Endogenous opiate peptides which are not transported by PTS-1 include dynorphin$_{1\text{-}17}$, ß-endorphin,

**Table 2.** Endogenous peptides transported by PTS-1.

| | | |
|---|---|---|
| Tyr-MIF-1 | Methionine enkephalin | Leucine enkephalin |
| Oxytocin* | Tyr-W-MIF-1* | $Dynorphin_{1-8}$# |
| ß-Casomorphin# | | |

*Less robustly transported. #Indirect evidence with competition studies.

dermorphin kyotorphin, and morphiceptin[15]. The peptide fragments Tyr, Tyr-Pro, Tyr-Pro-Leu, and MIF-1 (Pro-Leu-Gly) are not transported by PTS-1, showing that only intact peptide is transported[15]. Recovery of radioactivity from the vascular space also shows that Tyr-MIF-1 remains largely intact during the transport process[19].

The presence of an efflux system from the CNS for a substance often gives experimental results that appear as if the BBB is less permeable to the substance than is really the case. For example, potassium has an apparently low uptake by brain and the concentrations in the CSF do not correlate with concentrations in blood in large part as a consequence of brain to blood transport[20]. The presence of PTS-1 probably explains why the uptake by brain of some small peptides with N-terminal tyrosines is lower than predicted based on their physicochemical properties[21].

**Table 3.** Transport by PTS-1: Km and Vmax for Tyr-MIF-1 and Met-enkephalin.

| CONDITION | Km (nmol/g) | Vmax (nmol/g-min) | Reference |
|---|---|---|---|
| Fischer 344 Rats | | | 25 |
| Young | 84.9 | 3.22 | |
| Aged | 65.1 | 2.41 | |
| Mice (Blue Spruce ICR) | | | 15 |
| Method 1* | 15.2 | 0.266 | |
| Method 2* | 15.1 | 0.297 | |
| Met-enkephalin# | 24.9 | 0.630 | |
| Effects of L-leucine@ | | | 26 |
| Control | 9.73 | 0.317 | |
| + 100 nmol | 3.89 | 0.149 | |
| + 1 nmol | 11.5 | 0.367 | |
| Effects of Aluminum@ | | | 27 |
| Control@ | 3.17 | 0.187 | |
| + Al | 3.58 | 0.107 | |

*Method 1 kept the amount of radioactivity administered constant, but varied specific activity; Method 2 kept the specific activity constant, but varied the amount of radioactivity administered. All studies used Method 1 except as noted. #Km and Vmax for Met-enkephalin transport; all other studies were performed for Tyr-MIF-1. Leucine@ was given icv simultaneously with Tyr-MIF-1; aluminum chloride@ or acidified saline control@ was given about 1 h before icv injection of Tyr-MIF-1.

The kinetic parameters Vmax and Km have been measured for PTS-1 under a number of conditions (Table 3). During the period of study after icv injection of I-Tyr-MIF-1, radioactivity distributes in the periventricular area[22]. This means that I-Tyr-MIF-1 would be available for transport at both the choroid plexus and the endothelium near the ventricles. Brain endothelial cells in suspension do take up I-Tyr-MIF-1 by a saturable process[23]. However, this method cannot determine whether such uptake represents binding vs transport nor whether such uptake is occurring from the luminal or abluminal side. Recent work has shown that saturable efflux of I-Tyr-MIF-1 occurs after injection into the lateral ventricle, but not after injection into brain parenchyma, suggesting that PTS-1 may be located at the choroid plexus, but not at brain endothelia[24]. However, the location of PTS-1 remains to be confirmed. Administration of ouabain, theophylline, or reserpine does not interfere with PTS-1, suggesting that transport is not ATPase dependent and not coupled to the production of CSF[15]. Since both furosemide and acetazolamide affect transport, transport may be sodium and carbonic anhydrase dependent[15].

## REGULATORS AND PATHOPHYSIOLOGICAL IMPLICATIONS

Serotonin (5HT) inhibits PTS-1 while 5HT antagonists and high doses of histamine stimulate PTS-1[8]. Dopamine, acetylcholine, epinephrine, GABA, kainic acid, cAMP, cGMP, and numerous hormones are without effect[8,28]. Most of the activity of 5HT apparently occurs at the $5HT_1$ receptor subtype; compounds active at the $5HT_1$ receptor affect PTS-1 whereas compounds active at $5HT_2$ and $5HT_3$ receptors have little or no activity. The $5HT_1$ agonist 8-OH-DPAT is active after ip as well as icv injection[17].

Leucine acts as an allosteric regulator of PTS-1, influencing both Vmax and Km equally[26]; low doses stimulate and high doses inhibit PTS-1. When given icv, D-leucine is about 100 times more potent than L-leucine[29]. Both D- and L-leucine are more potent when given by the icv than the iv route. This suggests that the site at which leucine acts is stereospecific and located on the CNS side of the BBB. Levels achieved in brain after injection of the high dose of L-leucine (100 nmol/mouse icv) are unlikely to be attained except in pathological conditions such as maple syrup urine disease. This raises the possibility that some of the effects of this disease could be mediated through inhibition of PTS-1 and the accumulation within the brain of the peptides which it transports.

Aluminum, a neurotoxin known to cause dialysis dementia[30] and controversially implicated in Alzheimer's disease[31], acts as a non-competitive inhibitor of PTS-1[27]. Aluminum causes almost total inhibition of PTS-1 at doses which result in blood levels similar to those seen in dialysis dementia.

Circumstantial evidence suggests that brain levels of Met-enkephalin may be linked to regulation of PTS-1. In aged vs young rats, DBA vs C57 and C58 mice, and ethanol-addicted vs control mice the group with the higher level of Met-enkephalin in the brain also has the higher level of PTS-1 activity[25,32]. This suggests that the chronic enkephalinergic tone may set the level of PTS-1 activity. By transporting Met-enkephalin from the brain, PTS-1 would decrease the concentration of Met-enkephalin in the brain. A stimulatory effect of Met-enkephalin on PTS-1 and an inhibitory effect of PTS-1 on Met-enkephalin could together form a classic negative feedback loop that would serve to regulate the level of Met-enkephalin in the brain.

The setting of the PTS-1 activity level may occur during a critical period of brain development and be under opiate control. Mice given morphine during the first seven days of life have enhanced PTS-1 activity at age 23 days[33]. Adding Tyr-MIF-1 to the morphine further enhances this effect, increasing the rate at which I-Tyr-MIF-1 is transported to about three times that of mice treated with injections of vehicle.

Aging, as noted above, also affects PTS-1[25]. Both Vmax and Km are decreased proportionately. This suggests that in aged animals there are differences in the concentrations of either an allosteric regulator or a substance capable of acting in the manner of an uncompetitive inhibitor.

Ethanol administered acutely either peripherally or directly into the brain has little effect on PTS-1[32]. However, mice that are physically addicted to ethanol through ingestion of liquid feed have depressed PTS-1 activity. When ethanol is removed from the liquid feed and mice allowed to undergo physical withdrawal, PTS-1 levels return to normal; this results in an increase in efflux that may explain why brain concentrations of Met-enkephalin, already depressed in animals addicted to ethanol, show a further drop at the time of withdrawal. This additional fall in the Met-enkephalin may underlie alcohol seizures. PTS-1, therefore, may be a point at which therapy for alcoholism and alcohol withdrawal could be directed[34].

Genetic and species differences also have been described for PTS-1. The Km and Vmax are higher for rats than for mice[15,25]. PTS-1 activity also seems to vary among strains of mice[32]. Genetic differences in PTS-1 activity may relate to susceptibility to such conditions as the desire of mice to drink ethanol.

Some types of stress may alter PTS-1 activity[35]. Restraint of mice for 60 min inhibits PTS-1 while deprivation of food and water has no effect. This is consistent with the view that some types of stress, such as restraint, activate the enkephalinergic system while other types of stress, such as deprivation of food and water, activate other aspects of the opiate peptide system, such as ß-endorphin[36,37].

## SUMMARY

Work with PTS-1 has revealed much about its functional aspects. Taken together, this work suggests that peptide transport systems in the BBB are an integral aspect of the regulatory systems controlling concentrations of peptides in the CNS. PTS-1 is influenced by physiological and pathological modulators and may play a role in some disease processes.

## REFERENCES

1. D.J. Begley, Peptides and the blood-brain barrier. *in*: "Handbook of Experimental Pharmacology, vol 103, Physiology and Pharmacology of the Blood-Brain Barrier," M.W.B. Bradbury, Ed., Springer-Verlag. Berlin (1992).
2. W.A. Banks and A.J. Kastin, Editorial Review: Peptide transport systems for opiates across the blood-brain barrier. *Am. J. Physiol.* 259:E1 (1990)
3. T.Y. Terasaki, Y. Deguchi, H. Sato, K. Hirai, and A. Tsuji, In vivo transport of a dynorphin-like analgesic peptide, E-2078, through the blood-brain barrier: an application of brain microdialysis. *Pharm. Res.* 8:815 (1991).
4. W.A. Banks, K.L. Audus, and T.P. Davis, Permeability of the blood-brain barrier to peptides: an approach to the development of therapeutically useful analogs. *Peptides* 13:1289 (1992).
5. B.V. Zlokovic, D.J. Begley, M.B. Segal, H. Davson, L.J. Rakic, M.N. Lipovac, D.M. Mitrovic, and R.M. Jankov, Neuropeptide transport mechanisms in the central nervous system. *in*: "Peptide and Amino Acid Transport Mechanisms in the Central Nervous System," L.J. Rakic, D.J. Begley, H. Davson, and B.V. Zlokovic, Eds., Stockton Press, New York (1988).
6. W.A. Banks and A.J. Kastin, A brain-to-blood carrier-mediated transport system for small, N-tyrosinated peptides. *Pharmacol. Biochem. Behav.* 21:943 (1984).
7. C.M. Barrera, W.A. Banks, and A.J. Kastin, Passage of Tyr-MIF-1 from blood to brain. *Brain Res. Bull.* 23:439 (1989).
8. W.A. Banks and A.J. Kastin, Effect of neurotransmitters on the system that transports Tyr-MIF-1 and the enkephalins across the blood-brain barrier: a dominant role for serotonin. *Psychopharmacology* 98:380 (1989).

9. J.T. Huang and A. Lajtha, The accumulation of [$^3$H]enkephalinamide (2-D-alanine-5-methioninamide) in rat brain tissues. *Neuropharmacology* 17:1075 (1978).
10. H.M. Firemark, Choroid-plexus transport of enkephalins and other neuropeptides. *in*: "Neurobiology of Cerebrospinal Fluid 2." 2nd edit. J.H. Wood, Ed., Plenum Press. New York (1983).
11. D.J. Begley and D.G. Chain, The transport of encephalins from the cerebrospinal fluid of the rabbit. *in:* "Peptide and Amino Acid Transport Mechanisms in the Central Nervous System. LJ. Rakic, D.J. Begley, H. Davson, and B.V. Zlokovic, Eds., Stockton Press. New York (1988).
12. H. Davson, J.G. Hollingsworth, M.B. Carey, and J.D. Fenstermacher, Ventriculo-cisternal perfusionof twelve amino acids in the rabbit. *J. Neurobiol.* 13:293 (1982).
13. B.V. Zlokovic, J.B. Mackic, B.M. Djuricic, and H. Davson, Kinetic analysis of leucine-enkephalin cellular uptake at the luminal side of the blood-brain barrier of an in situ perfused guinea pig brain. *J. Neurochem.* 53:1333 (1989).
14. W.A. Banks, A.J. Kastin, and E.A. Michals, Tyr-MIF-1 and met-enkephalin share a saturable blood-brain barrier transport system. *Peptides* 8:899 (1987).
15. W.A. Banks, A.J. Kastin, A.J. Fischman, D.H. Coy, and S.L. Strauss, Carrier-mediated transport of enkephalins and N-Tyr-MIF-1 across blood-brain barrier. *Am. J. Physiol.* 251:E477 (1986).
16. D.A. Durham, W.A. Banks, and A.J. Kastin, Carrier-mediated transport of labeled oxytocin from brain to blood. *Neuroendocrinology* 53:447 (1991).
17. W.A. Banks, A.J. Kastin, and C.A. Ehrensing, Endogenous peptide Tyr-Pro-Trp-Gly-$NH_2$ (Tyr-W-MIF-1)is transported from the brain to the blood by peptide transport system-1. *J. Neurosci. Res.* 35:690 (1993).
18. W.A. Banks, and A.J. Kastin, Opposite direction of transport across the blood-brain barrier for Tyr-MIF-1 and MIF-1: comparison with morphine. *Peptides* 15:23 (1994).
19. W.A. Banks, A.J. Kastin, E.A. Michals, and C.M. Barrera, Stereospecific transport of Tyr-MIF-1 across the blood-brain barrier by peptide transport system-1. *Brain Res. Bull.* 25:589 (1990).
20. M.W.B. Bradbury, M.B. Segal, and J. Wilson, Transport of potassium at the blood-brain barrier. *J. Physiol. (London)* 221:617 (1972).
21. W.A. Banks and A.J. Kastin, Peptides and the blood-brain barrier: lipophilicity as a predictor of permeability. *Brain Res. Bull.* 15:287 (1985).
22. L.M. Maness, A.J. Kastin, W.A. Banks, and J.E. Zadina, *In vivo* translocation and selective removal of $^{125}$I-Tyr-MIF-1 from the rat brain after icv injection. *Soc. Neurosci. Abstr.* 17:240 (1991).
23. H.T. Pretorius, A.J. Kastin, and W.A. Banks, Binding of Tyr-MIF-1 to isolated brain capillaries. *Brain Res.Bull.* 17:829 (1986).
24. W.A. Banks, A.J. Kastin, H.M. Sam, V.T. Cao, B. King, L.M. Maness, and A.V. Schally, Saturable efflux of the peptides RC-160 and Tyr-MIF-1 by different parts of the blood-brain barrier. *Brain Res. Bull.* (in press).
25. W.A. Banks and A.J. Kastin, Ageing and the blood-brain barrier: changes in the carrier-mediated transport of peptides in rats. *Neurosci. Lett.* 61:171 (1985).
26. W.A. Banks and A.J. Kastin, Modulation of the carrier-mediated transport of the Tyr-MIF-1 across the blood-brain barrier by essential amino acids. *J. Pharmacol. Exp. Ther.* 239:668 (1986).
27. W.A. Banks, A.J. Kastin, and M.B. Fasold, Differential effect of aluminum on the blood-brain barrier transport of peptides, technetium and albumin. *J. Pharmacol. Exp. Ther.* 244:579 (1988).
28. W.A. Banks and A.J. Kastin, Twenty-one hormones fail to inhibit the brain to blood transport system for Tyr-MIF-1 and the enkephalins in mice. *J. Pharm. Pharmacol.* 40:289 (1988).
29. W.A. Banks and A.J. Kastin, Leucine modulates peptide transport system-1 across the blood-brain barrier at a stereospecific site within the central nervous system. *J. Pharm. Pharmacol.* 43:252 (1991).
30. A.C. Alfrey, J.M. Mishell, and J. Burks, Syndrome of dyspraxia and multifocal seizures associated with chronic hemodialysis. *Trans. Am. Soc. Artif. Inter.Organs* 18:257 (1972).
31. P. Zatta, R. Giordano, B. Corain, and G.G. Bombi, Alzheimer dementia and the aluminum hypothesis. *Med. Hypoth.* 26:139 (1988).
32. W.A. Banks and A.J. Kastin, Inhibition of the brain to blood transport system for enkephalins and Tyr-MIF-1 in mice addicted or genetically predisposed to drinking ethanol. *Alcohol* 6:53 (1989).

33. L.M. Harrison, J.E. Zadina, W.A. Banks, and A.J. Kastin, Effects of neonatal treatment with Tyr-MIF-1, morphiceptin, and morphine on development, tail flick, and blood-brain barrier transport. *Dev. Brain Res.* 75:207 (1993).
34. W.A. Banks, and A.J. Kastin The potential for alcohol to affect the passage of peptide and protein hormones across the blood-brain barrier: a hypothesis for a disturbance in brain-body communication. *in:* "NIAAA Research Monograph 23, Alcohol and the Endocrine System," S. Zakhari, Ed., 401 (1993).
35. W.A. Banks, A.J. Kastin, and B.J. Nager, Analgesia and the blood-brain barrier transport system for Tyr-MIF-1/enkephalins: evidence for a dissociation. *Neuropharmacology* 27:175 (1988).
36. K.K. Vaswani and G.A. Tejwani, Food deprivation-induced changes in the level of opioid peptides in the pituitary and brain of rat. *Life Sci.*38:197 (1986).
37. R. Greenberg and E.H. O'Keefe, Thiorphan potentiation of stress-induced analgesia in the mouse. *Life Sci.* 31:1185 (1982).

# LONG-CHAIN FATTY ACID TRANSPORT AT THE BLOOD-BRAIN BARRIER AND INCORPORATION INTO BRAIN PHOSPHOLIPIDS: A NEW *IN VIVO* METHOD FOR EXAMINING NEUROPLASTICITY, AND BRAIN SECOND MESSENGER SYSTEMS INVOLVING PHOSPHOLIPASE $A_2$ ACTIVATION

S. I. Rapoport[1] and P. J. Robinson[2]

[1]Laboratory of Neurosciences
National Institute on Aging
National Institutes of Health
Bethesda, Maryland 20892

and

[2]Miami Valley Laboratories
The Proctor and Gamble Company
Cincinnati, OH 45239-8707

## INTRODUCTION

Brain functional activity and structural integrity are related to rates of formation and turnover of phospholipids (Porcellati et al., 1983; Fisher and Agranoff, 1987; Axelrod et al., 1988). Until recently, these rates were estimated by injecting radiolabeled precursors intracerebrally in animals -- radiolabeled glucose, acetate, ethanolamine, glycerol, fatty acids (FAs) or ethanolamine -- and by killing the animals at different times thereafter to follow biochemical compartment radioactivities (Sun and Su, 1979; Porcellati et al., 1987). Unfortunately, intracerebral injections damage brain tissue, are not applicable to autoradiographic analysis or *in vivo* positron emission tomography (PET), and give half-lives of acyl groups within phospholipids which ignore effects of recycling.

To overcome these limitations, our laboratory has developed an *in vivo* method and model to quantify incorporation into brain phospholipids of intravenously (i.v.) injected radiolabeled long-chain FAs (Robinson et al., 1992). Brain metabolic pathways for FAs are multiple and complex. They include FA incorporation into phospholipids and triacylglycerols, elongation and desaturation, β-oxidation, lipogenesis and lipolysis (Fig. 1). Our FA model confirms that valid simplifications can be made to calculate unidirectional incorporation coefficients of FA tracers in plasma into brain phospholipids and, by means of an operational equation, the net rate of incorporation of a FA from the acyl-CoA brain precursor pool into brain phospholipids, and turnovers and half-lives of acyl groups within these phospholipids.

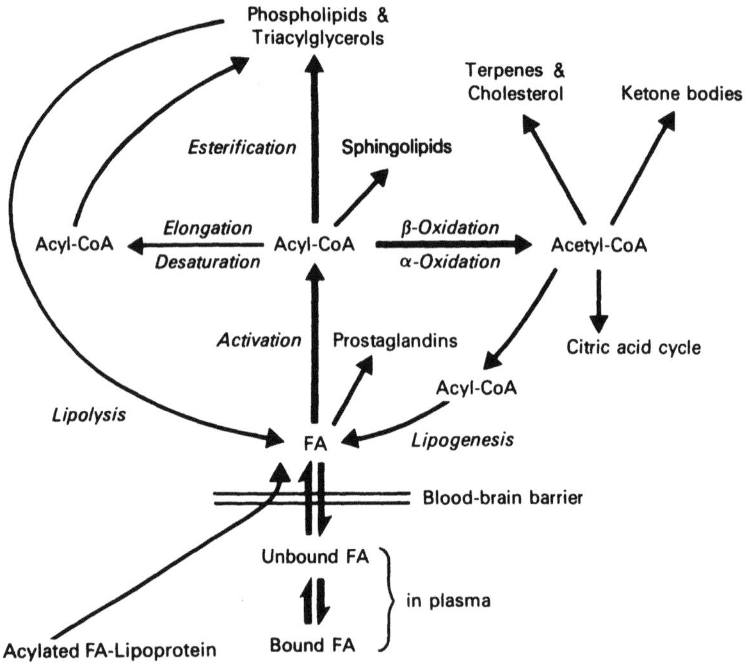


**Figure 1.** Brain metabolic pathways for fatty acids. Thick arrows indicate major pathways in the 15 min after intravenous injection of tracer FA. Thin arrows indicate alternate pathways that may be followed. Adapted from Robinson et al., (1992).

To apply the model, a radiolabeled albumin-bound unacylated FA is injected as a bolus or infused i.v. for 5 min into an animal. Labeled and unlabeled FAs in plasma are monitored until the animal is killed, when the brain is prepared for chemical analysis or quantitative autoradiography. Individual phospholipids in brain are rapidly 'pulse labeled' by this procedure. Our model can be used to evaluate this labeling because the following experimental conditions are satisfied: (i) short half-life of tracer in plasma (about 1 min); (ii) rapid equilibration between protein-bound and unbound FA in plasma; (iii) no effect of cerebral blood flow or of wide range of plasma cold FA concentrations on FA uptake and incorporation into brain from plasma; (iv) rapid equilibration (1 min or less) of labeled FAs between plasma and brain FA-CoA, the precursor pool for subsequent incorporation into brain phospholipids; (v) rapid entry of FA tracer into stable brain lipids from FA-CoA (80%-90% of brain radioactivity is within stable lipids within 1 min after label enters brain); and (vi) slow loss of FA label that has entered stable brain lipids.

In this paper, we discuss the FA model in detail. We consider the roles of protein binding and blood-brain barrier transport in FA entry into brain and show how experimental studies with the model have been used to examine brain phospholipid metabolism in relation to brain functional activity, neuroplasticity and cell growth.

## FATTY ACID ENTRY INTO BRAIN AND INCORPORATION INTO BRAIN STABLE COMPARTMENTS IS INDEPENDENT OF CEREBRAL BLOOD FLOW AND UNLABELED FA PLASMA CONCENTRATION

Long-chain FAs exist in blood as unacylated (unesterified) species in plasma; as acylated species within plasma phospholipids, esterified cholesterol and triglycerides; or as covalently bound species within erythrocytes and platelets (Staufenbiel, 1988; Bazán, 1990). The unacylated FAs normally are more than 99.99% bound to circulating plasma proteins (Wosilait and Soler-Argilaga, 1975). Their binding to albumin as compared to lipoproteins decreases as their chain length increases from 16 to 24 carbons, such that $C_{22}$ and longer chain saturated FAs bind more tightly to lipoproteins than to albumin (Shafrir et al., 1965).

Covalently bound FAs within erythrocytes slowly exchange with FAs in plasma, with half-lives of 30 min or longer (Staufenbiel, 1988). Acylated FAs in phospholipids, triglycerides and cholesterol esters are secreted into blood by the liver, as integral parts of lipoprotein complexes. They may enter brain as free FAs following their release by lipoprotein lipase at the cerebrovasculature (Fig. 1), or possibly as unacylated FAs through a receptor-mediated process involving the lipoprotein complex (not illustrated in Fig. 1) (Bazán, 1990). It is not known to what extent acylated FAs in blood contribute to brain FAs in well-fed adult animals, but they may do so in neonates (Scott and Bazán, 1989).

Net brain uptake $U$ of an unacylated FA is first order (Washizaki et al., 1994), characterized by a rate constant a ($s^{-1}$). Thus, the fraction $E$ of FA extracted from blood is given by:

$$E = 1 - e^{-\alpha\tau} = 1 - e^{-\alpha V / rCBF} \qquad \text{(Eqn. 1)}$$

where $\tau$ is transit time of blood in brain (1 s in the rat (Ohno et al., 1979)), $V$ is brain blood volume ($cm^3$ per g brain), and rCBF equals capillary blood flow ($cm^3.s^{-1}$ per g brain) (Robinson and Rapoport, 1986). Mean transit time $\tau = V/\text{rCBF}$. For small extraction fractions $E \ll 1$, Eqn. 1 gives:

$$E = \alpha\tau = \alpha V / rCBF \qquad \text{(Eqn. 2)}$$

Brain uptake $U$ ($\mu mol.s^{-1}$ per g brain) equals $E.c_{bl}.\text{rCBF}$, where $c_{bl}$ ($\mu$mol per $cm^3$) is the arterial end FA blood concentration:

$$U = \alpha V c_{bl} = \alpha \bar{c}_{bl} \qquad \text{(Eqn. 3)}$$

$\bar{c}_{bl}$ is the 'global' blood concentration normalized to 1 g brain, in units of μmol per g brain, and equals $V.c_{bl}$. Eqn. 3 indicates that FA uptake from blood into brain is independent of rCBF, provided there is no capillary recruitment so that $V$ is unchanged and $\alpha$ is independent of rCBF. Absence of capillary recruitment is suggested by observations in the rat that changes in rCBF due to hypercapnia or barbiturate anesthesia are unaccompanied by changes in $V$ (Robinson et al., 1990).

In general, $\alpha$ will depend on the binding properties of the unacylated FA to plasma protein, on FA blood-brain barrier permeability, and on the rate of incorporation of the FA into stable metabolic pools. Dissociation of the FA-palmitate complex is rapid, with rate $k_{off}$ = 0.12 $s^{-1}$ at 37°C for albumin-palmitate (Svenson et al., 1974). If brain uptake were limited solely by this dissociation ($\alpha = k_{off}$), by Eqn. 1 the extraction fraction $E=1-\exp(-k_{off}\tau) = 0.11$, assuming $\tau = 1$ s. This value for $E$ is somewhat greater than the measured extraction of 0.05 for several FA's (Pardridge and Mietus, 1980; Washizaki et al., 1991), implying that brain uptake of FAs is not solely dissociation-limited.

In view of rapid equilibration between bound and unbound forms of the unacylated FA in plasma, $\alpha$ reflects mainly transfer of the FA across the blood-brain barrier plus the kinetics of its brain incorporation (Robinson and Rapoport, 1986). Thus, the rate of change in brain concentration of unacylated FA in brain is:

$$\frac{dc_{br(unacylated)}}{dt} = k_1\bar{c}_{bl} - (k_2 + k_3)c_{br(unacylated)} \qquad \text{(Eqn. 4)}$$

where $c_{br(unacylated)}$ (μmol per g brain) equals unesterified brain FA concentration; $k_3$ is the rate constant ($s^{-1}$) for FA acylation from the free brain pool into FA-CoA *via* long chain acyl-CoA synthetase (Lehninger et al., 1993); $k_1$ and $k_2$ are rate constants ($s^{-1}$) for forward and backward transfer of unacylated FA between blood and brain ($k_1$ may include some effects of protein binding); and $t$ is time. At a steady state, $dc_{br(unacylated)}/dt = 0$ and by Eqn. 4:

$$\frac{c_{br(unacylated)}}{\bar{c}_{bl}} = \frac{k_1}{k_2+k_3} \qquad \text{(Eqn. 5)}$$

At steady-state, the rate of conversion of unacylated brain FA to FA-CoA is identical to its net brain uptake $U$ (Eqn. 3), and like $U$ is independent of rCBF:

$$k_3 . c_{br(unacylated)} = U = \alpha . \bar{c}_{bl} \qquad \text{(Eqn. 6)}$$

By Eqns. 5 and 6:

$$\alpha = \frac{k_3 . c_{br(unacylated)}}{\bar{c}_{bl}} = \frac{k_3 k_1}{k_2 + k_3} \qquad \text{(Eqn. 7)}$$

With rapid equilibration between unbound and bound forms in plasma, if $k_3$ increases, then

$k_1/(k_2+k_3) = c_{br(unacylated)} / \bar{c}_{bl}$ will not change provided capillary transfer is sufficiently rapid ($k_1$, $k_2 >> k_3$). This indeed is the case.

$k_1 = PA/D_{br}$, where $PA$ ($cm^3.s^{-1}$ per g brain) is the permeability-surface area product of the unacylated FA at the blood–brain barrier and $D_{br}$ ($cm^3$ per g brain) is the distribution volume of unacylated FA within parenchyma (Robinson et al., 1992). $PA >> 1.1\ cm^3.s^{-1}$ per g brain, as measured by *in situ* perfusion in the absence of plasma protein, at 4-fold normal blood flow rates (Smith et al., 1991). This value for $PA$ is higher than any previously measured value for any substance, 100 times a prior estimate (Spector, 1988).

From Eqns. 7 and 8, with $k_2 >> k_3$:

$$\alpha = k_3.k_1/k_2 \qquad \text{(Eqn. 8)}$$

A change in $k_3$ will be mirrored by a change in $\alpha$ and hence in $U$ (Eqn. 3), while $c_{br(unacylated)}$ will not change appreciably. Indeed, increased demand will be easily met because the brain unacylated FA concentration is in equilibrium with the large reservoir of unacylated FA in blood. Thus, $k_3$ can increase without substantially modifying $c_{br(unacylated)}$. As uptake $U$ is independent of rCBF (Eqn. 3), uptake in response to changed demand also is independent of fluctuations in rCBF (Eqn. 6). The result is a robust system in which unacylated FA in brain is maintained at a relatively constant concentration despite changes in blood flow or in demand for incorporation.

Experiments confirm that FA incorporation into brain is independent of rCBF. Incorporation rates of labeled palmitate and arachidonate from plasma are unaffected by elevating arterial $CO_2$ partial pressure from 35-45 mm Hg (during respiration with air) to 70-80 mm Hg, when rCBF varies 2-fold (Yamazaki et al., 1994; Chang et al., In preparation). Flow independence also is evidenced by the observation that muscarinic cholinergic activation by arecoline does not alter [9,10-$^3$H]palmitate incorporation into brain sites in which rCBF increases (see below) (DeGeorge et al., 1991).

Incorporation of radiolabeled FA into brain is independent of 4-fold differences in plasma unacylated cold concentrations (DeGeorge et al., 1989; Noronha et al., 1989; Noronha et al., 1990), reflecting the fact that plasma unacylated FA normally contributes to only 1-5% of the precursor acyl-CoA pool and thus does not influence its concentration markedly (Table III below). Consequently, unbound labeled unacylated plasma FA available for diffusion into brain is independent of $c_{pl}$ as well. Concentration-independence is consistent with the following analysis. If the unbound fraction $f$ of unacylated FA in plasma is small, $f << 1$ ($f = 0.0007$ for palmitate-albumin complex, (Wosilait and Soler-Argilaga, 1975)), the following approximation holds (Robinson and Rapoport, 1986):

$$f = k_{off}/[k_{on}(c_T - c_{pl})] \qquad \text{(Eqn. 9)}$$

where $c_T$ (μmol per $cm^3$ plasma) is plasma concentration of protein binding sites and $k_{on}$ ($s^{-1}.cm^3$ per μmol) the association rate constant for the FA-protein complex. For $c_{pl} \leq c_T$, Eqn. 9 becomes:

$$f = k_{off}/[k_{on}c_T]$$ (Eqn. 10)

Thus, $f$ is independent of $c_{pl}$ if the binding site concentration is far in excess of the FA concentration and if the unbound fraction is small. These conditions generally are met in our *in vivo* experiments but can be bypassed during *in situ* perfusion with protein-free solution (Washizaki et al., 1991).

## MODEL FOR FATTY ACID INCORPORATION IN BRAIN

In deriving equations to interpret experimental data, we further simplify our model according to Fig. 2. We identify two experimentally accessible brain compartments: (i) an 'unstable' final precursor FA-CoA pool (see above), whose specific activity can equilibrate rapidly, on the time-scale of the experiments, with plasma specific activity; and (ii) a 'stable' phospholipid/neutral lipid compartment in which FA turnover is slow compared to the time scale of the tracer incorporation experiments. Provided experiments are completed between 10 and 50 min after a 5-min infusion of a labeled FA, the assumptions with regard to these compartments are generally valid (Robinson et al., 1992), as the distribution of FA label in individual phospholipids, established within 5 min after a bolus injection, does not change significantly with one exception (see discussion of Fig. 4 below) in the following 50 min.

We also assume that unlabeled FAs in blood and brain are in a steady state during the experimental period. In this case (Fig. 2), net flux of the FA from the FA-CoA pool into the stable compartment, $J_{FA}$ ($\mu$mol.s$^{-1}$ per g brain) equals the sum of fluxes for the stable compartment into this pool: from unacylated FA in plasma ($J_1$), from acylated FA in blood ($J_2$), and from recycling and/or *de novo* synthesis within brain ($J_3$). A dilution factor $\lambda$ is defined as the ratio of $J_1$ to the sum of all fluxes into the unstable FA-CoA pool, equal to net flux from this compartment:

$$\lambda = \frac{J_1}{J_1 + J_2 + J_3} = \frac{J_1}{J_{FA}}$$ (Eqn. 11)

We have derived (Robinson et al., 1992) an expression for the net incorporation coefficient of an unacylated plasma FA into the stable brain lipid compartment, $k^*$ cm$^3$.s$^{-1}$.per g brain; as well as an 'operational equation' for the flux $J_{FA}$ of unlabeled FA from brain FA-CoA into this compartment. $k^*$ is the ratio of net brain radioactivity $c^*_{br}(T)$ at time $T$ of sampling, to the integral of plasma FA radioactivity from the beginning of infusion to $T$,

$$k^* = \frac{c^*_{br}(T)}{\int_0^T c^*_{pl}\,dt}$$ (Eqn. 12)

The 'operational equation' can be obtained for $J_{FA}$, taking into account dilution $\lambda$ of specific activity of unacylated FA in the FA-CoA pool due to the contributions of influxes $J_2$ and $J_3$ (Eqn. 11 and Fig. 3):

$$J_{FA} = \frac{\overset{*}{c}_{br}(T)}{\frac{\lambda}{c_{pl}}\int_0^T \overset{*}{c}_{pl}\,dt} = \frac{k^* c_{pl}}{\lambda} \qquad \text{(Eqn. 13)}$$

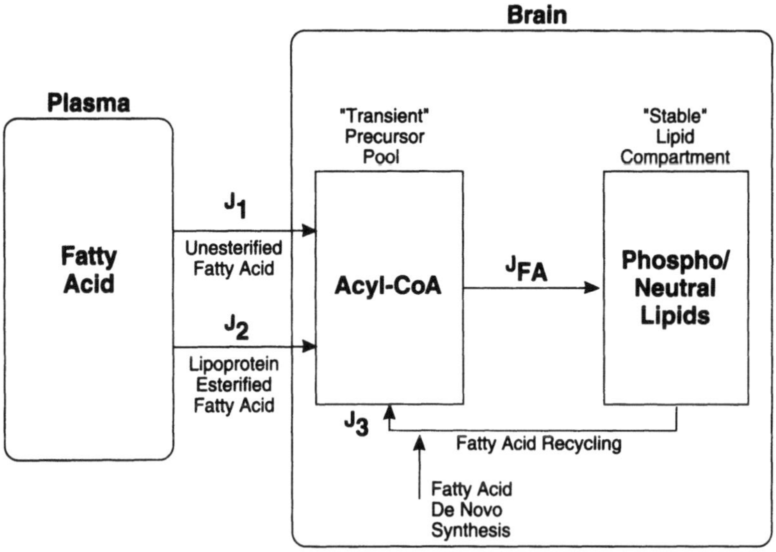


**Figure 2.** Diagram for uptake of a FA from plasma into brain acyl-CoA (final precursor pool) and from there into stable phospholipids and neutral lipids. $J_{FA}$, net flux of FA from FA-CoA to stable compartment; $J_1$, flux from plasma unesterified FA into FA-CoA pool; $J_2$,, flux of esterified FA on plasma lipoproteins into FA-CoA; $J_3$, flux from *de novo* synthesis and recycling from stable lipids. At steady state, $J_{FA} = J_1 + J_2 + J_3$

The second part of the right side of Eqn. 13 shows that net flux from the FA-CoA precursor pool into stable brain lipids is the flux of cold FA from plasma, $k^* c_{pl}$, corrected for dilution of the FA-CoA pool by the factor $\lambda$.

Expressions like Eqns. 12 and 13 can be derived for individual stable lipid compartments $i$ in brain and can be used when radioactive concentrations $\overset{*}{c}_{br,i}$ in these compartments are experimentally determined. As $\overset{*}{c}_{br} = \sum \overset{*}{c}_{br,i}$, the net incorporation coefficient $k^*$ can be represented as the sum of separate incorporation coefficients $k_i^*$ for brain lipids $i$, such that $k^* = \sum k_i^*$. Similarly, net flux from the precursor FA-CoA pool into stable lipid compartments, $J_{FA}$, is the sum of individual fluxes $J_{FA,i}$ from FA-CoA into individual lipid compartments $i$, where $J_{FA} = \sum J_{FA,i}$:

$$J_{FA,i} = \frac{\overset{*}{c}_{br,i}(T)}{\frac{\lambda_i}{c_{pl}}\int_0^T \overset{*}{c}_{pl}\,dt} = \frac{k_i^* c_{pl}}{\lambda_i} \qquad \text{(Eqn. 14)}$$

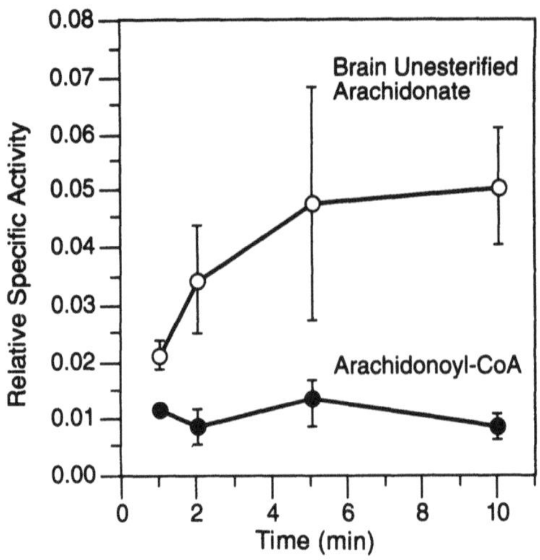


**Figure 3.** Specific activity of arachidonate in brain unacylated arachidonate and arachidonoyl-CoA pools related to specific activity of unacylated arachidonate in arterial plasma of pentobarbital-anesthetized rats, following step elevation in plasma [$^3$H]arachidonate. Each point represents mean ± S.E.M. for 3-4 animals, except for arachidonyl-CoA at 1 min when n = 1.(Washizaki et al., 1994).

Eqns. 13 and 14 were derived when assuming that specific activity in the FA-CoA pool of brain can equilibrate rapidly with FA specific activity in plasma (Robinson et al., 1992). This is the case. Fig. 3 illustrates the time course of this equilibration for arachidonyl-CoA and shows that equilibrium with plasma is established within 1 min and that $\lambda$ = 0.011 (Washizaki et al., In press). A similar time course has been noted for equilibration between plasma [9,10-$^3$H]palmitate and brain palmitoyl-CoA, with $\lambda$ = 0.017 (Grange et al., In press).

Fractional turnover $F_{FA,i}$ ($s^{-1}$) of a FA in a stable brain compartment $i$ equals the steady-state rate of incorporation into $i$ from plasma (Eqn. 14), divided by the concentration of unlabeled FA in $i$:

$$F_{FA,i} = \frac{J_{FA,i}}{c_{br,i}} = \frac{k_i^* c_{pl}}{\lambda_i c_{br,i}} \qquad \text{(Eqn. 15)}$$

Half-life $t_{1/2}$ of the FA within $i$ is defined as:

$$t_{1/2} = 0.693/F_{FA,i} \qquad \text{(Eqn. 16)}$$

## ENTRY OF FATTY ACIDS INTO SPECIFIC BRAIN PHOSPHOLIPIDS

### Entry into selected sn positions of specific phospholipids

The choice of an appropriate FA probe depends on the nature of the FA, how it is labeled, whether or not autoradiography is to be used, and which phospholipid and indeed which stereospecifically numbered (sn) position in that phospholipid is to be preferentially labeled. A labeled saturated palmitate tracer is not ideally suited for quantitative autoradiography, as about half of the label is converted *via* β-oxidation in brain mitochondria to aqueous metabolites (Noronha et al., 1989). This is not true for the unsaturated [1-$^{14}$C]arachidonate and [1-$^{14}$C]docosahexaenoate, which are preferentially incorporated into lipids compared to oxidation (Osmundsen et al., 1982); only about 15% of these labels enter the brain aqueous compartment. To overcome this limitation with regard to palmitic acid, [9,10-$^{3}$H]palmitate can be used in animals rather than [$^{14}$C]palmitate, as 75% of the tritium is converted to [$^{3}$H]$H_2O$ following oxidation (Greville and Tubbs, 1968) and is eliminated during chemical processing or autoradiography. However, if a carbon-label such as [$^{11}$C]palmitate is to be used in humans with PET, pretreatment with an inhibitor of mitochondrial oxidative metabolism is advised (Freed et al., 1994; Chang et al., 1994). Pretreatment changes the ratio of organic/aqueous brain radioactivity with carbon-labeled palmitate from 0.67 in the absence of treatment to 5.7.

Table I illustrates the distribution of radioactivity in brain lipid fractions (as percent of total dry weight radioactivity), 15 min after a 5-min i.v. infusion in awake rats of each of three radiolabeled FAs (Noronha et al., 1990; Nariai et al., 1994). Phospholipids are the most common lipid class labeled in each case, but the pattern of labeling depends on the tracer employed. [9,10-$^{3}$H]Palmitate mainly labels phosphatidylcholine (PC), [1-$^{14}$C]arachidonate mainly phosphatidylinositol (PI) and PC, and [1-$^{14}$C]arachidonate mainly phosphatidylethanolamine (PE) and PC . Labeling is maximal by 15 min. The net amount and distribution of label among individual phospholipids are largely unchanged between 15 min and 1 h following infusion of each tracer (e.g., Fig. 4, top), indicating little exchange of label among phospholipids in this time (DeGeorge et al., 1989; Noronha et al., 1990; Nariai et al., 1993). One exception follows pulse labeling of PI by [1-$^{14}$C]arachidonate, when 20% of label is lost within 1 h (Fig. 4, bottom).

Different FA tracers enter different sn sites within phospholipids. Of the phospholipids labeled most (Table I), Table II shows that [9,10-$^{3}$H]palmitate is incorporated mainly into sn1 positions of PC and PE; [1-$^{14}$C]arachidonate into sn2 positions of PI and PC; and [1-$^{14}$C]docosahexaenoate into sn2 positions of PE and PC (Nariai et al., 1994). Thus, combined use of the three FA tracers could provide estimates of *in vivo* FA metabolism of specific acyl groups within specific brain phospholipids.

### Half-lives of fatty acids within brain phospholipids

Half-lives of FAs within brain phospholipids were calculated by Eqns. 15-17 from data gathered by chemically analyzing brain after infusing awake rats with different radiolabels (Robinson et al., 1992; Nariai et al., 1994) using experimentally determined values for $\lambda_i$. As illustrated by Table III, estimated $t_{1/2}$'s are shortest in PI of all the phospholipids, consistent with participation by PI in numerous activation processes (Fisher and Agranoff, 1987; Axelrod et al., 1988; Bazán, 1990). Of the three FAs, arachidonate had the shortest half-lives within each phospholipid, likely reflecting its involvement in these processes and its role as a second messenger. Our calculated arachidonate $t_{1/2}$'s of 0.88 h and 5.47 h in PI and PE, respectively, are much less than published values of 0.4-0.54 days and 10.8-12.5 days, respectively, for mouse brain microsomes (Sun and Su, 1979; Porcellati et al., 1983). The latter were calculated without taking dilution into account.

**Table I.** Distribution of radioactivity within brain lipid fractions, in awake rats 15 min after a 5-min i.v. infusion of each of three labeled fatty acids.

| *Fraction* | *[9,10-³H]Palmitate* | *[1-¹⁴C]Arachidonate* | *[1-¹⁴C] Docosahexaenoate* |
|---|---|---|---|
| | *Percent Total Brain Radioactivity* | | |
| Lipid fraction | 89.2 ± 1.11 | 89.4 ± 0.49 | 87.1 ± 0.95 |
| Phospholipids | 51.7 ± 0.64 | 72.9 ± 0.80 | 64.7 ± 0.94 |
| Fatty acids | 9.19 ± 0.76 | 6.20 ± 0.41 | 3.37 ± 0.37 |
| Diglycerides | 8.37 ± 0.70 | 6.35 ± 0.41 | 5.10 ± 0.26 |
| Triglycerides | 19.9 ± 0.81 | 4.02 ± 0.53 | 13.8 ± 0.92 |
| Sphingomyelin | 1.21 ± 0.18 | -- | -- |
| Phosphatidylcholine | 31.7 ± 1.09 | 27.4 ± 0.22 | 17.2 ± 0.79 |
| Phosphatidylserine | 0.92 ± 0.20 | 2.09 ± 0.23 | 2.59 ± 0.30 |
| Phosphatidylinositol | 7.34 ± 0.20 | 35.6 ± 0.41 | 6.92 ± 0.68 |
| Phosphatidyl-ethanolamine | 10.5 ± 0.94 | 7.14 ± 0.22 | 37.6 ± 0.82 |

Values are means ± S.E.M. for 3-7 rats [Nariai, 1991; Noronha, 1990].

**Table II.** Distribution of radiolabeled fatty acids within sn1 and sn2 positions of brain phospholipids that receive most of the label 15 min after i.v. administration label in awake rats.

| *Fatty acid tracer* | *Phospholipid* | *Percent distribution of radiolabel* | |
|---|---|---|---|
| | | *sn1 position* | *sn2 position* |
| [9,10-³H]Palmitate | PC | 82.5 ± 1.9 | 17.5 ± 1.9 |
| | PE | 90.7 ± 0.5 | 9.3 ± 0.5 |
| [1-¹⁴C]Arachidonate | PC | 7.8 ± 1.9 | 92.2 ± 1.9 |
| | PI | 0.0 ± 0.0 | 100 ± 0 |
| [1-¹⁴C]Docosahexaenoate | PC | 22.9 ± 5.2 | 77.1 ± 5.2 |
| | PE | 0 ± 0 | 100 ± 0 |

Values are mean ± S.E.M. for 3-7 rats (Nariai et al., 1991a).

A half-life of less than 1 h for arachidonate in PI is consistent with the observed 20% decrease in PI label within 1 h after pulse labeling by [1-$^{14}$C]arachidonate (Fig. 4), allowing for recycling. It also is consistent with the observation (Fig. 5) that more than 90% of brain radioactivity is in stable lipids, 1 min after a step increment in plasma [1-$^{14}$C]arachidonate has been established in anesthetized rats(Washizaki et al., 1991). As arachidonoyl-CoA in the brain of the anesthetized rat equals 2.3 nmol.per g brain, and Fig. 5 indicates this compartment turns over within at least 60 sec, $J_{FA,i}$ << 0.038 nmol.s$^{-1}$. per g brain. Independently measured $J_1$ = 0.006 nmol.s$^{-1}$ per g brain, so that $\lambda_i$ << 0.16 by Eqn 17, compared with 0.011 measured directly by dilution of specific activity (Washizaki et al., In press). In view of the relations of Fig. 2 and Eqn. 11, low values for $\lambda_i$ and $t_{1/2}$ obtained in this way suggest that $J_3 \approx J_{FA,i}$ and thus that most dilution of arachidonoyl-CoA arises from recycling of arachidonic acid, not from unacylated arachidonate on plasma lipoproteins ($J_2$ ). This is consistent with evidence that ingested [9,10-$^3$H]palmitic acid, after being absorbed within triglycerides in the form of chlylomicrons, enter brain only after being released in blood from circulating lipoproteins by lipoprotein lipases (D. Purdon, unpublished observations).

**Table III.** Half-lives of fatty acids within individual phospholipids of the rat cerebral cortex.

| *Fatty Acid* | *Phospholipid Class* | | | | |
| --- | --- | --- | --- | --- | --- |
| | *SM* | *PC* | *PS* | *PI* | *PE* |
| | | | $t_{1/2}$, hours | | |
| Palmitate (16:0) | 12.4 | 9.38 | 9.45 | 1.24 | 2.68 |
| Arachidonate (20:4) | --- | 0.43 | 1.14 | 0.24 | 3.01 |
| Docosahexaenoate (22:6) | --- | 6.10 | 76.8 | 2.46 | 10.5 |

Data calculated with Eqn. 16 as described (Robinson et al., 1992) using $\lambda_i$ = 0.011 for arachidonate, 0.017 for palmitate and 0.02 for docosahexaenoate (Gazzah et al., 1993; Grange et al., In preparation, Washizaki et al., In press). SM, sphingomyelin; PC, phosphatidylcholine; PS, phosphatidylserine; PI, phosphatidylinositol; PE, phosphatidylethanolamine

## BRAIN FUNCTIONAL ACTIVITY, NEUROPLASTICITY AND CELL GROWTH

### Phospholipase $A_2$ activation and brain functional activity

Use of different tracers with the *in vivo* FA method allows us to identify roles of specific phospholipids with regard to changes in brain functional activity, development and neuroplasticity, and cell growth. The unsaturated FA labels can be used to examine changes involving acyl groups at sn2 positions of phospholipids, in relation to signal transduction and altered functional activity; whereas labeled saturated palmitate can be used to examine the sn1 acyl group on PC, as well as the entire PC molecule, in relation to membrane events distinct from signal transduction.

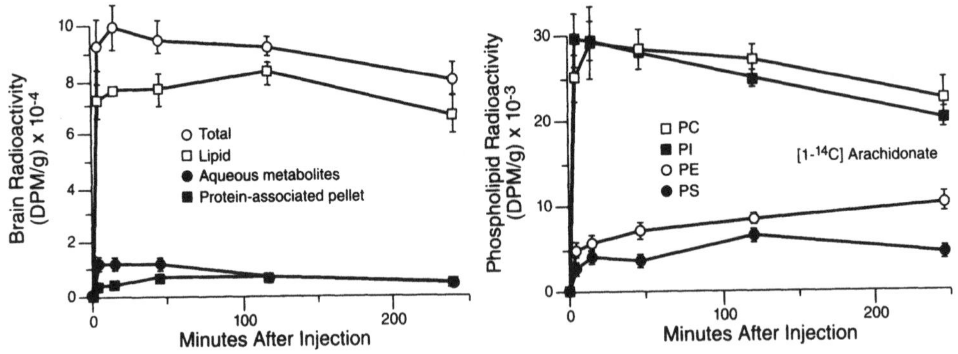


**Figure 4.** Radioactivity in different brain fractions, following an iv injection of [1-$^{14}$C]arachidonate (170 mCi.kg$^{-1}$) in awake rats. Values are means ± S.E.M. for 4 to 8 rats. PC, phosphatidylcholine; PI, phosphatidylinositol; PE, phosphatidylethanolamine; PS, phosphatidylserine (DeGeorge et al., 1989).

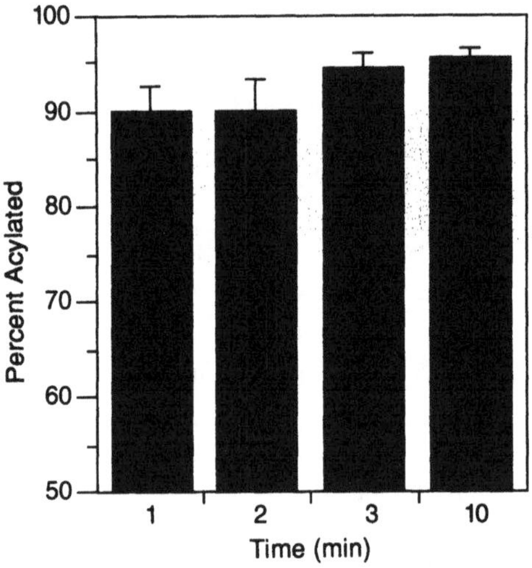


**Figure 5.** Percent acylated [$^3$H]arachidonate in brain of anesthetized rats at various times after establishing a steady state plasma specific activity by i.v. infusion. Each bar represents mean ± S.D. for 3-4 animals. From Washizaki et al., (1994).

A number of neurotransmitters can activate phospholipase $A_2$ *via* GTP binding proteins in neuronal membranes, thereby releasing arachidonate and other unsaturated acyl groups from sn2 positions of phospholipids (Axelrod et al., 1988). With prior injection of [1-$^{14}$C]arachidonate, the unacylated sn2 sites on the resultant lysophospholipids would then be re-esterified by labeled arachidonyl-CoA, through the action of arachidonyl-CoA synthetase-arachidonyl transferase (Fig. 6) (DeGeorge et al., 1987). Other neurotransmitters can induce activation of phospholipase C, with resultant cleavage of phosphatidylinositol 4,5-bisphosphate ($PIP_2$) and formation of diacylglycerol and inositol 1,4,5-triphosphate (1,4,5-

$IP_3$) (Fisher and Agranoff, 1987). The 1,4,5-$IP_3$ can mobilize calcium from intracellular stores (Berridge and Irvine, 1984), and thus promote arachidonate release by Ca-induced activation of phospholipases A (Bazán, 1989). The diacylglycerol also may release arachidonate if hydrolyzed by diglyceride lipase. Increased phospholipid labeling then would result from resynthesis of $PIP_2$ with labeled arachidonyl-CoA.

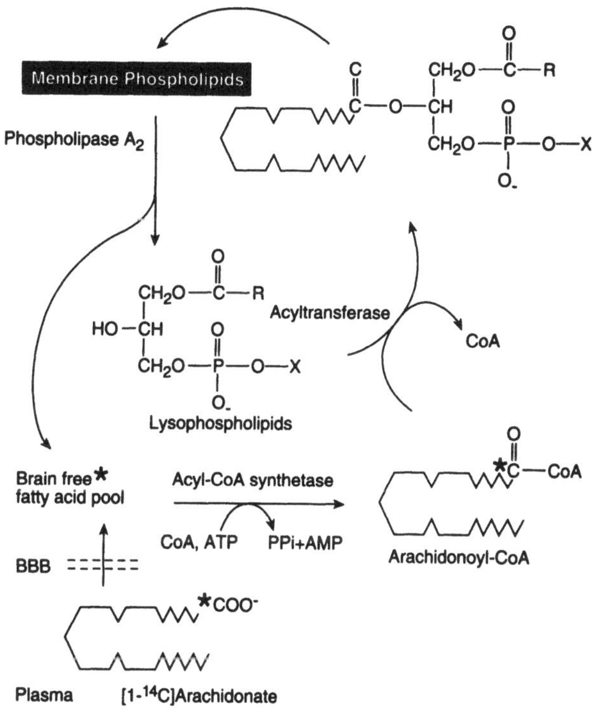


**Figure 6.** Recycling of arachidonate in brain, following activation of phospholipase $A_2$. Unlabeled arachidonate is released from a phospholipid (upper right) and is reacylated into the brain acyl-CoA pool. The latter also is labeled(*) *via* acylation of [1-$^{14}$C]arachidonate derived from blood. Tracer is incorporated into resultant lysophospholipids to label phospholipids, particularly phosphatidylinositol and phosphatidylcholine at the sn2 sites.

Phospholipases $A_1$ exist in the mammalian brain (Dawson, 1985), but their roles in signal transduction are less certain than those of phospholipases $A_2$ and C. It would not necessarily be expected that incorporation of this [9,10-$^3$H]palmitate would be augmented during changes in brain functional activity. Indeed, it is not (see below).

## Acutely increased functional activity: cholinergic stimulation

As illustrated by Table IV, brain incorporation of [9,10-$^3$H]palmitate is not increased by i.p. administration of 15 mg.kg$^{-1}$ arecoline, a cholinergic agonist. In contrast, incorporation rates of [1-$^{14}$C]arachidonate and of [1-$^{14}$C]docosahexaenoate rise by 41% and 48%, respectively (DeGeorge et al., 1991). Most brain regions with known $M_1$ muscarinic receptors (83%) show significant increments. Increased incorporation of [1-$^{14}$C]arachidonate can be prevented by prior administration of atropine to block central muscarinic action, but not of methylatropine, which does not cross the blood-brain barrier.

**Table IV.** Unidirectional whole-brain incorporation coefficients for fatty acid tracers, measured 15 min following 5-min i.v. infusion of radiolabel and i.p. administration of 15 $mg.kg^{-1}$ arecoline in awake rats. Atropine (4 mg/kg s.c.) was given prior to arecoline infusion in one set of experiments, whereas methylatropine (4 mg/kg s.c.) was administered when atropine was not given to prevent peripheral effects of arecoline.

| *Fattyacid/treatment* | *Unidirectional incorporation coefficient $k^*$, $cm^3 \cdot s^{-1}$ per g brain x $10^{-4}$* | | |
|---|---|---|---|
| | *Control* | *Arecoline (15 mg/kg)* | *Change(%)* |
| [9,10-$^3$H]Palmitate | 0.88 ± 0.06 | 0.94 ± 0.04 | +11 |
| [1-$^{14}$C]Docosahexaenoate | 3.08 ± 0.14 | 4.48 ± 0.26* | +48 |
| [1-$^{14}$C]Arachidonate | 3.13 ± 0.18 | 4.42 ± 0.27* | +41 |
| [1-$^{14}$C]Arachidonate ± atropine | 3.30 ± 0.15 | 2.54 ± 0.31 | -24 |

Mean ± S.E.M. for 3-9 animals (DeGeorge et al., 1991).
*differs significantly from control mean, $p < 0.05$

These observations suggest that arecoline promotes phospholipase $A_2$ activation to release unsaturated acyl groups from sn2 positions of PC, PI and PE, and that labeled unsaturated FA labels are transferred from FA-CoA to the resultant lysophospholipids. No arecoline effect on [9,10-$^3$H]palmitate incorporation suggests that sn1 acyl groups of PC are unaffected by arecoline, and that the increased incorporation of the unsaturated FA labels is not mediated by increased rCBF (Maiese et al., 1991).

**Acutely reduced functional activity: acute enucleation**

Contralateral-ipsilateral differences in incorporation of each of the three FA tracers into central visual areas were determined in awake adult rats from which one eye had been removed one day earlier, and were compared to differences in incorporation of [$^{14}$C]2-deoxy-D-glucose ([$^{14}$C]DG) as measures of regional cerebral metabolic rate for glucose ($rCMR_{glc}$) and functional activity (Wakabayashi et al., 1994). Compared with $k^*$ values ipsilateral to the enucleation, $k^*$ values for [1-$^{14}$C]arachidonate and [1-$^{14}$C]docosahexaenoate but not for [9,10-$^3$H]palmitate, and incorporation of [$^{14}$C]DG, were reduced in most of the 11 contralateral visual areas (layers of the superior colliculus, lateral geniculate body, primary and secondary visual cortex) that were examined. Contralateral reductions in the $k^*$ values of the unsaturated labels were proportional to reductions in [$^{14}$C]DG incorporation (Fig. 7). Thus, acutely reduced functional activity due to sensory deprivation results in reduced incorporation of the unsaturated FA labels but not of the saturated label into brain, suggesting reduced activity of phospholipase $A_2$ but not phospholipase $A_1$.

### Membrane remodeling and synthesis related to neuroplasticity and cell growth

As phospholipids are major constituents of cytoplasmic and plasma cell membranes, changes in membrane composition or in cell growth should be reflected in altered phospholipid turnover (Eqn. 15) and in incorporation of labeled FAs into phospholipids. This has been demonstrated with regard to Wallerian regeneration, chronic unilateral cochlea damage, transient cerebral ischemia (Tone et al., 1987; Tone et al., 1988; Yamazaki et al., 1989), as well as during brain development, in a brain tumor and following a chronic lesion to the nucleus basalis magnocellularis (NBM) (see below).

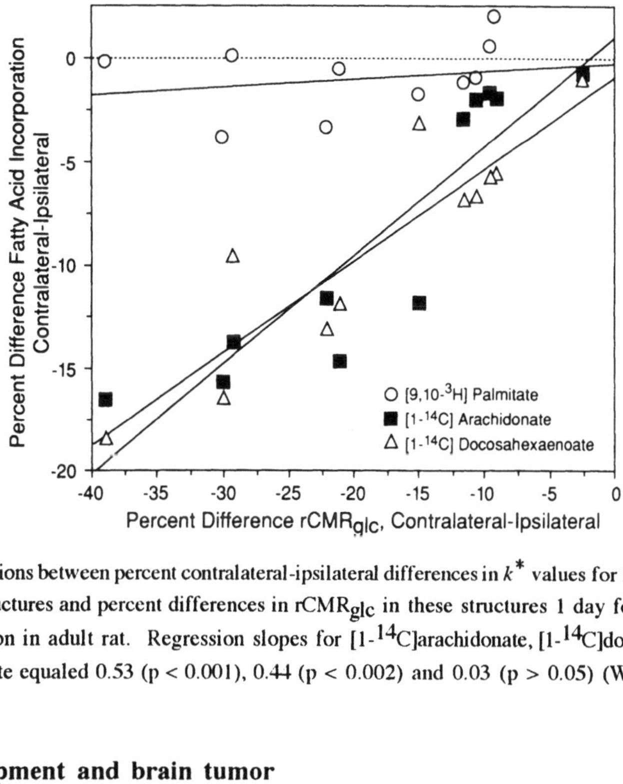


**Figure 7.** Relations between percent contralateral-ipsilateral differences in $k^*$ values for fatty acid tracers in central visual structures and percent differences in $rCMR_{glc}$ in these structures 1 day following unilateral orbital enucleation in adult rat. Regression slopes for [1-$^{14}$C]arachidonate, [1-$^{14}$C]docosahexaenoate and [9,10-$^{3}$H]palmitate equaled 0.53 ($p < 0.001$), 0.44 ($p < 0.002$) and 0.03 ($p > 0.05$) (Wakabayashi et al., 1994).

### Brain development and brain tumor

The rate of incorporation of labeled plasma palmitate into the rat brain rises between 15 and 20 days of age and then falls by up to 90% by adulthood (Tabata et al., 1986). This time course corresponds to the time course of myelination, which accelerates between 10 and 20 days of age, peaks at day 20 (at weaning), but is virtually over by 60 days (Norton and Poduslo, 1973). Furthermore, between 10 and 25 days of age, there is rapid proliferation of mitochondria in neurons and glia of gray matter, rapid growth of dendrites, axons and synaptic junctions, and marked increases in brain lipid content (Ford, 1973).

Autoradiographic studies show that Walker 256 carcinosarcoma cells, transplanted intracerebrally in rats, incorporate [9,10-$^{3}$H]palmitate 3-fold to 6-fold faster than does normal contralateral brain (Table V) (Nariai et al., 1991b). In brain-surrounding-tumor, $k^*$ for [9,10-$^3$H]palmitate is 30 % above normal, reflecting reactive gliosis. Necrotic areas within tumor have markedly reduced optical densities in autoradiographs and can be distinguished from tumor areas with high and low cell densities. Incorporation rates of [1-$^{14}$C]arachidonate and [1-$^{14}$C]docosahexaenoate also are elevated in tumor as compared to surrounding brain (Nariai et al., 1993).

**Table V.** Incorporation coefficients of [9,10-$^3$H]palmitate in control lesioned rats and in rats with intracerebral Walker 256 carcinosarcoma.

| *Tissue Specimen* | *Incorporation coefficient $k^*$* | *Ratio to contralateral brain* |
|---|---|---|
| | $cm^3.s^{-1}$ per g brain, x $10^{-4}$ | |
| *Tumor-bearing rats* | | |
| Tumor region† | | |
| highest | $11.82 \pm 1.76^{++}$ | $6.07 \pm 0.74^*$ |
| lowest | $6.23 \pm 1.02^+$ | $3.06 \pm 0.36^*$ |
| Adjacent to tumor | $2.55 \pm 0.30$ | $1.28 \pm 0.08^*$ |
| Contralateral to tumor | $2.03 \pm 0.22$ | -- |
| *Control rats* | | |
| Control lesion | $3.48 \pm 0.92$ | $1.33 \pm 0.07^*$ |
| Contralateral to lesion | $2.63 \pm 0.6$ | -- |

Means ± S.E.M. (calculated by Eqn. 12) for 10 tumor bearing animals and 6 control lesioned animals (Nariai et al., 1991a).

†Highest and lowest areas of radioactivity in the tumor were measured in each tumor-bearing rat, excluding necrotic areas

$^+p < 0.05$, $^{++}p < 0.01$, mean different from contralateral brain; $^*p < 0.01$, different from 1.0

## Neuroplasticity: Chronic unilateral lesion of nucleus basalis magnocellularis (NBM)

Modulation of brain structure or function in response to an imposed change or insult is termed 'neuroplasticity'. Neuroplasticity has been demonstrated *in vivo* with the FA method as increased membrane turnover involving PC and upregulation of receptor mechanisms involving phospholipase $A_2$.

Unilateral destruction of the NBM in the rat by ibotenic acid injection results in loss of cholinergic projections to, and in reduced choline acetyltransferase and acetylcholinesterase activities within, the ipsilateral frontal and parietal neocortices. Two weeks after a lesion, despite absence of axonal sprouting, incorporation of [$^{14}$C]DG is equal in denervated and unaffected sides (Orzi et al., 1988).

At 2 weeks, however, increased membrane turnover is evidenced by higher $k^*$ values for [9,10-$^3$H]palmitate in ipsilateral compared to contralateral neocortex, particularly upper cortical layers (Fig. 8) (De Micheli and Rapoport, In preparation). These changes are accompanied by recovery of normal phospholipid composition in ipsilateral neocortex, after

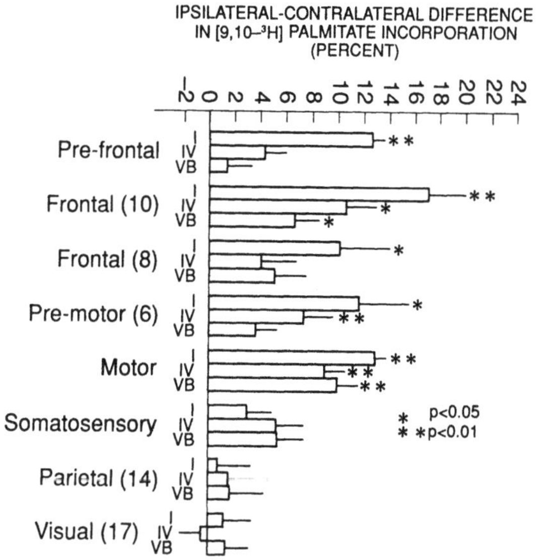


**Figure 8.** Ipsilateral-contralateral difference in $k^*$ values in awake rats for [9,10-$^3$H]palmitate, as percent contralateral $k^*$, in different neocortical brain regions 2 weeks after unilateral ablation of NBM. Cortical layers I, IV and VB are identified. Asterisks signify significant difference from 0, $^*p < 0.05$, $^{**}p < 0.01$ (De Micheli and Rapoport, In preparation).

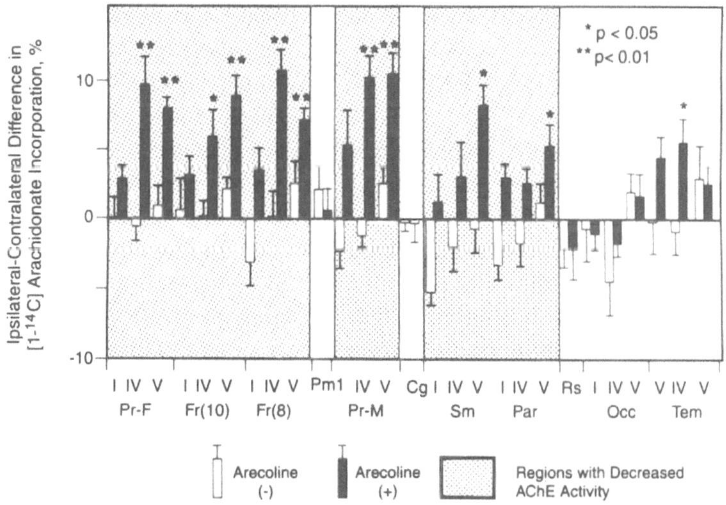


**Figure 9.** Ipsilateral-contralateral differences in $k^*$ for [1-$^{14}$C]arachidonate, as percent contralateral $k^*$, in brain regions of awake rats 2 weeks after unilateral ablation of the NBM. Studies were done with and without prior administration of 15 mg/kg i.p. arecoline, with 4 mg/kg s.c. methylatropine administered in both cases. Cortical layers I, IV and VB are identified. Filled area indicates regions with reduced acetylcholinesterase activity. Asterisks signify significant difference from 0, $^*p < 0.05$, $^{**}p < 0.01$. Abbreviations: Pr-F, prefrontal cortex; Fr(10), frontal area 10, Fr(8), frontal area 8; Pm, prefrontal medial; Pr-M, premotor ; Sm, sensorimotor ; Par, parietal ; Cg, cingulate; Occ, occipital; Rs, retrosplenial; Tem, temporal; Hipp, hippocampus CA1, CA2, CA3 and dentate regions (Nariai et al., 1991b).

an initial, post-lesion decline. Upregulation of post-synaptic cholinergic mechanisms involving phospholipase $A_2$ is evidenced as well, by increased incorporation of [1-$^{14}$C]arachidonate in the lesioned compared with the nonlesioned neocortex in response to arecoline, whereas without arecoline incorporation of [1-$^{14}$C]arachidonate is bilaterally symmetrical (Fig. 9) (Nariai et al., 1991b).

**Neuroplasticity: Chronic unilateral enucleation**

In contrast to no significant effect of acute removal of 1 eye in the rat (see above), at 3 mo after unilateral orbital enucleation in a 15-day old rat, incorporation of [9,10-$^3$H]palmitate is reduced significantly in primary visual areas contralateral enucleation: superficial gray of the superior colliculus and dorsal nucleus of the lateral geniculate body (Wakabayashi et al., In press). Incorporation rates of [1-$^{14}$C]arachidonate and [1-$^{14}$C]docosahexaenoate also are reduced in the contralateral superficial gray, whereas [$^{14}$C]DG incorporation is bilaterally symmetrical in central visual areas. These changes in contralateral primary visual areas correspond to morphological evidence of neuroplasticity: axonal sprouting and altered size and number of neurons, synapses and dendrites.

## SUMMARY

An experimental method and its associated mathematical model are described to quantitate *in vivo* incorporation rates into and turnovers of FAs within stable brain metabolic compartments, particularly phospholipids. A radiolabeled FA is injected i.v. in a rat, and arterial plasma FA radioactivities and unlabeled FA concentrations are sampled until the animal is killed after 15 min, when the brain is analyzed biochemically or with quantitative autoradiography. Unbound unacylated label in blood rapidly crosses the blood-brain barrier due to its very high permeability; rapidly equilibrates in the unacylated FA, FA-CoA and phosphatidate-diacylglycerol brain pools; then is incorporated into stable phospholipids and other stable compartments. Uptake from plasma and incorporation from the precursor pool of labeled FAs are independent of cerebral blood flow and of 4-fold differences in unlabeled plasma unacylated FA concentrations. Different labeled FAs enter different sn positions of different brain phospholipids, suggesting that a combination of labels can be used to investigate metabolism of these phospholipids. Thus, [9,10-$^3$H]palmitate preferentially labels the sn1 position of phosphatidylcholine; [1-$^{14}$C]arachidonate, the sn2 positions of phosphatidylinositol and phosphatidylcholine; and [1-$^{14}$C]docosahexaenoate, the sn2 positions of phosphatidylethanolamine and phosphatidylcholine.

The FA model leads to equations for rates of incorporation of radiolabeled FA tracers from blood into brain lipids. It also provides an operational equation for rates of incorporation of unlabeled FAs into brain phospholipids from the precursor acyl-CoA pool, when taking into account intracerebral recycling and *de novo* synthesis of the FA, as well as entry into brain of FA from acylated blood sources on lipoproteins. The equation is essentially independent of specific details of the proposed model and can be used to calculate rates of turnover and half-lives of FAs within different phospholipid classes. For the model to be entirely applicable, experiments should satisfy conditions for pulse-labeling of the phospholipids with brain sampling times short enough to minimize exchange of label between stable metabolic compartments. A 15-20 min sampling time satisfies these criteria.

The FA method can be used to elucidate brain phospholipid metabolism in relation to functional activity, and growth and turnover of cell membranes. Altered incorporation of the unsaturated arachidonate and docosahexaenoate labels reflects changes in sn2 positions of phospholipids, mediated by phospholipase $A_2$ and other enzymes. Such changes have been noted following acute cholinergic drug activation and acute sensory deprivation. Altered membrane growth or remodeling related to neuroplasticity, during development, brain tumor or chronic sensory deprivation, can be identified by examining incorporation of labeled saturated palmitate. Its incorporation involves the sn1 position of phosphatidylcholine and other lipid sites not directly related to signal transduction. Neuroplasticity also may be accompanied by upregulation of receptor mechanisms involving phospholipase $A_2$, leading to increased incorporation of unsaturated labels in response to drug activation.

Once $[^{11}C]$positron-emitting isotopes of appropriate long chain FAs are available (Channing and Simpson, 1993), it should be possible to extend the FA method to human studies using PET. Preliminary experiments on the rhesus monkey provide the basis for this extension (Chang et al., In preparation).

## REFERENCES


Axelrod, J., Burch, R. M., and Jelsema, C. L., 1988, Receptor-mediated activation of phospholipase $A_2$ via GTP-binding proteins: Arachidonic acid and its metabolites as second messengers, *Trends Neurosci.* 11: 117-123.

Bazán, N. G., 1989, Arachidonic acid in the modulation of excitable membrane function and at the onset of brain damage., *Ann. N.Y. Acad. Sci.* 559: 1-16.

Bazán, N. G., 1990, Supply of n-3 polyunsaturated fatty acids and their significance in the central nervous system, *in*: "Nutrition and the Brain," New York, Raven Press. 1-24.

Berridge, M. J., and Irvine, R. F., 1984, Inositol trisphosphate, a novel second messenger in cellular signal transduction, *Nature.* 312: 315-321.

Chang, M. C. J., Arai, T., Freed, L. M., Wakabayashi, S., Channing, M. A., Dunn, B. B., Der, M. G., Bell, J. M., Herscovitch, P., Eckelman, W. C., and Rapoport, S. I., In preparation, Brain incorporation of $^{11}C$-arachidonate in normocapnic and hypercapnic monkeys: a study using positron emission tomography.

Chang, M. C. J., Wakabayashi, S., and Bell, J. M., 1994, The effect of methyl palmoxirate on incorporation of $[U-^{14}C]$palmitate into rat brain, *Neurochem. Res.* 19: 1217-1223.

Channing, M. A., and Simpson, N., 1993, Radiosynthesis of $1-[^{11}C]$polyhomoallylic fatty acids, *J. Labelled Compds. Radiopharmacol.* 33: 541-546.

Dawson, R. M. C., 1985, Enzymatic pathways of phospholipid metabolism in the nervous system, *in*: "Phospholipids in nervous tissues," New York, John Wiley and Sons. 45-78.

De Micheli, E., and Rapoport, S. I., In preparation, Increased incorporation of labeled palmitic acid ipsilateral to chronic damage of the nucleus basalis signifies dendritic-synaptic neuroplasticity.

DeGeorge, J. J., Nariai, T., Yamazaki, S., Williams, W. M., and Rapoport, S. I., 1991, Arecoline-stimulated brain incorporation of intravenously administered fatty acids in unanesthetized rats, *J. Neurochem.* 56: 352-355.

DeGeorge, J. J., Noronha, J., Bell, J. M., Robinson, P., and Rapoport, S. I., 1989, Intravenous injection of $[1-^{14}C]$arachidonate to examine regional brain lipid metabolism in unanesthetized rats, *J. Neurosci. Res.* 24: 413-423.

DeGeorge, J. J., Ousley, A. H., McCarthy, K. D., Lapetina, E. G., and Morell, P., 1987, Acetylcholine stimulates selective liberation and re-esterification of arachidonate and accumulation of inositol phosphates and glycerophosphoinositol in C62B glioma cells, *J. Biol. Chem.* 262: 8077-8083.

Fisher, S. K., and Agranoff, B. W., 1987, Receptor activation and inositol lipid hydrolysis in neural tissues, *J. Neurochem.* 48: 999-1017.

Ford, D. H., 1973, Selected maturational changes observed in the postnatal rat brain, *Progr. Brain Res.* 40: 1-12.

Freed, L. M., Wakabayashi, S., Bell, J. M., and Rapoport, S. I., 1994, Effect of inhibition of β-oxidation on incorporation of [U-$^{14}$C]palmitate and [1-$^{14}$C]arachidonate into brain lipids, *Brain Res.* 645: 41-48.

Gazzah, N., Gharib, G., Bobillier, P., Lagarde, M., and Sarda, N., 1993, Evidence for docosahexaenoate recycling in the free-moving rat brain. 1st Intl. Congress ISSFAL, June 30-July 3, Lugano,

Grange, E., Deutsch, J., Chang, M., Smith, Q. R., Rapoport, S. I., and Purdon, A. D., In press, Brain palmitic acid incorporation in the awake rat: relation between specific activity of [9,10-$^{3}$H]palmitate in plasma and brain palmitoyl-CoA pool *J. Neurochem.*

Greville, G. D., and Tubbs, P. K., 1968, The catabolism of long-chain fatty acids in mammalian tissues, *Essays Biochem.* 4: 155-212.

Lehninger, A. L., Nelson, D. L., and Cox, M. M., 1993, "Principles of Biochemistry," New York, Worth.

Maiese, K., Holloway, H. W., Larson, D. M., Reis, D. J., and Soncrant, T. T., 1991, Coupling of regional cerebral metabolism and blood flow in the conscious rat is maintained after treatment with the muscarinic agonist arecoline, *Abstr. Soc. Neuroscience* 17: 26.

Nariai, T., DeGeorge, J. J., Greig, N. H., Genka, S., Rapoport, S. I., and Purdon, A. D., 1994, Differences in rates of incorporation of intravenously injected radiolabeled fatty acids into phospholipids of intracerebrally implanted tumor and brain in awake rats, *Clin. Exptl. Metast.* 12: 213-225.

Nariai, T., DeGeorge, J. J., Greig, N. H., and Rapoport, S. I., 1991a, In vivo incorporation of [9,10-$^{3}$H]palmitate into a rat metastatic brain-tumor model, *J. Neurosurg.* 74: 643-649.

Nariai, T., DeGeorge, J. J., Lamour, Y., and Rapoport, S. I., 1991b, In vivo brain incorporation of [1-$^{14}$C]arachidonate in awake rats, with or without cholinergic stimulation, following unilateral lesioning of nucleus basalis magnocellularis, *Brain Res.* 559: 1-9.

Nariai, T., Greig, N. H., DeGeorge, J. J., Genka, S., and Rapoport, S. I., 1993, Intravenously injected radiolabelled fatty acids image brain tumor phospholipids in vivo: differential uptakes of palmitate, arachidonate and docosahexaenoate, *Clin. Exptl. Metastasis* 11: 141-149.

Noronha, J. G., Bell, J. M., and Rapoport, S. I., 1990, Quantitative brain autoradiography of [9,10-$^{3}$H]palmitic acid incorporation into brain lipids, *J. Neurosci. Res.* 26: 196-208.

Noronha, J. G., Larson, D. M., and Rapoport, S. I., 1989, Regional cerebral incorporation of plasma [$^{14}$C]palmitate, and cerebral glucose utilization in water-deprived Long-Evans and Brattleboro rats, *Exp. Neurol.* 103: 267-276.

Norton, W. T., and Poduslo, S. E., 1973, Myelination in rat brain: changes in myelin composition during brain maturation, *J. Neurochem.* 21: 759-773.

Ohno, K., Pettigrew, K. D., and Rapoport, S. I., 1979, Local cerebral blood flow in the conscious rat, as measured with $^{14}$C-antipyrine, $^{14}$C-iodoantipyrine and $^{3}$H-nicotine, *Stroke.* 10: 62-67.

Orzi, F., Diana, G., Casamenti, F., Palombo, E., and Fieschi, C., 1988, Local cerebral glucose utilization following unilateral and bilateral lesions of the nucleus basalis magnocellularis in the rat, *Brain Res.* 462: 99-103.

Osmundsen, H., Cervenka, J., and Bremer, J., 1982, A role for 2,4-enoyl-CoA reductase in mitochondrial b-oxidation of polyunsaturated fatty acids. Effects of treatment with clofibrate on oxidation of polyunsaturated acylcarnitines by isolated rat liver, *Biochem. J.* 208: 749-757.

Pardridge, W. M., and Mietus, L. J., 1980, Palmitate and cholesterol transport through the blood-brain barrier, *J. Neurochem.* 34: 463-466.

Porcellati, G., Goracci, G., and Arienti, G., 1983, Lipid turnover, *in*: "Modern Methods in Pharmacology, Vol. 4, Methods for Studying Platelets and Megakaryocytes," New York, Alan R. Liss. 229-265.

Robinson, P., Villacreses, N., Rapoport, S., and Smith, Q. R., 1990, Brain capillary perfusion and blood volume during acute hypoxia, hypercapnia or pentobarbital anesthesia, *Abstr. Soc. Neurosci.* 16: 741.

Robinson, P. J., Noronha, J., DeGeorge, J. J., Freed, L. M., Nariai, T., and Rapoport, S. I., 1992, A quantitative method for measuring regional in vivo fatty-acid incorporation into and turnover within brain phospholipids: review and critical analysis, *Brain Res. Rev.* 17: 187-214.

Robinson, P. J., and Rapoport, S. I., 1986, Kinetics of protein binding determine rates of uptake of drugs by brain., *Am. J. Physiol.* 251: R1212-R1220.

Scott, B. L., and Bazán, N. G., 1989, Membrane docosahexaenoate is supplied to the developing brain and retina by the liver, *Proc. Natl. Acad. Sci. U.S.A.* 86: 2903-2907.

Shafrir, E., Gatt, S., and Khasis, S., 1965, Partition of fatty acids of 20-24 carbon atoms between serum albumin and lipoproteins, *Biochim. Biophy. Acta.* 98: 365-371.

Smith, Q. R., Nagura, H., Washizaki, K., DeGeorge, J., Robinson, P. J., and Rapoport, S. I., 1991, Kinetics of fatty acid dissociation from albumin and transport into brain, *Abstr. Soc. Neurosci.* 17: 239.

Spector, R., 1988, Fatty acid transport through the blood-brain barrier, *J. Neurochem.* 50: 639-643.

Staufenbiel, M., 1988, Fatty acids covalently bound to erythrocyte proteins undergo a differential turnover in vivo, *J. Biol. Chem.* 263: 13615-13622.

Sun, G. Y., and Su, K. L., 1979, Metabolism of arachidonoyl phosphoglycerides in mouse brain subcellular fractions, *J. Neurochem.* 32: 1053-1059.

Svenson, A., Holmer, E., and Andersson, L. O., 1974, A new method for the measurement of dissociation rates for complexes between small ligands and proteins as applied to the palmitate and bilirubin complexes with serum albumin, *Biochim. Biophys. Acta.* 342: 54-59.

Tabata, H., Bell, J. M., Miller, J. C., and Rapoport, S. I., 1986, Incorporation of plasma palmitate into the brain of the rat during development, *Dev. Brain Res.* 29: 1-8.

Tone, O., Miller, J. C., Bell, J. M., and Rapoport, S. I., 1987, Regional cerebral palmitate incorporation following transient bilateral carotid occlusion in awake gerbils, *Stroke.* 18: 1120-1127.

Tone, O., Miller, J. C., Bell, J. M., and Rapoport, S. I., 1988, Regional cerebral palmitate incorporation after unilateral auditory deprivation in immature and adult Fischer-344 rats, *Exp. Neurol.* 100. 491-505.

Wakabayashi, S., Freed, L. M., Bell, J. M., and Rapoport, S. I., 1994, In vivo cerebral incorporation of radiolabeled fatty acids after acute unilateral orbital enucleation in adult hooded Long-Evans rats, *J. Cereb. Blood Flow Metab.* 14: 312-323.

Wakabayashi, S., Freed, L. M., Chang, M., and Rapoport, S. I., In press, In vivo imaging of brain incorporation of fatty acids and of 2-deoxy-D-glucose suggests distinct remodeling and functional neuroplastic effects of chronic unilateral visual deprivation in rats, *Brain. Res.*

Washizaki, K., Purdon, D., DeGeorge, J., Robinson, P. J., Rapoport, S. I., and Smith, Q. R., 1991, Fatty acid uptake and esterification by the in situ perfused rat brain, *Abstr. Soc. Neuroscience* 17: 864.

Washizaki, K., Smith, Q. R., Rapoport, S. I., and Purdon, A. D., 1994, Brain arachidonic acid incorporation and precursor pool specific activity during intravenous infusion of unesterified [$^3$H]arachidonate in the anesthetized rat, *J. Neurochem.* 63: 727-736.

Wosilait, W. D., and Soler-Argilaga, C., 1975, A theoretical analysis of multiple binding of palmitate by bovine serum albumin: the relationship to uptake of free fatty acids by tissues, *Life Sci.* 17: 159-166.

Yamazaki, S., DeGeorge, J. J., Bell, J. M., and Rapoport, S. I., 1994, Effect of pentobarbital on incorporation of plasma palmitate into rat brain, *Anesthesiology* 80: 151-158.

Yamazaki, S., Noronha, J. G., Bell, J. M., and Rapoport, S. I., 1989, Incorporation of plasma [U-$^{14}$C]palmitate into the hypoglossal nucleus following unilateral axotomy of the hypoglossal nerve in adult rat, with and without regeneration, *Brain Res.* 477: 19-28.

# HAEMOSTATIC FUNCTIONS OF THE BLOOD-BRAIN BARRIER: POSSIBLE IMPLICATIONS IN THE PATHOGENESIS OF STROKE

Berislav V. Zlokovic, Liang Wang,
Steven S. Schreiber and Mark J. Fisher

Departments of Neurosurgery and Neurology
and Division of Neurosurgery Childrens Hospital
Los Angeles
CA 90033

The balance between clot-forming (procoagulant) and clot-dissolving (fibrinolytic) pathways in the brain is closely controlled, but the mechanisms responsible for its maintenance are not known. Alterations in this balance may ultimately result in cerebral infarction or stroke. Understanding haemostatic mechanisms in the brain ultimately requires the delineation of the haemostatic functions of the blood-brain barrier (BBB).

## ANTICOAGULANT FUNCTION OF NORMAL BBB

The most important factors for preventing clotting in the cerebral microcirculation are most probably: (1) the barrier which prevents tissue factor (TF) exposure and activation of the extrinsic clotting mechanism, (2) the anticoagulant properties which include an integral endothelial membrane protein, thrombomodulin (TM), heparan-sulfate and (3) the physico-chemical properties of the luminal endothelial membrane, which prevents the attachment of platelets and the contact phase amplification of the intrinsic clotting mechanism.

TM is expressed at the luminal surface of endothelial cells and normally serves as a binding site for thrombin[1]. Circulating thrombin promotes formation of a fibrin clot and activates platelets, while bound thrombin is deprived of these procoagulant activities. The thrombin-TM complex activates protein C which in turn inactivates factors Va, VIIIa, and plasminogen activator inhibitor type 1 (PAI-1)[2,3]. These events shift the haemostatic balance toward anticoagulation. In human brain, TM is present in nearly all vessels in the neocortex, but is absent in 30-50% of the vessels in the pons, putamen and mesencephalon[4]. This non-uniform distribution is intriguing since both pons and putamen are amongst the most common sites for brain infarction, particularly small "lacunar" infarcts. The mechanism responsible for the regional distribution of TM is unknown.

Recently, by using quantitative competitive PCR, TM mRNA has been demonstrated in bovine cerebral endothelial cells (Fisher et al., unpublished observations).

Other endothelial mechanisms may include production of prostacyclin, a potent inhibitor of platelet aggregation and adherence, or complexing of antithrombin III to vascular heparan sulfate. These mechanisms, however, may be less significant at the intact BBB since endothelial cells isolated from microvessels secrete barely detectable amounts of prostacyclin, in contrast to cells isolated from large vessels[5]. Heparan-sulfate from cerebral microvessels exhibits a small amount of anticoagulant activity *in vitro*, but its location *in situ* is exclusively on the abluminal side of brain capillaries[6].

## BLOOD COAGULATION IN BRAIN *IN VIVO*

Despite major advances in our understanding of blood coagulation *in vitro*, remarkably little is known about functions of coagulation mechanisms *in vivo*[7]. Current concepts suggest an interaction between extrinsic and intrinsic mechanisms[8]. The key for initiation of the coagulation process is TF, an integral membrane protein expressed on the surface of non-vascular cells. Under normal conditions, TF is not expressed by endothelium at the BBB as we recently reported[9]. When the cerebral endothelial membrane is damaged, the barrier which prevents TF exposure is lost. The extracellular domain of TF is the factor VII receptor, and the TF-activated factor VIIa complex activates factors IX and X[7]. Factor IXa forms a complex with factor VIIIa and activates factor X, while factor Xa forms a complex with factor Va and activates prothrombin[7]. Thrombin cleaves fibrinogen, yielding monomeric fibrin, which then polymerizes to form a fibrin clot. It is suggested that the components of the extrinsic pathway activate the intrinsic pathway which is dormant *in vivo* with regard to the generation of factor Xa or thrombin[10]. Exposure of the subendothelial basement membrane activates the contact phase which amplifies the intrinsic mechanism[7,10].

In parallel with the coagulation process, cerebral vascular injury prompts circulating platelets to marginate against the vessel wall[7]. Their adherence is most likely mediated by von Willebrand factor (vWF) released from endothelial cells. vWF binds to glycoprotein Ib on the platelet membrane, and activated platelets degranulate and rapidly express factor VIII and V receptors, and also P-selectin (leukocyte receptor on platelets) on their surfaces. Platelets aggregate in an interaction mediated by fibrinogen and the fibrinogen receptor (glycoprotein IIb-IIIa complex) and form a platelet plug. Circulating leukocytes interact with the microvascular endothelium and the platelets via different receptor adhesion molecules (selectins)[11].

## POSSIBLE FIBRINOLYTIC FUNCTION OF THE BBB

When a cerebral clot is formed, a large amount of plasminogen is trapped in the clot along with other plasma proteins. Plasminogen requires activation to become plasmin[12]. Plasmin is a trypsin-like enzyme with fibrinolytic activity which digests fibrin, fibrinogen, and factors V, VIII and XII. Fibrinolysis promotes slow clearing of clotted blood, and eventually reopening of cerebral microvessels. The tissue plasminogen activator (tPA) is a key enzyme in the control of fibrinolysis. tPA activity is measured in brain capillary-like structures *in vitro*[13], and we recently reported that tPA is present in cerebral microvessels in the guinea-pig brain[9]. We have confirmed this finding by a quantitative Western blot study, in which tPA band of 67 kDa was demonstrated only in brain microvessels, but not in capillary-depleted brains[14]. Initial immunohistochemical study has revealed that in brain microvessels, tPA is preferentially distributed in endothelial cells of small precapillary arterioles and venules[15], while endothelial cells of larger cerebral vessels were negative for both tPA protein and tPA mRNA. More recent study confirmed tPA expression in brain capillaries[14]. These findings give a rise to the hypothesis that the normal BBB may have a role in promoting plasmin-dependent fibrinolysis in the brain microcirculation.

The injured cerebral endothelium releases tPA which then preferentially activates fibrin bound plasminogen, so that plasmin is generated primarily within the fibrin clot[12]. Much of the tPA in normal plasma is inactive and complexed with PAI-1. Thrombin adjacent to microclots may also stimulate tPA and PAI-1 release from the intact endothelium. Thrombin-TM activated protein C simultaneously inactivates PAI-1[2]. It is suggested that tPA may have important functions during development by regulating the migration/invasion of astrocytes[16] and neurons[17], and by modulating astroglial-induced microvascular brain morphogenesis[13]. tPA is also rapidly induced by neuronal activity which suggests a role in synaptic plasticity[18].

## BBB MECHANISMS AND STROKE

Recent studies have provided strong experimental evidence in support of the hypothesis that stroke may develop as a result of changes in the brain microcirculation[19,20]. It has been demonstrated that stroke risk factors such as hypertension, diabetes, advanced age, and genetic predisposition, may prepare rat brainstem tissue for a local Shwartzman reaction. The reaction is provoked by either intracisternal or intravenous injection of endotoxin. Endotoxin (lipopolysaccharide, LPS) is derived from the cell wall of Gram-negative bacteria, and is a natural substance to which animals and humans are repeatedly exposed. Endotoxin has powerful effects on the host humoral and cellular systems which mediate haemostasis and inflammation. Affected rats exhibit profound neurological impairment caused primarily by microthrombi and petechial haemorrhages in the brainstem. These studies demonstrate that stroke risk factors may predispose to brain ischemia by local effects on haemostatic systems in the cerebral microcirculation.

## STROKE-RISK FACTORS, PROCOAGULANT CHANGES, AND BBB MECHANISMS

Smoking, diabetes and hypertension are the major factors which increase the risk of cerebrovascular disease and ischemic stroke.

### Smoking

Nicotine, a primary constituent in cigarette smoke, is responsible for the increased platelet aggregability, and decreased fibrinolysis found in habitual smokers[21]. Several studies have demonstrated that chronic exposure to nicotine produces morphological alterations in various endothelial cells of non-BBB origin. In our recent study on nicotine effects in guinea pig brain, we demonstrated ultrastructural changes in neurohypophyseal capillaries (outside the BBB), but not at the BBB itself. *In vitro* studies have suggested that the exposure of endothelial cells to nicotine impairs fibrinolytic activity (Hofman et al., unpublished observations). Our preliminary studies have shown that chronic treatment of rats with nicotine results in complete depletion of tPA expression at the BBB[22].

## Diabetes

Recent studies suggest that advanced glycosylated end-products (AGE) may have a central role in the pathogenesis of vascular changes which occur in diabetic patients[23]. AGE are proteins modified by non-enzymatic covalent glycosylation. In diabetes these molecules accumulate progressively on long-lived proteins in plasma and on subendothelial tissues including vascular basement membranes. AGE proteins could modulate the expression of endothelial coagulation receptors *in vitro* by inducing TF expression, and suppressing the activity of the TM pathway[23]. We have recently analyzed expression of tPA in cerebral microvessels isolated from streptozotocin-treated rats. After 14 days, quantitative Western blot analysis indicated a complete tPA depletion at the BBB in these rats and this was corroborated at the level of mRNA[24]. These procoagulant changes could then lead to an imbalance of endothelial haemostatic functions.

## Hypertension

Prolonged clot lysis time, lower tPA activity, increased platelet aggregation, elevated fibrinogen, increased PAI-1, and a link between cerebral venous thrombosis and hereditary protein C deficiency has been demonstrated in patients with mild essential hypertension[25]. These changes suggest impaired fibrinolysis and a procoagulant haemostatic blood profile. Experimental studies indicate that the spontaneous dissolution of a cerebral embolus is significantly lower in hypertensive compared with normotensive rats[26]. Recent studies have also demonstrated that rats which are genetically prone to hypertension, or spontaneously hypertensive rats, with a mean arterial blood pressure of 175 mmHg, develop brain microthrombi and petechial hemorrhages only after a provocative dose of LPS[19].

# CONCLUSIONS AND FUTURE DIRECTIONS

There is vast body of clinical evidence that emphasizes the importance of haemostasis in pathophysiology of ischemic stroke[27]. However, little is known of the mechanisms which regulate haemostasis in the brain. We present evidence that the BBB may be an important mediator of brain regulation of haemostasis, and that this role is enhanced in the presence of stroke risk factors. Delineation of the molecular mechanisms of BBB-mediated haemostasis will likely contribute to future efforts in stroke prevention.

## Acknowledgments

This work was supported by funds provided by the Cigarette and Tobacco Surtax Fund of the State of California through the Tobacco-Related Disease Research Program of the University of California, Grant 2RT0071, the Hoover Foundation and the NIH Grant PO1NS31945.

# REFERENCES

1. E.G. Nabel, Biology of the impaired endothelium, *Am. J. Cardiol.* 68:6C (1991).
2. V.W.M. Van Hinsberg, R.M. Bertina, A. Van Wijngaarden, N.H. Van Tilburg, J.J. Emeis, and F. Haverkate, Activated protein C decreases plasminogen activator inhibitor activity in endothelial cell-conditioned medium, *Blood* 65:444 (1985).
3. W.L. Chandler, S.L. Trimble, S.C. Loo, and D. Mornin, Effect of PAI-1 Levels on the molar concentrations of active tissue plasminogen activator (t-PA) and t-PA/PAI-1 complex in plasma, *Blood* 76:930 (1990).

4. V.L.Y. Wong, F.M. Hofman, H. Ishii, and Fisher M., Regional distribution of thrombomodulin in human brain, *Brain Res.* 556:1 (1991).
5. O. Durieu-Trautmann, N. Foignant-Chaverot, F. Roux, A.D. Strosberg, and P.O. Couraud, Immortalized rat brain microvessel endothelial cells: II-pharmacological characterization, *Adv. Exp. Med. Biol.* 331:205 (1993).
6. A.W. Vorbrodt, Ultracytochemical characterization of anionic sites in the wall of brain capillaries, *J. Neurocyto.* 18:359 (1989).
7. B. Furie and B. Furie, Molecular and cellular biology of blood coagulation, *N. Engl. J. Med.* 800 (1992).
8. K.A. Bauer, B.L. Kass, H. Cate, J.J. Hawiger, and R.D. Rosenberg, Factor IX is activated in vivo by the tissue factor mechanism, *Blood* 76:731 (1990).
9. M. Fisher, M. Lipovac, A. Martin, V. Wong, and B.V. Zlokovic, Hemostasis regulation in guinea pig brain, *Soc. Neurosci.* 18:179 (1992).
10. K.A. Bauer, P.M. Mannucci, A. Gringeri, F. Tradati, S. Barzegar, B.L. Kass, H. Cate, A. Kestin, D.B. Brettler, and R.D. Rosenberg, Factor IXa-Factor VIIIa-cell surface complex does not contribute to the basal activation of the coagulation mechanism in vivo, *Blood* 79:2039 (1992).
11. R.P. McEver, Leukocyte Interactions mediated by selectins, *Thromb. Haemost.* 66:80 (1991).
12. R.B. Francis Jr., Clinical disorders of fibrinolysis: a critical review, *Blut* 59:1 (1989).
13. J. Laterra, R.R. Indurti, and G.W. Goldstein, Plasminogen activation and astroglial-induced neural microvessel morphogenesis, *Adv. Exp. Med. Biol.* 331:189 (1993).
14. B.V. Zlokovic, L. Wang, N. Sun, S, Haffne, S. Verral, N.W. Seeds, and M. Fisher, Expression of tissue plasminogen activator in cerebral capillaries: possible fibrinolytic function of the blood-brain barrier, *Neurosurgey* 36: (1995) in press.
15. E.G. Levin and G.J. del Zoppo, Localization of tissue plasminogen activator in the endothelium of a limited number of vessels, *Am. J. Pathol.* 144(5):855 (1994).
16. N. Kalderon, K. Ahonen, and S. Fedoroff, Developmental transition in plasticity properties of differentiating astrocytes: Age-related biochemical profile of plasminogen activators in astroglial cultures, *Glia* 3:413 (1990).
17. N.W. Seeds, S.P. Haffke, R.L. Hawkins, A. Krystosek, P.G. McGuire, and S. Verrall, Neuronal growth cones: battering rams or lasers? *in*: "The Nerve Growth Cone," P.C. Letourneau, S.B. Kater, and E.R. Macagno, eds., Raven Press, Ltd., New York (1991).
18. Z. Quian, M.E. Gilbert, M.A. Colicos, E.R. Kandel, and D. Kuhl, Tissue plasminogen activator is induced as an immediate early gene during seizure, kindling and long-term potentiation, *Nature* 361:453 (1993).
19. J.M. Hallenbeck, A.J. Dutka, P.M. Kochanek, A. Siren, G.H. Pezeshkpour, and G. Feuerstein, Stroke risk factors prepare rat brainstem tissues for modified local Shwartzman reaction, *Stroke* 19:863 (1988).
20. J.M. Hallenbeck, A.J. Dutka, S.N. Vogel, E. Heldman, D.A. Doron, and G. Feuerstein, Lipopolysaccharide-induced production of tumor necrosis factor activity in rats with and without risk factors for stroke, *Brain Res.* 541:115 (1991).
21. W.D. Haire, J.C. Goldsmith, and J. Rasmussen, Abnormal fibrinolysis in healthy male cigarette smokers: role of plasminogen activator inhibitors, *Am. J. Hematol.* 31:36 (1989).
22. B.V. Zlokovic, L. Wang, S. Schreiber, and M. Fisher, Nicotine downregulates tissue plasminogen activation at the blood-brain barrier, *J. Neurochem.* 64:591 (1995).
23. C. Esposito, H. Gerlach, J. Brett, D. Stern, and H. Vlassara, Endothelial receptor-mediated binding of glucose-modified albumin is associated with increased monolayer permeability and modulation of cell surface coagulant properties, *J. Exp. Med.* 170:1387 (1989).
24. B.V. Zlokovic, L. Wang, S.S. Schreiber, and M. Fisher, Downregulation of tissue plasminogen activator at the blood-brain barrier in diabetic rats, *Stroke* 26:11 (1995).
25. K. Landin, L. Tengborn, and U. Smith, Elevated fibrinogen and plasminogen activator inhibitor (PAI-1) in hypertension are related to metabolic risk factors for cardiovascular disease, *J. Intern. Med.* 227:273 (1990).
26. H. Kimata, K. Nakajima, H. Suzuki, T. Koide, S. Yamamoto, and T. Narita, Fibrinolytic effect of tissue plasminogen activator on cerebral embolism in stroke-prone spontaneously hypertensive rats. *Chem. Pharm. Bull.* 39:3327 (1991).
27. M. Fisher and R. Francis, Altered coagulation in cerebral ischemia: platelet, thrombin, and plasmin activity, *Arch. Neurol.* 47:1075 (1990).

# AUTACOIDS AS MEDIATORS OF VASOGENIC BRAIN OEDEMA

M. Wahl, L. Schilling,
A. Unterberg[1], and A. Baethmann[2]

Department of Physiology, Univ. Munich
Department of Neurosurgery[1], Free Univ. Berlin
Institute of Surgical Research[2], Univ. Munich

## INTRODUCTION

Under several pathological conditions such as trauma, incomplete ischemia, reperfusion after complete ischemia, tumor, and allergic or inflammatory diseases an opening of the blood-brain barrier (BBB) may occur. The extravasation inducing vasogenic oedema is obviously induced by the release of several autacoids as will be discussed in an updated review of items published previously.[1-3] The mainly discussed mediators are histamine, bradykinin, arachidonic acid, free radicals, leukotrienes, and serotonin. Five criteria will be discussed for some favoured mediator candidates: 1) the effect on BBB permeability, 2) the vasomotor action which may affect driving force for transmural bulk flow, 3) the influence on oedema formation, 4) the actual concentration of the mediator in the tissue or in the interstitial space under pathological conditions and 5) the therapeutic results after treatment with specific inhibitors or antagonists.

## HISTAMINE (HA)

HA may influence cerebral microcirculation from the intravascular and parenchymal side. A gross release from the latter source, i.e. mast cells, histaminergic neurons and possibly perivascular nerves[4] can be simulated by cortical superfusion. In a cat model with an open cranial window and an intact BBB (3-4h) under control conditions,[5-7] cortical superfusion of HA induced an extravasation of tracers which was measured by intravital fluorescence microscopy.[7] Extravasation of $Na^+$-fluorescein (MW, 376; Stokes' radius, 5.5 Å) started at $10^{-9}$ M HA whereas extravasation of FITC (fluoresceinisothiocyanate)-dextran 70,000 (Stokes' radius, 60 Å) and 150,000 (Stokes' radius, 87 Å) was first detected at $10^{-8}$ M and $10^{-6}$ M HA, respectively. The leakage started at venular sites which have been found to have a lower transmural electrical resistance than arteries.[8] In fact, the transmural electrical resistance could be reduced by HA.[9-11]

Similarly, extravasation of various tracers (MW, 103 - 67,000) has frequently been found with intravascular administration of HA which acts mainly via $H_2$ receptors.[12-16] Some

*New Concepts of a Blood–Brain Barrier*
Edited by J. Greenwood *et al.*, Plenum Press, New York, 1995

negative results of others may be explained by methodical or species differences as discussed elsewhere.[1-4] The positive results demonstrate that HA induces an unspecific opening (i.e. allowing escape of low and high molecular weight tracers) of the BBB via $H_2$ receptors located at the luminal and abluminal membrane of endothelial cells.[3,4,7] The mechanisms of HA-induced BBB opening are still unclear. Some experimental data support the hypothesis of increased vesicular transport[4,13,14] while other data are in favour of an opening of tight junctions.[4,9]

Perivascular microapplication[17,18] or cortical superfusion of HA[7] revealed that the dilatation of feline pial arteries was concentration dependent. In most species the vasomotor effect is mediated mainly by $H_2$ receptors of the smooth muscle.[4,17,18] Furthermore, intraparenchymal arteries were dilated by HA.[19,20] Venous diameters increased during cortical superfusion of HA[7] but not during perivascular microapplication.[19]

Thus, a mediator role of HA in the formation of brain oedema may be deduced from its vasomotor and permeability enhancing effects. Indeed, an increase of brain water content and a swelling of astrocytic endfeet has been found after intracarotid infusion of HA.[14]

An increase of the HA concentration in the cortical tissue after cold lesion, stab injury, or ischemia has been reported recently.[21-24] However, quantitative measurements of the HA concentration in cerebral interstitial space are still lacking.

Application of antihistamines may exert therapeutic results, but discrepant results with pretreatment and treatment using antagonists have been reported. Schilling and Wahl[25] using a cold lesion model did not find a reduction of brain swelling with application of the $H_1$ and $H_2$ antagonists mepyramine and zolantidine, respectively. Similarly, pretreatment with the $H_2$ antagonist cimetidine was ineffective in an ischemia model.[26] However, a beneficial effect of the $H_2$ antagonists metiamide, cimetidine, and ranitidine has been found in models with stab wound[22], heat stress[27], hypoxia due to pneumothorax,[28] ischemia,[28] irradiation trauma[29,30] and seizure.[31] Thus the therapeutic results with antihistamines appear to depend on the model used.

## BRADYKININ (BK)

Cortical superfusion with BK induced a leakage of $Na^+$-fluorescein from feline pial vessels starting at $4 \times 10^{-7}$ to $4 \times 10^{-6}$ M.[32,33] Since extravasation of FITC-albumin (MW, 67,000) or FITC-dextran (MW, 19,400 - 62,000) was never observed, a selective increase of BBB permeability by opening of functional pores with a diameter of at least 11 Å can be estimated. Similarly, intracarotid infusion of BK revealing concentrations of $10^{-6}$ to $10^{-5}$ M BK in the sagittal sinus blood induced extravasation of $Na^+$-fluorescein, but not of FITC-dextran.[32] The leakage is mediated by $B_2$-kininergic receptors[32] which have been located at the luminal and abluminal membrane of the endothelial cell.[34] Phosphoinositide turnover in cerebral microvessels is stimulated by $B_2$ kininergic receptors.[35] This may result in an activation of protein kinase C which is involved in the permeability increasing effect of BK in microvessels of the hamster cheek pouch.[36] In accordance with such a mechanism are findings obtained in cultured cerebral endothelial cells in which BK or phorbolester stimulated the fluid-phase endocytosis of a fluorescent tracer.[37] However some evidence is now accumulating supporting the hypothesis that BK opens tight junctions by inducing contraction of endothelial cells[34,38] which would be compatible with the increased transmural ionic conductance in pial vessels of the frog and rat during intravascular or cortical administration of BK, respectively.[10,39]

The BK-induced extravasation is primarily due to an increased permeability of the vessel wall since intra- and extravascular BK caused a comparable leakage without and with concomitant vasomotor responses,[32] respectively, which is also true for the HA-evoked leakage.[7,14] Arterial vasodilatation which increases the driving force for transmural bulk flow and the vascular surface area is per se not sufficient to elicit penetration of tracers.

However, there is no doubt that arterial dilatation increases extravasation if vascular permeability is enhanced.[33]

Bradykinin is a potent dilator of pial arteries when applied by perivascular microapplication[34,40,41] or by cortical superfusion.[32,33,42-44] The vasomotor effect of BK is mediated by $B_2$-kininergic receptors[41] which appear to be located at the abluminal membrane of the endothelial cell.[34] Whether the signal transmission from the endothelium to the smooth muscle cell is mediated by the endothelium-derived relaxing factor (nitric oxide), prostanoids, free radicals, $H_2O_2$ or leukotrienes depends on the species and the location of the artery.[34] Similarly, BK dilates intraparenchymal arteries.[45]

The increase of vascular permeability and diameter can explain the formation of brain oedema after ventriculo-cisternal perfusion or interstitial infusion of about $10^{-6}$ M BK, respectively.[46,47]

Bradykinin can be formed in the brain which contains all components of an intraparenchymal kallikrein-kinin system as reviewed elsewhere.[34,48,49] After cold lesion an uptake of plasma kininogen into the brain has been found[50] from which BK concentrations of $10^{-7}$ to $10^{-6}$ M in the interstitial space can be calculated. Thus, after cold lesion the BK concentration raises to values which increase permeability and arterial diameter and induce oedema formation. Similarly, a seven- to eightfold increase of brain BK concentration was found after concussive brain injury.[51] In addition, tissue kininogen and kinin content were increased after traumatic spinal cord[52] and ischemic brain[53] injury.

Further support for a mediator role of BK in brain oedema is provided by therapeutic results. The cold lesion[54] and ischemia[53] induced brain swelling could be significantly diminished by treatment with kallikrein-inhibitors. Furthermore, the dilatation of cerebral arterioles after concussive brain injury could be diminished by pretreatment with a $B_2$-kininergic receptor antagonist.[55] In conclusion, all criteria favour BK as one mediator of vasogenic oedema.

## ARACHIDONIC ACID (AA)

Cortical superfusion with AA ($10^{-5}$ - $10^{-3}$ M) evoked an unspecific opening of the BBB for small ($Na^+$-fluorescein) and large (FITC-dextran 62,000) tracers.[33,56] This is in accordance with investigations using Evans Blue-albumin as tracer.[57-59] Since cortical superfusion[56] or intraparenchymal application[57] of oleic acid were without effect the opening due to AA was believed to be specific. However, recently intracarotid oleic acid was found to open BBB for albumin.[60] Pretreatment with inhibitors of cyclooxygenase and lipoxygenase did not prevent the barrier opening for $Na^+$-fluorescein by AA. Therefore, it was concluded that this effect was mediated by AA itself and not by its metabolites.[56] This is in contradiction to findings of others.[61,62] The extravasation due to AA obviously results from an interaction between polymorphonuclear leukocytes and the vascular endothelium. Employing electron microscopy, sticking of leukocytes and their penetration through endothelial cells has been found after cortical superfusion with AA.[56] This might explain why leukopenia reduced extravasation of $\alpha$-aminoisobutyric acid due to acute cerebral inflammation.[63]

A weak constriction (5 - 10%) of feline pial arteries due to $3 \times 10^{-6}$ to $3 \times 10^{-4}$ M AA, but no significant response to $3 \times 10^{-3}$ M was observed during cortical superfusion.[56] In contrast, others[61,64] found dilatation of pial arteries exposed to AA which was explained by the generation of free radicals.[61] Since arachidonate did not change venular diameter as well,[56] the weak vasomotor effects obviously do not participate in the development of vasogenic oedema. Indeed, AA induced cytotoxic and vasogenic oedema[57]. Correspondingly, an intraparenchymal infusion-induced brain oedema was markedly increased by addition of AA to the infusion fluid.[65,66]

Release of AA from brain tissue was detected in several pathological states such as hypoxia, ischemia, seizure, and head injury.[67-69] The AA concentration in the interstitial space increased up to $2 \times 10^{-4}$ M after cold lesion,[67] a concentration which is effective to induce tracer extravasation.[56] However, convincing therapeutic results are still lacking since a complete pharmacological inhibition of the AA release is not yet possible. The beneficial effect of steroid therapy on vasogenic oedema formation might be explained by an inhibition of phospholipase $A_2$ and a reduction of AA release. Thus, most of the criteria favour AA as a mediator of vasogenic oedema, but convincing therapeutic results have yet to be demonstrated.

## FREE RADICALS (FR)

Cortical superfusion with oxygen FR induced endothelial lesions and arterial dilatation.[70] Since the extravasation of Evans Blue-albumin and of horseradish peroxidase during cortical superfusion with AA could be prevented by concomitant application of superoxide dismutase (SOD) and catalase, a mediation by FR was concluded.[59] Similarly, FR reduced the transmural electrical resistance of frog pial venules indicating an increase of ionic permeability which could be inhibited by SOD and catalase.[71] Furthermore, extravasation of intravenously applied Evans Blue from intraparenchymal vessels was demonstrated after injection of a FR generating system into the brain.[72] In contrast to these findings, cortical superfusion with xanthine-oxydase and hypoxanthine induced leakage of $Na^+$-fluorescein only in 30% of the cats investigated, whereas extravasation of FITC-dextran (MW, 62,000) was never detected.[73] Extravasation was increased by an additional hypertonicity of the superfusion fluid which also caused leakage of FITC-dextran (MW 62,000) in some cases.[73]

However, discrepant results concerning the vasomotor effects of FR have been reported.[70,73,74] Unterberg et al.[73] found only weak dilatation of pial arteries of up to 10% when feline cortex was superfused with normotonic hypoxanthine and xanthine-oxydase, but pronounced arteriolar dilation of up to 70% when undialyzed, hypertonic xanthine-oxydase and hypoxanthine were applied.

FR are generated in the brain tissue during hypertensive insults and after ischemia or trauma.[75-78] However, recent findings failed to show an increase in the generation of FR after brief ischemia without and with acidosis in the brain parenchyma.[79] In contrast, a pronounced increase of oxygen FR production was detected in rats 1 h after cold lesion using the nitroblue tetrazolium assay.[80] Thus, one may speculate that the vessel wall is a more important source of FR than brain parenchyma.

Controversial therapeutic results on treatment of oedema with free radical scavengers have been reported.[81] However, the cold lesion induced vasogenic oedema was reduced by treatment with liposome-entrapped SOD[80] or a SOD derivative bound to albumin with a half life of 6h.[82] Similarly, pretreatment with SOD coupled to polyethylene-glycol, which increased the half life, reduced $O_2^-$ production and extravasation of urea in the reperfusion period after ischemia in the piglet.[83] Thus, it is not surprising that cold induced brain oedema was markedly reduced in transgenic mice which overexpressed human CuZn-SOD.[84,85] Recently, treatment with an inhibitor of nitric oxide synthase (NOS) was found to reduce cold induced brain oedema.[85] This can be explained by a diminished formation of NO which reacts with $O_2^-$ to peroxinitrite ($ONOO^-$) from which $OH^-$ and $NO_2$ are released at acidic pH.[86] However, treatment with NOS-inhibitors is crucial since it has a dual effect, reduction of peroxinitrite formation and of perfusion in the ischaemic border zone.[87]

Taken together, FR appear to be involved in the generation of the vasogenic oedema, but to be of less importance than the mediators discussed previously.

## INTERACTION BETWEEN BK, AA AND FR

An interaction between the kallikrein-kinin system, AA metabolism and FR has been suggested by Ellis et al.[55] According to this hypothesis brain injury induces either directly or by activation of the kallikrein-kinin system a release of AA. Free radicals are generated when AA is further metabolized via the cyclooxygenase pathway. Therefore, Ellis et al.[55] concluded that the arterial dilatation and endothelial lesions after concussive brain injury are finally mediated by FR. However, studies of Unterberg et al.[32,56,73] and Wahl et al.[33] do not support this hypothesis. These authors found divergent vasomotor and permeability effects of BK, AA and FR in their experimental model, which are not compatible with the concept of Ellis et al.[55]

## LEUKOTRIENES (LT)

Conflicting results on the effects of LTs on BBB permeability have been published. $LTC_4$ induced an increase of transmural ionic conductance of frog pial venules.[39] In addition, a brain uptake of intravenously applied Evans Blue after intraparenchymal injection of $LTC_4$ dissolved in distilled water was reported first by Black and Hoff[62] for the rat. However, Unterberg et al.[65,66,88] were unable to detect extravasation of Evans Blue or $Na^+$ fluorescein in cats after intraparenchymal injection of $LTB_4$ or $LTC_4$ dissolved in artificial cerebrospinal fluid. Correspondingly, using the cortical superfusion model in cats, an increase of BBB permeability of pial vessels for $Na^+$-fluorescein was not observed, even when $LTC_4$, $B_4$, $D_4$ were tested up to 2$\mu$M.[66,88-90] Similarly, no increase of BBB permeability for Evans Blue was detected in guinea pigs after intravenous infusion of LTs.[91] Later on, Black's group[92-94] found in a rat model that intracarotid $LTC_4$ induced extravasation of $\alpha$-aminoisobutyric acid but only in vessels of ischaemic or tumor brain areas. Thus, they now conclude that LTs do not open BBB of "normal" vessels but only of those from ischaemic or tumor regions.[92-94] $LTC_4$, $D_4$, $E_4$ constricted feline pial arterioles and venules up to 25% without affecting BBB permeability[88] and constriction of pial arteries was found in other species such as rabbits, mice, and piglets.[42,95,96]

Oedema formation is not evoked by $LTC_4$ and $B_4$.[65,66,88] In a cat model intraparenchymal infusion of artificial cerebrospinal fluid induced an oedema of white matter which was not increased by addition of LTs. So it was concluded that LTs neither promote oedema nor act as mediators of brain oedema.[88]

Brain concentrations of LTs were increased after ischaemia, subarachnoid hemorrhage, concussive brain injury, and in tumors.[97-102] From these data one may calculate LT concentrations of the interstitial space in the range of $10^{-9}$ to $10^{-8}$ M. Therapeutic studies demonstrated that pretreatment and treatment with BW755C, an inhibitor of lipoxygenase and cyclooxygenase, did not reduce cold lesion induced brain swelling.[88-90] In contrast, treatment with BW755C reduced extravasation of $\alpha$-aminosobutyric acid in vessels of brain tumor regions,[94] which was explained by the therapeutic prevention of an accumulation of $LTC_4$ due to a lack of $\gamma$-glutamyl-transpeptidase in the wall of tumor vessels.

Taken together, with conflicting data on tracer extravasation and therapeutic results, further convincing studies have to be performed to support the mediator role of LTs.

## CYTOKINES (CK)

Few papers indicate that cytokines are also involved in the opening of BBB besides the above mentioned classic mediators of inflammation. Thus, it is likely that some of the above mentioned autacoids act, at least in part, via the release of CK.

Tumor necrosis factor alpha (TNF$\alpha$) has recently been found to induce extravasation and brain uptake of $Na^+$ fluorescein and pial arterial constriction in the piglet during cortical superfusion.[103] In the rat intracisternal application of TNF$\alpha$ induced extravasation of albumin and emigration of leukocytes.[104] Similarly, intracerebral injection of TNF$\alpha$ or interleukin-1ß (IL-1ß) activated metalloproteinases and plasminogen activators and increased extravasation of the tracer sucrose.[105] The involvement of IL-1ß in the development of ischaemic brain oedema is also supported by the beneficial effect with treatment of the inhibitor zinc-protoporphyrin.[106]

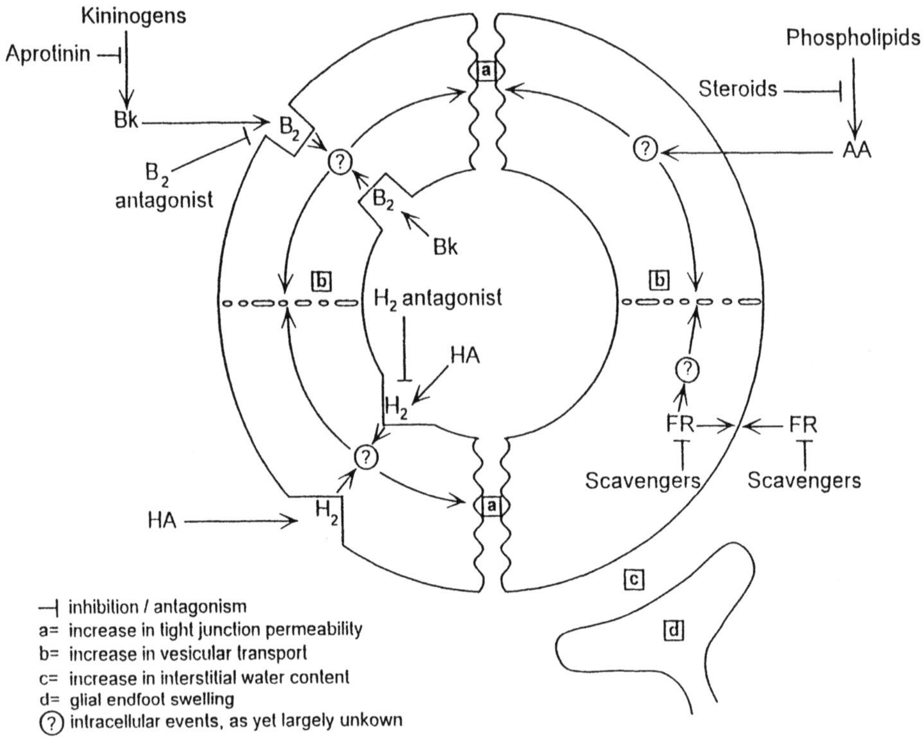


**Figure 1.** The role of various autocoids in the development of brain oedema.

## CONCLUSIONS

Taken together, vasogenic oedema appears to be mediated by several mediators (Fig. 1). All criteria favor BK as a mediator of brain oedema. Much evidence also supports the mediator role of HA, AA, and FR. Recent data indicate the involvement of CK also.

## ACKNOWLEDGEMENT

The authors greatly acknowledge the secretarial assistance of E. Held. Supported by the Deutsche Forschungsgemeinschaft (DFG) and Wilhelm-Sander Foundation.

## REFERENCES


1. M. Wahl, A. Unterberg, Baethmann, and L. Schilling L, Mediators of blood-brain-barrier dysfunction and formation of vasogenic brain oedema, *J. Cereb. Blood Flow Metab.* 8:621 (1988).
2. M. Wahl, L. Schilling, A. Unterberg, and A. Baethmann, Mediators of vascular and parenchymal mechanisms in secondary brain damage, *Acta Neurochir.* 57, Suppl.:64 (1993).
3. M. Wahl, L. Schilling, A. Unterberg, and A. Baethmann, Opening of the blood-brain barrier by autacoids, *in*: "CNS Barriers and Modern CSF Diagnostics. Centennial of Quincke's Lumbar Puncture", K. Felgenhauer, M. Holzgraefe, and H.W. Prange, eds., VCH, Weinheim (1993).
4. L. Schilling and M. Wahl, Histaminergic effects on cerebral hemodynamics, *in*: "The Regulation of Cerebral Blood Flow", J.W. Phillis, ed., CRC Press, Boca Raton (1993).
5. L. Schilling, A. Bultmann, and M. Wahl, Lack of effect of topically applied nicotine on pial arteriole diameter and blood brain barrier integrity in the cat, *Clin. Invest.* 70:210 (1992).
6. A. Findling, L. Schilling, A. Bultmann, and M. Wahl, Computerised image analysis in conjunction with fluorescence microscopy for the study of blood-brain barrier permeability in vivo, *Pflügers Arch.* 427:86 (1994).
7. L. Schilling and M. Wahl, Opening of the blood-brain barrier during cortical superfusion with histamine. *Brain Res.* 653:289 (1994).
8. A.M. Butt, H.C. Jones, and N.J. Abbott, Electrical resistance across the blood-brain barrier in anaesthetized rats: a developmental study, *J. Physiol.* 429:47 (1990).
9. A.M. Butt and H.C. Jones, Effect of histamine and antagonists on electrical resistance across the blood-brain barrier in rat brain-surface microvessels, *Brain Res.* 569:100 (1992).
10. A.M. Butt, Electrical resistance measurements of blood-brain barrier permeability, *This issue* (1994).
11. A.S. Easton and P.A. Fraser, Regulation of cerebral microvascular brain permeability by histamine and CGRP, *Proc. Int. Union Physiol. Sci. XXXII* 88:17/P (1993).
12. F.R. Domer, S.B. Boertje, E.G. Bing, and I. Reddix, Histamine- and acetylcholine-induced changes in the permeability of the blood-brain barrier of normotensive and spontaneously hypertensive rats *Neuropharmacology* 22:615 (1983).
13. E. Dux, and F. Joó, Effects of histamine on brain capillaries. Fine structural and immunohistochemical studies after intracarotid infusion, *Exp. Brain Res.* 47:252 (1982).
14. P.M. Gross, G.M. Teasdale, D.I. Graham, W.J. Angerson, and A.M. Harper, Intra-arterial histamine increases blood-brain transport in rats, *Am. J. Physiol.* 243:H307 (1982).
15. A. Gulati, K.N. Dhawan, R. Shukla, R.C. Srimal, and B.N. Dawan, Evidence for the involvement of histamine in the regulation of blood-brain barrier permebility, *Pharmacol. Res. Commun.* 17:39 (1985).
16. R. Oishi, M. Baba, M. Nishibori, Y. Itoh, and K. Saeki, Involvement of central histaminergic and cholinergic systems in the morphine-induced increase in blood-brain barrier permeability to sodium fluorescein in mice, *Naunyn-Schmiedeberg's Arch. Pharmacol.* 339:159 (1989).
17. P.M. Gross, A.M. Harper, and G.M.Teasdale, Cerebral circulation and histamine: 2. Responses of pial veins and arterioles to receptor agonists, *J. Cereb. Blood Flow Metab.* 1:219 (1981).
18. M. Wahl, and W. Kuschinsky, The dilating effect of histamine on pial arteries of cats and its mediation by $H_2$ receptors, *Circ. Res.* 44:161 (1976).
19. R.G. Dacey, and J.E. Bassett, Histaminergic vasodilatation of intracerebral arterioles in the rat, *J. Cereb. Blood Flow Metab.* 7:327 (1987).
20. G. DeLey, J. Weyne, G. Demeester, and I. Leusen I, Response of local blood flow in the caudate nucleus of the cat to intraventricular administration of histamine, *Stroke* 13:499 (1982).
21. E.L. Orr, Cryogenic lesions induce a mast cell-dependent increase in cerebral histamine levels in the mouse, *Neurochem. Pathol.* 8:43 (1988)
22. S. Mohanty, P.K. Dey, H.S. Sharma, S. Singh, J.P.N. Chansouria, and Y. Olsson, Role of histamine in traumatic brain edema. An experimental study in the rat, *J. Neurol. Sci.* 90:87 (1989).
23. N. Adachi, R. Oishi, and K. Saeki, Changes in the metabolism of histamine and monoamines after occlusion of the middle cerebral artery in rats, *J. Neurochem.* 57:61 (1991).
24. N. Adachi, Y. Itoh, R. Oishi, and K. Saeki, Direct evidence for increased continuous histamine release in the striatum of conscious freely moving rats produced by middle cerebral artery occlusion, *J. Cereb. Blood Flow Metab.* 12:477 (1992).
25. L. Schilling and M. Wahl, Effects of antihistaminics on experimental brain edema, *Acta Neurochir. Suppl.* 60:79 (1994).
26. H.P.J.M. Leistra and W.D. Dietrich, Effect of histamine antagonist cimetidine on infarct size in the rat, *J. Neurotrauma* 10:83 (1993).

27. H.S. Sharma, F. Nyberg, J. Cervos-Navarro, and P.K. Dey, Histamine modulates heat stress-induced changes in blood-brain barrier permeability, cerebral blood flow, brain oedema and serotonin levels: an experimental study in conscious young rats, *Neuroscience.* 50:445 (1992).
28. F. Joó, J. Kovacs, P. Szerdahelyi, P. Temesvari, and A. Tosaki, Treatment with histamine receptor antagonists of ischemic brain edema, *in*: "Pharmacology of Cerebral Ischemia 1992", J. Krieglstein and H. Oberpichler-Schwenk, eds., Wissenschaftliche Verlagsgesellschaft, Stuttgart (1992).
29. E. Csanda, Radiation brain edema, *in*: "Brain Edema," J. Cervos-Navarro, and R. Ferszt, eds., Raven Press, New York (1980).
30. F. Joó, A. Zücs, and E. Csanda, Metiamide-treatment of brain oedema in animals exposed to $^{90}$yttrium irradiation, *J. Pharm. Pharmacol.* 28:162 (1976).
31. L. Sztriha, F. Joó, and P. Szerdahelyi, Histamine $H_2$-receptors participate in the formation of brain edema induced by kainic acid in rat thalamus, *Neurosci. Lett.* 75:334 (1987).
32. A. Unterberg, M. Wahl, and A. Baethmann, Effects of bradykinin on permeability and diameter of pial vessels in vivo, *J. Cereb. Blood Flow Metab.* 4:574 (1984).
33. M. Wahl, A. Unterberg, and A. Baethmann, Intravital fluorescence microscopy for the study of blood-brain-barrier function, *Int. J. Microcirc. Clin. Exp.* 4:3 (1985).
34. M. Wahl and L. Schilling, Effects of bradykinin in the cerebral microcirculation, *in*: "The Regulation of Cerebral Blood Flow", J.W. Phillis, ed., CRC Press, Boca Raton (1993).
35. P. Homayoun, and S.I. Harik, Bradykinin receptors of cerebral microvessels stimulate phosphoinositide turnover, *J. Cereb. Blood Flow Metab.* 11:557 (1991).
36. M.A. Murray, D.D. Heistad, and W.G. Mayhan, Role of protein kinase C in bradykinin-induced increases in microvascular permeability, *Circ. Res.* 68:1340 (1991).
37. F.L. Guillot, and K.L. Audus, Angiotensin peptide regulation of fluid-phase endocytosis in brain microvessel endothelial cell monolayers, *J. Cereb. Blood Flow Metab.* 10:827 (1990).
38. N.J. Abbott and P.A. Revest, Control of brain endothelial permeability, *Cerebrovasc. Brain Metab. Rev.* 3:39 (1991).
39. S.-P. Olesen and C. Crone, Substances that rapidly augment ionic conductance of endothelium in cerebral venules, *Acta Physiol. Scand.* 127:233 (1986).
40. M. Wahl, A.R. Young, L. Edvinsson, and F. Wagner, Effects of bradykinin on pial arteries and arterioles *in vitro* and *in situ*, *J. Cereb. Blood Flow Metab.* 3:231 (1983).
41. E.T. Whalley, and M. Wahl, Analysis of bradykinin receptor mediating relaxation of cat cerebral arteries *in vivo* and *in vitro*, *Naunyn-Schmiedeberg's Arch. Pharmacol.* 323:66 (1983).
42. T. Kamitani, M.H. Little, and E.F. Ellis, Effect of leukotrienes, 12-HETE, histamine, bradykinin, and 5-hydroxytryptamine on *in vivo* rabbit cerebral arteriolar diameter, *J. Cereb. Blood Flow Metab.* 5:554 (1985).
43. H.A. Kontos, E.P. Wei, J.T. Povlishock, and C.W. Christman, Oxygen radicals mediate the cerebral arteriolar dilatation from arachidonate and bradykinin in cats, *Circ. Res.* 55:295 (1984).
44. H.A. Kontos, E.P. Wei, R.C.Kukreja, E.F. Ellis, and M.L. Hess, Differences in endothelium-dependent cerebral dilation by bradykinin and acetylcholine, *Am. J. Physiol.* 258:H1261 (1990).
45. R.G. Dacey, J.E. Bassett, and M. Takayasu, Vasomotor responses of rat intracerebral arterioles to vasoactive intestinal peptide, substance P, neuropeptide Y, and bradykinin, *J. Cereb. Blood Flow Metab.* 8:254 (1988).
46. A. Unterberg, A. Baethmann, The kallikrein-kinin system as mediator on vasogenic brain edema. Part 1: Cerebral exposure to bradykinin and plasma, *J. Neurosurg.* 61:87 (1984).
47. I.R. Whittle, I.R. Piper, and J.D. Miller, The role of bradykinin in the etiology of vasogenic brain edema and perilesional brain dysfunction, *Acta Neurochir.* 115:53 (1992).
48. M. Wahl, A. Unterberg, E.T. Whalley, A. Baethmann, A.R. Young, L. Edvinsson, and F.F.W. Wagner, Cerebrovascular effects of bradykinin, *in*: "Neural Regulation of Brain Circulation", C. Owman, J.E. Hardebo, eds., Elsevier, New York (1986).
49. M. Wahl, A. Unterberg, E.T. Whalley, A. Baethmann, A.R. Young, L. Edvinsson, and F. Wagner, Effect of bradykinin on cerebral hemodynamics and blood-brain-barrier function *in*: "Peptidergic Mechanisms in the Cerebral Circulation", L. Edvinsson, and J. McCulloch, eds., Horwood, Chichester (1987).
50. K. Maier-Hauff, A.J. Baethmann, M. Lange, L. Schürer, and A. Unterberg, The kallikrein-kinin system as mediator in vasogenic brain edema. Part 2: Studies on kinin formation in focal and perifocal brain tissue, *J. Neurosurg.* 61:97 (1984).
51. E.F. Ellis, J. Chao, and M.L. Heizer, Brain kininogen following experimental brain injury: evidence for a secondary event, *J. Neurosurg.* 71:437 (1989).
52. J. Xu, C.Y. Hsu, H. Junker, S. Chao, E.L. Hogan, and J. Chao, Kininogen and kinin in experimental spinal cord injury, *J. Neurochem.* 57:975 (1991).

53. T. Kamiya, Y. Katayama, F. Kashiwagi, and A. Terashi, The role of bradykinin in mediating ischemic brain edema in rats, *Stroke* 24:571 (1993).
54. A. Unterberg, C. Dautermann, A. Baethmann, and W. Müller-Esterl, The kallikrein-kinin system as mediator in vasogenic brain edema, Part 3: Inhibition of the kallikrein-kinin system in traumatic brain swelling, *J. Neurosurg.* 64:269 (1986).
55. E.F.Ellis, S.A. Holt, E.P. Wei, and H.A. Kontos HA, Kinins induce abnormal vascular reactivity, *Am. J. Physiol.* 255:H397 (1988).
56. A. Unterberg, M. Wahl, F. Hammersen, and A. Baethmann, Permeability and vasomotor response of cerebral vessels during exposure to arachidonic acid, *Acta Neuropathol.* 73:209 (1987).
57. P.H. Chan, R.A. Fishman, J. Caronna, J.W. Schmidley, G.Priolieau, and J. Lee, Induction of brain edema following intracerebral injection of arachidonic acid, *Ann. Neurol.* 13:625 (1983).
58. S. Wakai, K. Aritake, T. Asano, and K. Takakura K, Selective destruction of the outer leaflet of the capillary endothelial membrane after intracerebral injection of arachidonic acid in the rat, *Acta Neuropathol.* 58:303 (1982).
59. E.P. Wei, M.D. Ellison, H.A. Kontos, and J.T. Povlishock, $O_2$ radicals in arachidonate-induced increased blood-brain barrier permeability to proteins. *Am. J. Physiol.* 251:H693 (1986).
60. L. Sztriha and A.L. Betz, Oleic acid reversibly opens the blood brain barrier, *Brain Res* 550:25 (1991).
61. H.A. Kontos, E.P. Wei, J.T. Povlishock, W.D. Dietrich, C.J. Magiera, and E.P. Ellis, Cerebral arteriolar damage by arachidonic acid and prostaglandin $G_2$, *Science* 209:1242 (1980).
62. K.L. Black, and J.T. Hoff, Leukotrienes increase blood-brain barrier permeability following intraparenchymal injections in rats, *Ann. Neurol.* 18:349 (1985).
63. W. Lo and P. Clair, Leukopenia reduces blood-brain barrier (BBB) damage in acute cerebral inflammation, *Soc. Neurosci. Abst.* 19:225 (1993).
64. D.W. Busija, and D.D. Heistad, Effects of indomethacin on cerebral blood flow during hypercapnia in cats, *Am. J. Physiol.* 244:H519 (1983).
65. A. Unterberg, A. Baethmann, M. Wahl, L. Schürer, and A. Marmarou, New aspects in the formation of vasogenic brain edema, *in:* "Surgical Research. Recent Concepts and Results", A. Baethmann, and U. Messmer, eds., Springer, Heidelberg (1987).
66. A. Unterberg, T. Polk, E. Ellis, and A. Marmarou, Enhancement of infusion induced brain edema by mediator compounds, *in:* "Brain Edema. Pathogenesis, Imaging and Therapy", Adv Neurol 52, D.M. Long, ed., Raven Press, New York (1990).
67. A. Baethmann, K. Maier-Hauff, L. Schürer, M. Lange, C. Guggenbichler, W. Vogt, K. Jacob, and O. Kempski, Release of glutamate and of free fatty acids in vasogenic brain edema, *J. Neurosurg.* 70:578 (1989).
68. N.G. Bazan, and E.B. Rodriguez de Turco, Membrane lipids in the pathogenesis of brain oedema. Phospholipids and arachidonic acid, the earliest membrane components changed at the onset of ischemia, *in:* "Brain Edema", J. Cervos-Navarro, R. Ferszt, eds., Raven Press, New York, (1980).
69. K.K. Bhakoo, H.A. Crockard, and P.T. Lascelles, Regional studies of changes in brain fatty acids following experimental ischaemia and reperfusion in the gerbil. *J. Neurochem.* 43:1025 (1984).
70. E.P. Wei, C.W. Christman, H.A. Kontos, and J.T. Povlishock, Effects of oxygen radicals on cerebral arterioles, *Am. J. Physiol.* 248:H157 (1985).
71. S.P. Olesen, Free oxygen radicals decrease electrical resistance of microvascular endothelium in brain, *Acta Physiol. Scand.* 129:181 (1987).
72. P.H. Chan, J.W. Schmidley, R.A. Fishman, and S.M. Longar, Brain injury, edema, and vascular permeability changes induced by oxygen-derived free radicals, *Neurology* 34:315 (1984).
73. A. Unterberg, M. Wahl, and A. Baethmann, Effects of free radicals on permeability and vasomotor response of cerebral vessels, *Acta Neuropathol.* 76:238 (1988).
74. W.I. Rosenblum, Effects of free radical generation on mouse pial arterioles: probable role of hydroxylradicals, *Am. J. Physiol.* 245:H139 (1983).
75. H.B. Demopoulos, E.S. Flamm, M.L. Seligman, J.A. Mitamura, and J. Ransohoff, Membrane perturbations in central nervous system injury: theoretical basis for free radical damage and a review of the experimental data, *in:* "Neural Trauma", A.J. Popp, R.S. Bourke, L.R. Nelson, and H.K. Kimelberg, eds., Raven Press, New York (1979).
76. H.A. Kontos and E.P. Wei, Superoxide production in experimental brain injury, *J. Neurosurg.* 64:803 (1986).
77. B.K. Siesjö, G. Bendek, T. Koide, E. Westerberg, and T. Wieloch, Influence of acidosis on lipid peroxidation in brain tissues in vitro, *J. Cereb. Blood Flow Metab.* 5:253 (1985).
78. O. Suzuki, and K. Yagi, Formation of lipoperoxide in brain edema induced by cold injury, *Experientia* 30:248 (1974).

79. J. Lundgren, H. Zhang, C.D. Agardh, M.L. Smith, P.J. Evans, B. Halliwell, and B.K. Siesjö, Acidosis-induced ischemic brain damage: Are free radicals involved? *J. Cereb. Blood Flow Metab.* 11:587 (1991).
80. P.H. Chan, S. Longar, and R.A. Fishman, Protective effects of liposome-entrapped superoxide dismutase on posttraumatic brain edema, *Ann. Neurol.* 21:540 (1987).
81. D.M. Long, "Brain Edema. Pathogenesis, Imaging and Therapy", Adv Neurol 52, Raven Press, New York (1990).
82. Y. Ando, M. Inoue, M. Hirota, Y. Morino, and S. Araki, Effect of superoxide dismutase derivative on cold-induced brain edema, *Brain Res.* 477:286 (1989).
83. W.M. Armstead, R. Mirro, O.P. Thelin, M. Shibata, S.L. Zuckerman, D.R. Shanklin, D.W. Busija, and C.W. Leffler, Polyethylene glycol superoxide dismutase and catalase attenuate increases in blood-brain barrier permeability after ischemia in piglets, *Stroke* 23:755 (1992).
84. P.H. Chan, G.Y. Yang, S.F. Chen, E. Carlson, and C.J. Epstein, Cold-induced brain edema and infarction are reduced in transgenic mice overexpressing CuZn-superoxide dismutase, *Ann. Neurol.* 29:482 (1991).
85. T.D. Oury, C.A. Piantadosi, and J.D. Crapo, Cold-induced brain edema in mice, *J. Biol. Chem.* 268:15394 (1993).
86. J.S. Beckmann, J. Chen, H. Ischiropoulos, and K.A. Conger, Inhibition of nitric oxide synthesis and cerebral protection, *in*: "Pharmacology of Cerebral Ischemia 1992", J. Krieglstein and H. Oberpichler-Schwenk, eds., Wissenschaftliche Verlagsgesellschaft, Stuttgart (1992).
87. D.A. Pelligrino, Saying NO to cerebral ischemia, *J. Neurosurg. Anesth.* 5:221 (1993).
88. A. Unterberg, W. Schmidt, M. Wahl, E.F. Ellis, A. Marmarou, and A. Baethmann, Evidence against leukotrienes as mediators of brain edema, *J. Neurosurg.* 74:773 (1991).
89. A. Unterberg, W. Schmidt, T. Polk, M. Wahl, E. Ellis, A. Marmarou, and A. Baethmann, Evidence against leukotrienes as mediators of brain edema, *J. Cereb. Blood Flow Metab.* 7, Suppl. 1:S625 (1987).
90. A. Unterberg, W. Schmidt, M. Wahl, and A. Baethmann, "Role of Leukotrienes as Mediator Compounds in Brain Edema", *in*: "Brain Edema. Pathogenesis, Imaging and Therapy", Adv Neurol 52, D.M. Lond, ed., Raven Press, New York (1990).
91. X.-Y. Hua, S.-E. Dahlen, J.M. Lundberg, S. Hammarström, and P. Hedqvist, Leukotrienes $C_4$, $D_4$ and $E_4$ cause widespread and extensive plasma extravasation in the guinea pig, *Naunyn-Schmiedeberg's Arch. Pharmacol.* 330:136 (1985).
92. T. Baba, K.L. Black, K. Ikezaki, K. Chen, and D.P. Becker, Intracarotid infusion of leukotriene $C_4$ selectively increases blood-brain barrier permeability after focal ischemia in rats, *J. Cereb. Blood Flow Metab.* 11:638 (1991).
93. C.C. Chio, T. Baba, and K.L. Black, Selective blood-tumor barrier disruption by leukotrienes, *J. Neurosurg.* 77:407 (1992).
94. T. Baba, C.C. Chio, and K.L. Black, The effect of 5-lipoxygenase inhibition on blood-brain barrier permeability in experimental brain tumors, *J. Neurosurg.* 77:403 (1992).
95. W.I. Rosenblum, Constricting effect of leukotrienes on cerebral arterioles of mice, *Stroke* 16:262 (1985).
96. D.W. Busija, C.W. Leffler, and D.G. Beasley, Effects of leukotrienes $C_4$, $D_4$ and $E_4$ on cerebral arteries of newborn pigs, *Pediatr. Res.* 20:973 (1986).
97. A. Dembinska-Kiec, T. Simmet, and B.A. Peskar, Formation of leukotriene $C_4$-like material by rat brain tissue, *Eur. J. Pharmacol.* 99:57 (1984).
98. M.A. Moskowitz, K.J. Kiwak, K. Hekimian, and L. Levine, Synthesis of compounds with properties of leukotrienes $C_4$ and $D_4$ in gerbil brains after ischemia and reperfusion, *Science* 224:886 (1984).
99. K.J. Kiwak, M.A. Moskowitz, and L. Levins, Leukotriene production in gerbil brain after ischemic insult, subarachnoid hemorrhage, and concussive injury, *J. Neurosurg.* 62:865 (1985).
100. K.L. Black, J.T. Hoff, J.E. McGillicuddy, and S.S. Gebarski, Increased leukotriene $C_4$ and vasogenic edema surrounding brain tumors in humans, *Ann. Neurol.* 19:592 (1986).
101. R.J. Dempsey, M.W. Roy, K. Meyer, D.E. Cowen, and M.E. Maley, Development of cyclooxygenase and lipoxygenase metabolites of arachidonic acid after transient cerebral ischemia, *J. Neurosurg.* 64:118 (1986).
102. S.T. Chen, C.Y. Hsu, E.L. Hogan, P.V. Halushka, O.I. Linet, and F.M. Yatsu, Thromboxane, prostacyclin, and leukotrienes in cerebral ischemia, *Neurology* 36:466 (1988).
103. P. Megyeri, C.S. Abraham, P. Temesvari, J. Kovacs, T. Vas, and C.P. Speer, Recombinant human tumor necrosis factor $\alpha$ constricts pial arterioles and increases blood brain barrier permeability in newborn piglets, *Neurosci. Lett.* 148:137 (1992).

104. K.S. Kim, C.A. Wass, A.S. Cross, and S.M. Opal, Modulation of blood-brain barrier permeability by tumor necrosis factor and antibody to tumor necrosis factor in the rat, *Lymphokine Cytokine Res.* 11:293 (1992).
105. G.A. Rosenberg, P.G. McGuire, J.E. Dencoff, E.Y. Estrada, A.N. Clark, and W.G. Stetler-Stevenson, Metalloproteinases and serine proteinases link cytokines to blood-brain barrier injury in rat, *Soc. Neurosci. Abst.* 19:692 (1993).
106. Y. Yamasaki, T. Suzuki, H. Yamaya, N. Matsuura, H. Onodera, and K. Kogure, Possible involvement of interleukin-1 in ischemic brain edema formation, *Neurosci. Lett.* 142:45 (1992).

# BLOOD-BRAIN BARRIER SODIUM TRANSPORT AND BRAIN EDEMA FORMATION

A. Lorris Betz[1,2], Steven R. Ennis[1], Xiao-dan Ren[1], Gerald P. Schielke[3], and Richard F. Keep[1]

[1]Department of Surgery (Neurosurgery),
[2]Departments of Pediatrics, and Neurology,
University of Michigan
[3]Parke-Davis Pharmaceutical Research,
Division of Warner-Lambert Company
Ann Arbor, MI USA

## INTRODUCTION

In most forms of brain edema, increased brain water results from a net increase in brain sodium and chloride contents (Betz et al., 1989). In vasogenic edema, the blood-brain barrier (BBB) is disrupted and sodium and chloride enter brain in a relatively unimpeded fashion. However, in cytotoxic edema, sodium and chloride must traverse an intact BBB. Thus, identification of the processes which mediate sodium and chloride transport across the brain capillary endothelial cells could lead to new therapy for cytotoxic brain edema through inhibition of BBB ion transport.

It has long been recognized that the BBB permeability to sodium is quite low. Davson was among the first to determine that the transport rate constant is approximately 1.2 to 1.5 $\mu$l/gm/min (Davson, 1955; Davson and Welch, 1971). This low permeability led to the speculation that sodium crosses the BBB by diffusion through the tight junctions (Rapoport, 1976). However, we now recognize that sodium ions cross the BBB, in part, through specific ion transporters and channels in the brain capillary endothelial cells (Schielke and Betz, 1993).

In this chapter, we will summarize our current understanding of the mechanisms by which sodium enters brain via the BBB and then discuss the importance of this process in brain edema formation. The transport of sodium across the choroid plexus is also mediated through transporters and channels, however, these are reviewed elsewhere (Spector and Johanson, 1989) and will not be described here.

## MECHANISMS OF BBB SODIUM TRANSPORT

Early studies of BBB ion transporters focused on Na,K-ATPase since this active sodium pump plays a central role in ion transport in most tissues. Bradbury and Stulcová (1970) first implicated Na,K-ATPase in ion transport across the BBB by demonstrating that potassium efflux from brain is saturable and sensitive to ouabain, a specific inhibitor of Na,K-ATPase. Subsequently, Goldstein (1979) reported that the kinetics of ouabain-inhibitable potassium transport into isolated brain capillaries are similar to the kinetics for potassium efflux from brain. Interestingly, the rate of potassium transport by the BBB is sensitive to changes in extracellular potassium in the physiological range ($K_m$ of about 3 mM) both *in vivo* (Bradbury and Stulcová, 1970) and *in vitro* (Goldstein, 1979; Chaplin et al., 1981; Schielke et al., 1990). A similar relationship exists for ouabain-inhibitable sodium efflux from isolated brain capillaries (Schielke et al., 1990). In contrast, the $K_m$ for ouabain-sensitive potassium uptake into synaptosomes is approximately 0.5 mM (Schielke et al., 1990) indicating that Na,K-ATPase activity in neurons is not significantly affected by physiological changes in the extracellular potassium concentration.

The finding of Bradbury and Stulcová (1970) that Na,K-ATPase was involved in potassium efflux from brain suggested that this ion pump must be present on the abluminal (brain facing) membrane of the brain capillary endothelial cell which forms the BBB. This prediction was directly demonstrated using cytochemical staining for Na,K-ATPase (Firth, 1976; Betz et al., 1980) and purification of the luminal and abluminal membranes (Betz et al., 1980). These studies concluded that Na,K-ATPase is found only on the abluminal and not the luminal membrane of the endothelial cells although more recent reports suggest that, under some conditions, the enzyme may also be present on the luminal membrane (Vorbrodt et al., 1982; Nag, 1990).

Na,K-ATPase is a heterodimer consisting of one $\alpha$ and one $\beta$ subunit. To date, 3 different $\alpha$ and 2 different $\beta$ isoforms have been identified. The $\alpha$ subunit determines the catalytic activity, the effectiveness of ouabain inhibition, and the affinities for sodium and potassium (Sweadner, 1989). Availability of the $\beta$ subunit regulates the assembly of the $\alpha\beta$ heterodimers and the number of pumps inserted in the membrane (McDonough et al., 1990). The brain contains all of the isoforms, but the regional distribution varies (Sweadner, 1989). All three $\alpha$ and both $\beta$ subunits have also been detected in isolated brain microvessels using Western blots (Zlokovic et al., 1993). However, it is important to recognize that isolated microvessels contain endothelial cells, pericytes, and occasionally, smooth muscle cells and membranes from glial foot processes. Thus, specific localization of the Na,K-ATPase isoforms in brain capillary endothelial cells is still needed.

The recognition that Na,K-ATPase is largely restricted to one side of the brain capillary endothelial cell lead to the realization that these endothelial cells are structurally and functionally similar to epithelial cells (Betz, 1985; Crone, 1986). Other features that the brain capillary endothelium has in common with a typical epithelium include the presence of continuous tight junctions, a transcellular electrical resistance of more than 2000 $\Omega \cdot cm^2$ (Butt and Jones, 1992), and a high mitochondrial content (Oldendorf et al., 1977) which is believed to be related to a high capacity for active transport. In epithelia, the polar distribution of transporters provides the basis for fluid secretion. Thus, transendothelial ion transport may lead to fluid secretion by the BBB and account for interstitial fluid flow through brain (Szentistványi et al., 1984) and extrachoroidal CSF production (Milhorat et al., 1971). Na,K-ATPase in the brain capillary may also contribute to the electrical potential difference between blood and brain (Abbott et al., 1986) and it may play a role in brain potassium homeostasis (see Keep et al. in this volume).

More recent studies have focused upon the sodium transporters and channels that are present on the luminal membrane through which sodium enters the endothelial cell before being pumped out of the cell and into brain by the Na,K-ATPase in the abluminal membrane. Our current working model is shown in Figure 1. This model is derived primarily from *in vivo* studies on the effects of various sodium transport and channel inhibitors on the unidirectional movement of sodium from blood to brain.

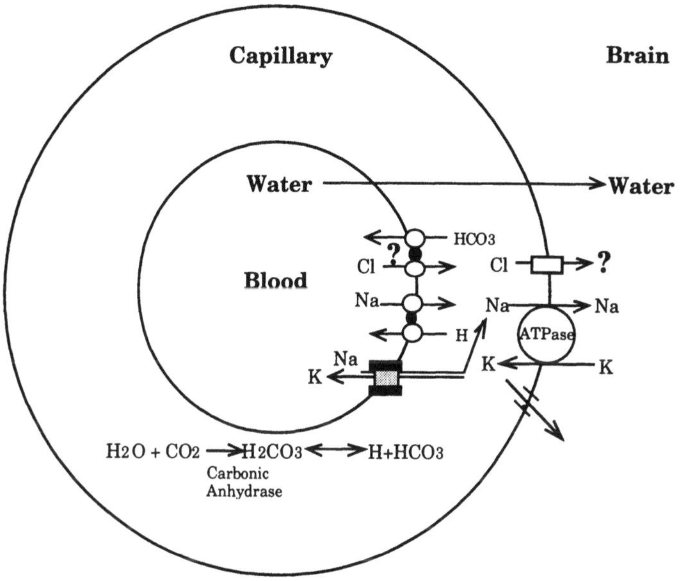


**Figure 1.** Schematic representation of the distribution of ion transporters and channels in the brain capillary endothelial cell. The presence of Na,K-ATPase, a Na/H exchanger, a nonselective cation channel, and an abluminal K channel have been documented experimentally. The Cl/$HCO_3$ exchanger and abluminal exit mechanism for Cl are, at this time, hypothetical.

Using intracarotid bolus injections to study single-pass sodium uptake, Betz (1983) observed two saturable sodium entry systems in the BBB. One was inhibited by furosemide suggesting the presence of a Na-Cl cotransporter. The other was inhibited by amiloride but its properties were not clearly identical to either the known amiloride-sensitive sodium channel or the Na/H exchanger. More recent studies confirm that brain capillary endothelial cells contain a unique amiloride-sensitive cation channel which is permeable to both sodium and potassium (Vigne et al., 1989). Furthermore, we have recently used *in situ* brain perfusion to demonstrate that phenamil, an amiloride analogue which is relatively selective for sodium channels, inhibits BBB sodium transport by about 25% (Figure 2). On the other hand, there is also now evidence that the BBB contains an amiloride-sensitive Na/H exchanger as well (Murphy and Johanson, 1989). We were able to confirm this finding using *in situ* brain perfusion and dimethylamiloride, an amiloride analogue with specificity for the Na/H exchanger (Figure 2). Under normal conditions, the Na/H exchanger appears to account for about one third of BBB sodium transport which is also the fraction of BBB sodium transport that is saturable (data not shown). The activity of this antiporter is dependent upon intracellular generation of hydrogen ions by carbonic anhydrase since, in the presence of the carbonic anhydrase inhibitor acetazolamide, BBB sodium permeability is reduced by about 40%. Although one study at reduced sodium concentrations suggested the presence of a furosemide-sensitive Na-Cl cotransporter (Betz, 1983) and this system has also been detected in isolated capillaries (Ibaragi et al., 1989), our recent work indicates that it is not a significant pathway for brain sodium uptake at a normal plasma sodium concentration since a more specific inhibitor (bumetanide) did not block BBB sodium transport (Figure 2).

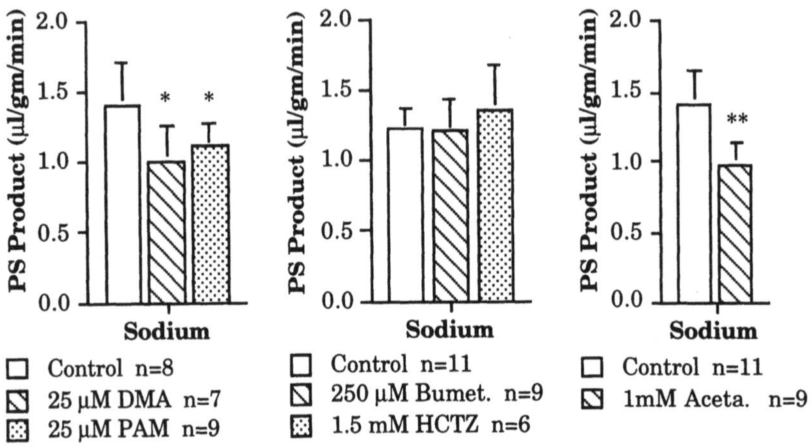


**Figure 2.** BBB sodium transport studied by *in situ* perfusion. Significant inhibition of sodium uptake by dimethylamiloride (DMA), a Na/H exchange inhibitor, phenamil (PAM), a sodium channel inhibitor, and acetazolamide (Aceta.), a carbonic anhydrase inhibitor, are apparent. The Na-Cl cotransport inhibitors bumetanide (Bumet.) and hydrochlorothiazide (HCTZ) did not have a significant effect on sodium transport. Values are means±SD. * $p<0.025$, ** $p<0.001$ compared to control.

Additional studies suggest the presence of yet other types of ion transporters and channels in the brain capillary endothelial cell. Abluminal potassium channels have been identified by Hoyer et al. (Hoyer et al., 1991) using patch clamp studies of isolated cerebral microvessels. By studying $^{86}$Rb efflux from isolated brain microvessels, we have identified components which are sensitive to Ca, pH and ouabain (Keep et al., 1993). Although BBB chloride transport is known to be saturable, the transporter has not yet been identified. Our model (Figure 1) predicts the presence of a Cl/$HCO_3$ exchanger as this antiporter is usually found in epithelial cells which utilize a Na/H exchanger for sodium transport.

Overall, therefore, it appears that 55-60% of blood to brain sodium transport occurs by identified transport pathways (30-35% by Na/H exchange, 25% via cation channel). The remaining sodium flux occurs through non-saturable mechanisms, either another sodium channel or simple diffusion.

## REGULATION OF BBB SODIUM TRANSPORT

The rate of ion transport is affected by the concentration of the ions themselves. For example, Na,K-ATPase is regulated by the intracellular sodium concentration (Gadsby, 1984) and this is one mechanism by which the activity of luminal sodium entry mechanisms are coupled to the activity of the abluminal Na,K-ATPase (Figure 1). Na,K-ATPase in the BBB is also uniquely regulated by the extracellular fluid potassium concentration (Goldstein,

1979; Chaplin et al., 1981; Schielke et al., 1990). Although the luminal transporters in the BBB are saturated at plasma sodium concentrations above 80 mM (Ennis et al., unpublished results), they are influenced by the availability of intracellular hydrogen ion (see above).

There are many potential hormonal regulators in brain, but there is very little information concerning their effects on BBB sodium transport. Brain capillaries contain atrial natriuretic peptide (ANP) receptors (Steardo and Nathanson, 1987) some of which are on the luminal side of the BBB (Ermisch et al., 1991). ANP activates guanylate cyclase, but not adenylate cyclase (Steardo and Nathanson, 1987) and it inhibits amiloride-sensitive sodium transport in isolated brain capillaries (Ibaragi et al., 1989). Intravenous ANP increases the BBB permeability to water, but not leucine and glucose (Brust et al., 1991). Intraventricular administration of ANP attenuates brain uptake of sodium and water following cerebral ischemia (Nakao et al., 1990), however, this might be due to an effect of ANP on brain cells (Kanda et al., 1992).

Brain capillary endothelial cells contain the $V_1$ type of receptor for vasopressin (Pearlmutter et al., 1988). Davson and Segal (1970) showed that intracarotid infusion of arginine vasopressin (AVP) increases the rate of brain sodium uptake by 16-18%. In isolated brain capillaries, AVP inhibits an abluminal potassium channel (Hoyer et al., 1991). Vasopressin deficiency in the Brattleboro rat is associated with reduced brain sodium uptake in response to acute systemic hypernatremia (DePasquale et al., 1989). Intraventricular administration of AVP has been shown to increase BBB permeability to $^3H_2O$ (Raichle and Grubb, 1978) and the rate of water exchange between CSF and brain (Rosenberg et al., 1986). Nevertheless, the significance of the observed change in water permeability is uncertain because the rate of exchange of $^3H_2O$ across the BBB is rapid, at least 1000 times greater than the net flux of water into brain. Thus, the overall effect on fluid and electrolyte balance in the brain of a change in BBB permeability to water is uncertain.

Chaplin et al. (1981) reported that high concentrations of steroids inhibit ouabain-sensitive potassium transport in isolated brain capillaries. The $IC_{50}$ for DOCA was 0.5 mM and for aldosterone it was > 5.0 mM.

Brain capillaries contain α- and β-adrenergic receptors (Harik et al., 1981) and show increases in intracellular cAMP when exposed to adrenergic agents such as norepinephrine (NE) (Herbst et al., 1979). The brain capillary receives adrenergic innervation from the locus ceruleus (Kalaria et al., 1989). Stimulation of the locus ceruleus causes an increase in the BBB permeability to $^3H_2O$ (Raichle et al., 1975) while ablation leads to an up-regulation of β-adrenergic receptors in brain capillaries (Kalaria et al., 1989) and a reduction in the Na,K-ATPase content of isolated brain capillaries (Harik, 1986).

Thus, a variety of data suggest that ion and water permeabilities of the brain capillary endothelial cell are under hormonal and/or neural regulation; however, with the exception of the effects of ANP on the amiloride-sensitive sodium channel (Ibaragi et al., 1989) and AVP on a potassium channel (Hoyer et al., 1991), the specific effects of hormones on the ion transporters of these cells have not been characterized.

## ROLE OF BBB SODIUM TRANSPORT IN ISCHEMIC BRAIN EDEMA

An acute interruption of blood flow to the brain results in the initiation of a series of events which produce brain injury. Among these are a sudden loss of potassium and uptake of sodium by brain cells which causes the cells to swell (Hansen and Nedergaard, 1988). If some degree of blood flow continues (i.e. incomplete or focal ischemia), water is drawn into the brain from the blood and brain edema develops. Numerous studies have shown that ischemic brain edema is characterized by increases in total brain sodium and reductions in total brain potassium (Gotoh et al., 1985; Young et al., 1987; Betz et al., 1989; Yang et al.,

1992; Menzies et al., 1993; Betz et al., 1994). However, the increase in sodium is greater than the reduction in potassium and, therefore, there is a net increase in brain cations. This change in cations is matched by a similar increase in brain chloride content (Betz et al., 1994). Assuming that the brain tissue remains isotonic with plasma, the change in ions explains the observed increase in brain water content (Betz et al., 1989; Menzies et al., 1993; Betz et al., 1994). Thus, the increase in brain sodium and chloride provides the osmotic driving force for brain edema formation.

In ischemia, the direct relationship between brain ion and water contents is seen whether or not the BBB is disrupted (Menzies et al., 1993). In the rat model of permanent focal ischemia produced by occlusion of the middle cerebral artery, barrier permeability is normal for approximately the first 6 hours. Once the BBB is disrupted, ion movements between plasma and brain probably occur by simple diffusion down concentration gradients. However, prior to barrier disruption, changes in brain sodium and potassium must be mediated, at least in part, by transport of these ions across the BBB. Interestingly, the unidirectional rate constant for sodium movement into brain is increased relative to that of a passive permeability tracer such as α-aminoisobutyric acid (AIB) (Betz and Coester, 1990; Ennis et al., 1990; Schielke et al., 1991; Dickinson and Betz, 1992; Betz et al., 1994). We speculate that this increase in the rate of sodium transport results from stimulation of brain capillary Na,K-ATPase by the increase in extracellular potassium concentration that is seen in ischemia (Schielke et al., 1991). Further support for the idea that brain capillary Na,K-ATPase activity plays an important role in these shifts of sodium and potassium is found in the ratio of sodium content increase to potassium content decrease. When the BBB is intact, this ratio is 1.4 to 1.6 (Betz et al., 1989; Betz and Coester, 1990; Dickinson and Betz, 1992; Menzies et al., 1993; Betz et al., 1994) a value essentially identical to the ratio of 3 sodium ions for 2 potassium ions that are exchanged by Na,K-ATPase.

In one study, the rate of sodium transport across the BBB approximated the rate of increase in total brain sodium content (Betz et al., 1989). In other studies, this relationship was not quite as close (Ishimaru et al., 1993; Betz et al., 1994), but the spread of edema fluid in models of focal ischemia (Menzies et al., 1993) can lead to an underestimation of the total sodium gained in the ischemic tissue (Betz et al., 1994). Thus, before the BBB is disrupted by ischemia, it is possible that the rate of sodium transport across the BBB determines the rate at which edema forms.

If BBB sodium transport is rate-limiting for edema formation, then it should be possible to reduce edema by reducing BBB sodium transport. This could be accomplished either by blocking one of the specific transport pathways (Figure 1) or reducing the diffusion of sodium across the BBB by tightening the barrier.

We have recently used the BBB Na/H exchange inhibitor dimethylamiloride (DMA) to reduce brain edema formation during the early stages of focal ischemia (Ennis et al., unpublished results). Focal ischemia was produced in rats using the intracarotid suture model of middle cerebral artery occlusion (MCAO) (Longa et al., 1989). DMA (53 mg/kg) or vehicle were administered by intravenous infusion over 40 minutes immediately prior to the MCAO. Control animals received an infusion of vehicle at the same rate. After 4 hours of ischemia, the brains were removed and the water, sodium, and potassium contents were measured. Plasma osmolality and pH were also measured and were not significantly different between the 2 groups.

As shown in Figure 3, DMA treatment significantly reduced edema formation in the ischemic cortex. Increases in water and sodium content and loss of potassium were all attenuated. The effect of DMA on sodium content was greater than its effect on water content. This is because loss of potassium was also prevented and, consequently, the net change in brain cations was only about one third of the change in sodium content.

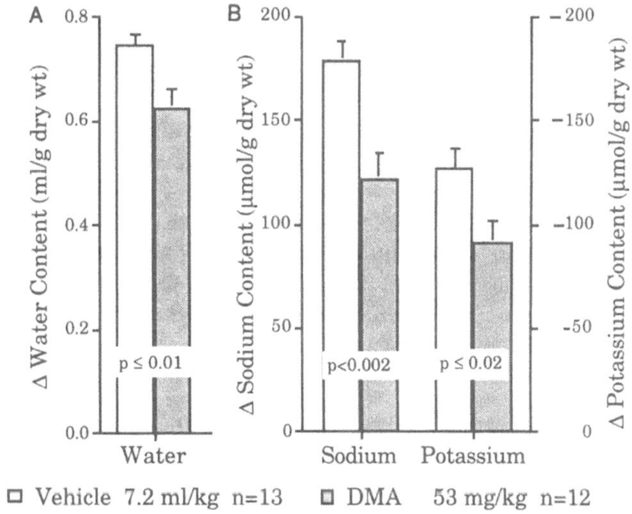


**Figure 3.** Effect of dimethylamiloride (DMA) on brain edema formation following MCAO. Rats were pretreated with either DMA or vehicle given intravenously. Brains were removed 4 hr after MCAO and samples were obtained from the ischemic and nonischemic cortex. The difference in water content (**A**) and ion content (**B**) between the ischemic and nonischemic tissue is a measure of edema formation. The sodium change represents an increase in total brain sodium while the potassium change is a decrease in total brain potassium. The effect of DMA Values are means±SE

This is the approximate magnitude of the water content change. The effect on sodium content was about 1.5 times the effect on potassium content, suggesting involvement of Na,K-ATPase in BBB ion exchange. Thus, inhibition of sodium entry into the endothelial cell may reduce Na,K-ATPase-mediated cation exchange by limiting the availability of sodium. The reduction in brain edema that resulted from DMA pretreatment was not due to an effect on cerebral blood flow as this was similar in the drug- and vehicle-treated groups (data not shown). Direct measurement of BBB permeability to sodium in this model will be required to prove that DMA reduces edema by inhibiting BBB sodium transport rather than by an effect on sodium transport in other brain cells.

Another strategy for reducing BBB sodium transport and ischemic brain edema formation is to alter the regulatory processes that control BBB sodium transport. Since AVP may be such a regulator (see above), we studied ischemic edema formation and BBB permeability in the vasopressin-deficient (Brattleboro) rat (Dickinson and Betz, 1992). After 4 hours of ischemia, the Brattleboro rat showed 36% less water accumulation and 20% less sodium accumulation than did the control Long-Evans strain, despite having a similar level of ischemia. Interestingly, there was no change in potassium loss. This is different than the effect seen with DMA treatment and suggests that AVP may act by a mechanism which does not involve Na,K-ATPase. Further support for this proposal was obtained by measuring BBB permeability in the ischemic Brattleboro rats. While blood to brain sodium transport was reduced by 36% in the ischemic cortex of the Brattleboro compared to Long-Evans rat, the BBB permeability to AIB showed a similar decrease. Thus, the effect of AVP appears to be on the passive permeability of the BBB. Nevertheless, reduction in passive uptake of sodium was associated with attenuation of edema formation. It appears that this effect of AVP is mediated from the brain side of the BBB since intraventricular but not systemic AVP therapy normalized edema formation in the Brattleboro rat (Dickinson and Betz, 1992).

## SUMMARY

These studies demonstrate that transporters and channels in the BBB are important for the movement of sodium from blood to brain. Furthermore, uptake of sodium by brain plays a primary role in the formation of ischemic brain edema. Our preliminary studies indicate that pharmacological inhibition of BBB sodium transporters and channels may provide a new approach to brain edema therapy in stroke.

## ACKNOWLEDGMENTS

Supported in part by grants from the National Institutes of Health (NS-23870, HL-18575, and NS-17760) and by the Section of Neurosurgery, University of Michigan.

## REFERENCES

Abbott N. J., Butt A. M., and Wallis W., 1986, The Na-K ATPase of the blood-brain barrier: a microelectrode study, *Ann New York Acad Sci* 481:390-391.

Betz A. L., 1983, Sodium transport from blood to brain: inhibition by furosemide and amiloride, *J Neurochem* 41:1158-1164.

Betz A. L., 1985, Epithelial properties of brain capillary endothelium, *FedProc* 44:2614-2615.

Betz A. L. and Coester H. C., 1990, Effect of steroids on edema and sodium uptake of the brain during focal ischemia in rats, *Stroke* 21:1199-1204.

Betz A. L., Ennis S. R., and Schielke G. P., 1989, Blood-brain barrier sodium transport limits development of brain edema during partial ischemia in gerbils, *Stroke* 20:1253-1259.

Betz A. L., Firth J. A., and Goldstein G. W., 1980, Polarity of the blood-brain barrier: distribution of enzymes between the luminal and antiluminal membranes of brain capillary endothelial cells, *Brain Res* 192:17-28.

Betz A. L., Iannotti F., and Hoff J. T., 1989, Brain edema: a classification based on blood-brain barrier integrity, *Cerebrovasc Brain Metab Rev* 1:133-154.

Betz A. L., Keep R. F., Beer M. E., and Ren X.-d., 1994, Blood-brain barrier permeability and brain content of sodium, potassium and chloride during focal ischemia, *J Cereb Blood Flow Metab* 14:29-37.

Bradbury M. W. B. and Stulcová B., 1970, Efflux mechanism contributing to the stability of the potassium concentration in cerebrospinal fluid, *J Physiol (Lond.)* 208:415-430.

Brust P., Baethmann A., Gjedde A., and Ermisch A., 1991, Atrial natriuretic peptide augments the blood-brain transfer of water but not leucine and glucose, *Brain Res* 564:91-96.

Butt A. M. and Jones H. C., 1992, Effect of histamine and antagonists on electrical resistance across the blood-brain barrier in rat brain-surface microvessels, *Brain Res* 569:100-105.

Chaplin E. R., Free R. G., and Goldstein G. W., 1981, Inhibition by steroids of the uptake of potassium by capillaries isolated from rat brain, *Biochem Pharmacol* 30:241-245.

Crone C., 1986, The blood-brain barrier as a tight epithelium: where is information lacking?, *Ann New York Acad Sci* 481:174-185.

Davson H., 1955, A comparative study of the aqueous humour and cerebrospinal fluid in the rabbit, *J Physiol* 129:111-133.

Davson H. and Segal M. B., 1970, The effects of some inhibitors and accelerators of sodium transport on the turnover of $^{22}$Na in the cerebrospinal fluid and the brain, *J Physiol* 209:131-153.

Davson H. and Welch K., 1971, The permeation of several materials into the fluids of the rabbit's brain, *J Physiol* 218:337-351.

DePasquale M., Patlak C. S., and Cserr H. F., 1989, Brain ion and volume regulation during acute hypernatremia in Brattleboro rats, *Am J Physiol* 256:F1059-F1066.

Dickinson L. D. and Betz A. L., 1992, Attenuated development of ischemic brain edema in vasopressin-deficient rats, *J Cereb Blood Flow Metab* 12:681-690.

Ennis S. R., Keep R. F., Schielke G. P., and Betz A. L., 1990, Decrease in perfusion of cerebral capillaries during incomplete ischemia and reperfusion, *J Cereb Blood Flow Metab* 10:213-220.

Ermisch A., Rühle H. J., Kretzschmar R., and Baethmann A., 1991, On the blood-brain barrier to peptides: Specific binding of atrial natriuretic peptide in vivo and in vitro, *Brain Res* 554:209-216.

Firth J. A., 1976, Cytochemical localization of the $K^+$ regulation interface between blood and brain, *Experientia* 33:1093-1094.

Gadsby D. C., 1984, The Na/K pump of cardiac cells, *Ann Rev Biophys Bioeng* 13:373-398.

Goldstein G. W., 1979, Relation of potassium transport to oxidative metabolism in isolated brain capillaries, *J Physiol* 286:185-195.

Gotoh O., Asano T., Koide T., and Takakura K., 1985, Ischemic brain edema following occlusion of the middle cerebral artery in the rat. I: The time courses of the brain water, sodium and potassium contents and blood-brain barrier permeability to $^{125}$ I-albumin, *Stroke* 16:101-109.

Hansen A. J. and Nedergaard M., 1988, Brain ion homeostasis in cerebral ischemia, *Neurochem Path* 9:195-209.

Harik S. I., 1986, Blood-brain barrier sodium/potassium pump: modulation by central noradrenergic innervation, *Proc Natl Acad Sci USA* 83:4067-4070.

Harik S. I., Sharma V. K., Wetherbee J. R., Warren R. H., and Banerjee S. P., 1981, Adrenergic and cholinergic receptors of cerebral microvessels, *J Cereb Blood Flow Metab* 1:329-338.

Herbst T. J., Raichle M. E., and Ferrendelli J. A., 1979, β-Adrenergic regulation of adenosine 3',5'-monophosphate concentration in brain microvessels, *Science* 204:330-332.

Hoyer J., Popp R., Meyer J., Galla H.-J., and Gögelein H., 1991, Angiotensin II, vasopressin and GTP[γ -S] inhibit inward-rectifying $K^+$ channels in porcine cerebral capillary endothelial cells, *J Membrane Biol* 123:55-62.

Ibaragi M.-A., Niwa M., and Ozaki M., 1989, Atrial natriuretic peptide modulates amiloride-sensitive $Na^+$ transport across the blood-brain barrier, *J Neurochem* 53:1802-1806.

Ishimaru S., Okada Y., Mies G., and Hossmann K.-A., 1993, Relationship between blood flow and blood-brain barrier permeability of sodium and albumin in focal ischaemia of rats: a triple tracer autoradiographic study, *Acta Neurochir (Wien)* 120:72-80.

Kalaria R. N., Stockmeier C. A., and Harik S. I., 1989, Brain microvessels are innervated by locus ceruleus noradrenergic neurons, *Neurosci Lett* 97:203-208.

Kanda F., Sarnacki P., and Arieff A. I., 1992, Atrial natriuretic peptide inhibits amiloride-sensitive sodium uptake in rat brain, *Am J Physiol Regul Integr Comp Physiol* 263:R279-R283.

Keep R. F., Xiang J., and Betz A. L., 1993, Potassium transport at the blood-brain and blood-CSF barriers, *in:* "Frontiers in Cerebrovascular Biology. Transport and its Regulation," L. R. Drewes and A. L. Betz, eds., New York, Plenum, pp. 43-54.

Longa E. Z., Weinstein P. R., Carlson S., and Cummins R., 1989, Reversible middle cerebral artery occlusion without craniectomy in rats, *Stroke* 20:84-91.

McDonough A. A., Geering K., and Farley R. A., 1990, The sodium pump needs its β subunit, *FASEB J* 4:1598-1605.

Menzies S. A., Betz A. L., and Hoff J. T., 1993, Contributions of ions and albumin to the formation and resolution of ischemic brain edema, *J Neurosurg* 78:257-266.

Milhorat T. H., Hammock M. K., and Fenstermacher J. D., 1971, Cerebrospinal fluid production by the choroid plexus and brain, *Science* 173:330-332.

Murphy V. A. and Johanson C. E., 1989, Acidosis, acetazolamide, and amiloride: effects on $^{22}Na$ transfer across the blood-brain and blood-CSF barriers, *J Neurochem* 52:1058-1063.

Nag S., 1990, Ultracytochemical localisation of $Na^{+}$, $K^{+}$-ATPase in cerebral endothelium in acute hypertension, *Acta Neuropathol (Berl)* 80:7-11.

Nakao N., Itakura T., Hideyoshi Y., Nakai K., and Komai N., 1990, Effect of atrial natriuretic peptide on ischemic brain edema: changes in brain water and electrolytes, *Neurosurg* 27:39-44.

Oldendorf W. H., Cornford M. E., and Brown W. J., 1977, The large apparent work capacity of the blood-brain barrier: a study of the mitochondrial content of capillary endothelial cells in brain and other tissues of the rat, *Ann Neurol* 1:409-417.

Pearlmutter A. F., Szkrybala M., Kim Y., and Harik S. I., 1988, Arginine vasopressin receptors in pig cerebral microvessels, cerebral cortex and hippocampus, *Neurosci Lett* 87:121-126.

Raichle M. E. and Grubb R. L. Jr., 1978, Regulation of brain water permeability by centrally-released vasopressin, *Brain Res* 143:191-194.

Raichle M. E., Hartman B. K., Eichling J. O., and Sharpe L. G., 1975, Central noradrenergic regulation of cerebral blood flow and vascular permeability, *Proc Nat Acad Sci USA* 72:3726-3730.

Rapoport S. I., 1976, "Blood-Brain Barrier in Physiology and Medicine," Raven Press, New York.

Rosenberg G. A., Kyner W. T., Fenstermacher J. D., and Patlak C. S., 1986, Effect of vasopressin on ependymal and capillary permeability to tritiated water in cat, *Am J Physiol* 251:F485-F489.

Schielke G. P. and Betz A. L., 1993, Electrolyte transport, *in:* "Physiology and Pharmacology of the Blood-Brain Barrier," M. W. B. Bradbury, ed., Heidelberg, Springer-Verlag, pp. 221-244.

Schielke G. P., Moises H. C., and Betz A. L., 1990, Potassium activation of the Na,K-pump in isolated brain microvessels and synaptosomes, *Brain Res* 524:291-296.

Schielke G. P., Moises H. C., and Betz A. L., 1991, Blood to brain sodium transport and interstitial fluid potassium concentration during focal ischemia in the rat, *J Cereb Blood Flow Metab* 11:466-471.

Spector R. and Johanson C. E., 1989, The mammalian choroid plexus, *Sci Am* 261:68-74.

Steardo L. and Nathanson J. A., 1987, Brain barrier tissues: end organs for atriopeptins, *Science* 235:470-473.

Sweadner K. J., 1989, Isozymes of the $Na^{+}/K^{+}$-ATPase, *Biochim Biophys Acta* 988:185-220.

Szentistványi I., Patlak C. S., Ellis R. A., and Cserr H. F., 1984, Drainage of interstitial fluid from different regions of rat brain, *Am J Physiol* 246:F835-F844.

Vigne P., Champigny G., Marsault R., Barbry P., Frelin C., and Lazdunski M., 1989, A new type of amiloride-sensitive cationic channel in endothelial cells of brain microvessels, *J Biol Chem* 264:7663-7668.

Vorbrodt A. W., Lossinsky A. S., and Wisniewski H. M., 1982, Cytochemical localization of ouabain-sensitive, $K^{+}$-dependent P-nitro-phenylphosphatase (transport ATPase) in the mouse central and peripheral nervous systems, *Brain Res* 243:225-234.

Yang G. Y., Chen S. F., Kinouchi H., Chan P. H., and Weinstein P. R., 1992, Edema, cation content, and ATPase activity after middle cerebral artery occlusion in rats, *Stroke* 23:1331-1336.

Young W., Rappaport Z. H., Chalif D. J., and Flamm E. S., 1987, Regional brain sodium, potassium, and water changes in the rat middle cerebral artery occlusion model of ischemia, *Stroke* 18:751-759.

Zlokovic B. V., Mackic J. B., Wang L., McComb J. G., and McDonough A., 1993, Differential expression of Na,K-ATPase alpha and beta subunit isoforms at the blood-brain barrier and the choroid plexus, *J Biol Chem* 268:8019-8025.

# MECHANISMS AND CONTROL OF BLOOD–BRAIN POTASSIUM TRANSPORT

Richard F. Keep,[1] Walter Stummer,[1] Susan A. Klarr,[1]
Philip B. Furspan,[2] and A. Lorris Betz[1,3]

[1]Department of Surgery (Neurosurgery)
[2]Department of Physiology
[3]Departments of Pediatrics and Neurology
University of Michigan
Ann Arbor, MI 48109-0532

## INTRODUCTION

Cerebrospinal fluid (CSF), brain interstitial fluid and total brain potassium concentrations ([K]) are all very tightly regulated during fluctuations in plasma [K]. The exact mechanisms involved in this regulation remain uncertain although there is evidence that the modulation of both potassium influx (Bradbury and Kleeman, 1967) and efflux (Bradbury and Stulcova, 1970) at the blood-brain and blood-CSF barriers can be important. The aim of the initial part of this study was to determine the relative role of K influx and efflux in brain [K] regulation during acute and chronic changes in plasma [K] and to examine the potential role of blood-brain barrier (BBB) Na/K ATPase in this regulation. We then examined the effect of mineralocorticoids on brain K homeostasis and choroid plexus K transport as Atarashi et al. (1988) showed that intracerebro-ventricular infusion of aldosterone lowers CSF [K] in the rat.

## MATERIALS AND METHODS

### Animals

Chronic hypo-, normo- and hyperkalemia was induced by feeding Sprague Dawley rats with diets containing potassium at 0.13, 130 or 2700 mmol/Kg. Chronic hyperkalemic animals were also uninephrectomized to produce a more consistent rise in plasma [K]. Acute hyperkalemia was induced by intravenous infusion (0.233 ml/kg/min) of 150 mM KCl for 40 min. Animals were rendered acutely hypokalemic by giving the same volume of a potassium free solution containing 150 mM NaCl, 2.5 % glucose and 857 U/l insulin. Acute controls received 146.5 mM NaCl and 3.5 mM KCl.

These rats were used for three sets of experiments. The first involved determining the extent of brain K homeostasis by measuring both CSF and total brain [K]. The second involved measurement of influx rate constant for K in brain and CSF (using $^{86}Rb$ as a marker for K) to assess the role of K influx modulation in brain K homeostasis. The last set of experiments examined the potential role of the BBB Na/K ATPase in K homeostasis. This pump has an abluminal distribution at the BBB (Betz et al., 1980) and, therefore, moves K from brain towards the blood. It appears that this pump may be a major route for Na entry into brain (Schielke et al., 1991) and, therefore, we examined whether the influx rate constant for $^{22}Na$ was altered by changes in plasma [K] as an indirect measurement of Na/K ATPase activity.

The role of mineralocorticoids in brain K homeostasis was examined in uninephrectomized rats given a deoxycorticosterone acetate (DOCA) silastic implant (200mg/kg) and salt (1%) in the drinking water for two weeks. Such DOCA-salt rats are a common model of hypertension. Sham rats were just subjected to the uninephectomy.

These rats were used for two sets of experiments. The first involved testing whether brain K homeostasis was altered by the DOCA-salt treatment. The second examined whether this treatment altered whole-cell potassium currents recorded from epithelial cells of the 4th ventricle choroid plexus.

### Potassium Concentrations

Blood was obtained from a femoral artery cannula and CSF by cisterna magna puncture. Plasma and CSF [K] were measured by flame photometry. Brain [K] was measured after desiccation at 100 °C and extraction in $HNO_3$ for 4 days.

### Influx Rate Constants

The influx rate constant for $^{22}Na$ was determined using a modification of the method of Ohno et al. (1978) with a $^{22}Na$ intravenous bolus injection, a 10 min circulation time and a continuous arterial withdrawal to determine the integral of the plasma isotope curve. For $^{86}Rb$, which enters red blood cells, individual plasma samples were taken at intervals during the circulation time to determine the integral of the plasma curve. For both isotopes the influx rate constant ($K_i$) was calculated as $K_i = C_{br} / \int C_p dt$, where $C_{br}$ is the extravascular concentration of isotope in the brain and $\int C_p dt$ is the integral of the plasma isotope curve. The calculation of $C_{br}$ requires the subtraction of a vascular correction from the total isotope concentration in the brain. For $^{22}Na$ the vascular correction is $VC_t$ where V is the brain plasma volume and $C_t$ is the final plasma concentration of isotope. For $^{86}Rb$, the correction is $VW_t / (1\text{-hct})$ where $W_t$ is the terminal whole blood concentration of isotope and hct is the hematocrit. The brain plasma volume (V) was determined using $^3H$-dextran which was given intravenously 3 min prior to sacrifice in the $^{22}Na$ experiments. For tracers such as dextran that remain intravascular, $V = C_{br} / C_t$. In the acute hypo- and hyperkalemic experiments $^{22}Na$ or $^{86}Rb$ were given 30 min after the onset of infusion in order that sacrifice would be at 40 min, the time point used in the experiments determining CSF and brain [K].

### Patch Clamp

Patch clamp experiments were performed on freshly isolated 4th ventricle choroid plexuses in artificial CSF. Giga-ohm seals were formed between the patch pipettes, containing K acetate, and the apical membrane of choroid plexus epithelial cells. The whole-cell configuration was then obtained by rupturing the membrane within the pipette. The holding potential was maintained at -60 mV. Outward K currents were elicited by progressive depolarization (10 mV steps). The current activated slowly and showed rectification at potentials above 0 mV. The results presented here are of the current elicited by stepping the command potential to +30 mV and the inhibition of this current by dibutryl cAMP and isoproterenol (all 10 $\mu$M) applied to the bathing medium.

## RESULTS

### Acute and Chronic Brain Potassium Homeostasis

The acute and chronic protocols used in this study produced similar degrees of hypokalemia, 2.9 ± 0.1 and 2.8 ± 0.1 mM, respectively (means ± S.E.) and hyperkalemia, 6.9 ± 0.2 and 7.2 ± 0.4 mM, respectively. Plasma [K] was 5.0 ± 0.2 and 4.2 ± 0.1 mM in acute and chronic controls. In neither acute or chronic experiments was there a significant relation between plasma and total brain [K]. Similarly, there was no significant increase in CSF [K] with increasing plasma [K] in either acute or chronic experiments. Indeed, in the acute experiments there was a small but significant decrease ($p<0.05$).

Although brain [K] did not increase with plasma [K] in both acute and chronic experiments, the mechanisms involved in K regulation appear to differ. In chronic experiments, the influx rate constant, $K_i$, for $^{86}$Rb uptake into anterior cortex decreased as plasma [K] increased ($p<0.001$; Fig. 1A). Because of this change in $K_i$, the unidirectional K influx ($J_{in}$), which equals $K_i$ x plasma [K], did not significantly change as plasma [K] increased (Fig. 1B). In the acute experiments, however, there was no decline in $K_i$ as plasma [K] increased and, therefore, $J_{in}$ increased with increasing plasma [K] ($p<0.001$; Fig. 1). This difference in influx regulation between acute and chronic experiments was found for CSF as well as brain. Thus, regulation of K influx can account for most of brain and CSF K homeostasis in the chronic but not the acute experiments.

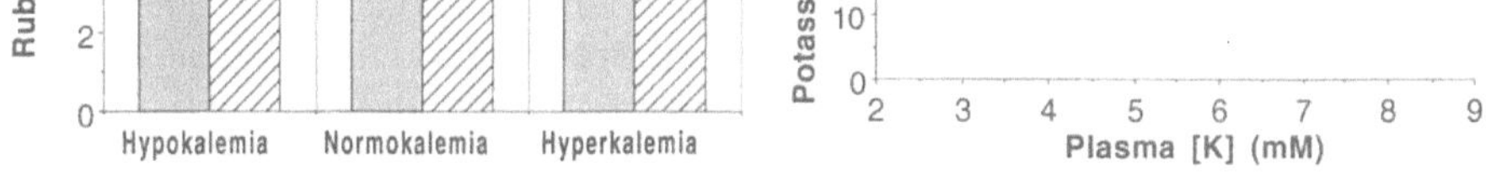


**Figure 1**. The effect of plasma [K] on (A) the influx rate constant, $K_i$, for $^{86}$Rb and (B) the unidirectional influx, Jin, of potassium in the anterior cortex. Measurements were either made in acute or chronic experiments (A) Values are means ± S.E.; n= 6-13; ***=p<0.001 vs normokalemia. (B) Acute y=4.6 + 5.60x, r=0.87, p<0.001; chronic y=24.3 + 0.91x, r=0.30, N.S.

In contrast to $^{86}$Rb, chronic changes in plasma [K] did not alter the influx rate constant for $^{22}$Na at the BBB. However, acute hyperkalemia resulted in a significant increase to 2.30 ± 0.21 $\mu$l/g/min compared to 1.65 ± 0.07 $\mu$l/g/min in normokalemic controls (Fig. 2). If this change in influx rate constant reflects an increase in BBB Na/K ATPase activity, brain K homeostasis in acute hyperkalemia may be achieved by increased K efflux to compensate for increased influx.

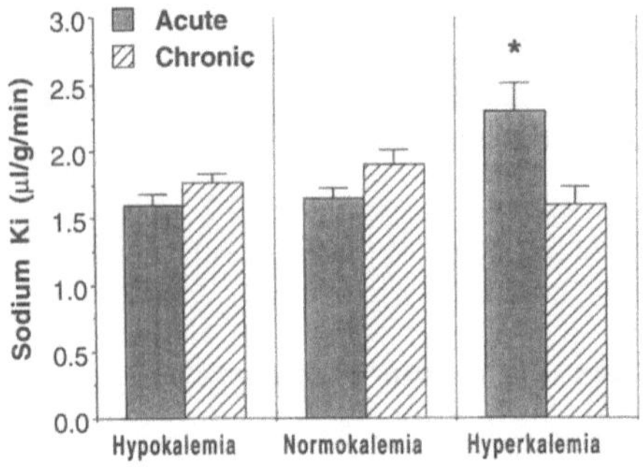


**Figure 2.** The effect of changes in plasma [K] on the influx rate constant, $K_i$, for $^{22}Na$ in the anterior cortex. Measurements were either made in acute or chronic experiments All values are means ± S.E.; n= 5-10; *=p<0.05 vs. acute hypo- and normokalemic animals.

## Potential Role of Mineralocorticoids in Brain Potassium Homeostasis

The difference in BBB and blood-CSF barrier potassium transport in response to acute and chronic changes in plasma [K] could either be due to a direct effect of potassium on these barrier tissues or an indirect effect via either a peripheral or a central potassium sensor. Atarashi et al. (1988) found that chronic intraventricular administration of aldosterone lowered CSF [K] without reducing plasma [K], a result we have confirmed. This suggests that this mineralocorticoid may have a role in brain K homeostasis as well as its known role in plasma K homeostasis.

We have examined the possible role of mineralocorticoids in brain K homeostasis using a common model for hypertension, the DOCA-salt rat. Similar to rats given aldosterone intraventricularly, DOCA-salt rats after two weeks had a significantly reduced CSF [K] compared to sham-operated rats (2.62 ± 0.02 vs 2.83 ± 0.06 mM; n= 12; p<0.01), although in contrast to the centrally infused aldosterone, this model also displays lowered plasma [K] than controls (2.83 ± 0.10 vs 3.69 ± 0.06 mM; n= 12; p<0.001).

The 4th ventricle choroid plexus was isolated from DOCA-salt and control rats for measurement of epithelial potassium currents using the whole-cell patch clamp technique. In control rats, we identified a K current, a delayed rectifier, that was inhibited in the presence of dibutryl cAMP (57 ± 4 %). It was also inhibited by the β-adrenergic agent isoproterenol, (31 ± 3%) which increases intracellular cAMP in choroid plexus (Crook et al., 1984). In DOCA-salt rats, the inhibition of the potassium current by isoproterenol (p< 0.05) was less than found in controls. In contrast, inhibition by dibutryl cAMP was not significantly different between the two groups. Thus it appears that hormonal regulation of choroid plexus potassium transport is changed in mineralocorticoid hypertension as a result of an alteration in the signal transduction pathway that links the hormone receptor to the generation of cAMP.

## DISCUSSION

Stability of the potassium gradient between the extracellular and intracellular compartments of the brain is important for neuronal function (Sykova, 1983). There are several mechanisms involved in maintaining this gradient. Glial cells have a role in buffering changes in extracellular [K] by either temporarily accumulating K or by redistributing K to other areas of the brain (Walz and Hertz, 1983; Kimelberg and Norenberg, 1989). However, alone, these mechanisms appear to have a limited role during hypo- and hyperkalemia as prolonged changes in glial cell K accumulation would involve changes in cell volume and because redistribution of K is of limited benefit during an insult that affects the whole brain. Potassium homeostasis during hypo- and hyperkalemia appears to primarily involve changes in potassium influx and efflux at the blood-brain and blood-CSF barriers ((Bradbury and Kleeman, 1967; Bradbury and Stulcova, 1970) and above). Whereas alterations in influx can account for most of the regulation during chronic changes in plasma [K], it appears that these alterations only occur slowly and that during acute hyperkalemia, at least, potassium regulation also involves changes in K efflux at the BBB.

The exact mechanisms involved in potassium transport at the BBB have proved elusive (Keep et al., 1993). A knowledge of such mechanisms is important not only for a general understanding of the function of the BBB and the brain but also because alterations in brain K may be important in several disease states. In particular, an imbalance between potassium loss and sodium gain across the BBB is associated with edema formation during cerebral ischemia (Betz et al., 1994). Our knowledge of K transport mechanisms at the choroid plexus is greater (Keep et al., 1993), although just how they are involved in CSF potassium regulation has still to be fully elucidated. The present study suggests that the delayed rectifier potassium channel may be involved in CSF K regulation although, until the channel is localized to either the basolateral or apical membrane, it is uncertain whether it is involved in reducing or increasing CSF [K]. As with the BBB, changes in choroid plexus K transport may be important in disease. In particular, the reduction in CSF [K] in the DOCA-salt rat may be linked to the genesis of the hypertension found in that model. We have found that chronic intraventricular infusions of small amounts of K-enriched CSF prevent the onset of hypertension in these rats (Klarr et al., 1993).

The BBB and the choroid plexus are innervated and they also have receptors for many hormones (Betz, 1992; Nilsson et al., 1992). The exact functions of these inputs are not fully understood. It appears that one function of a number of hormones which increase choroid plexus epithelial cAMP (Crook et al., 1984) is to inhibit a specific potassium conductance. Thus, as suggested by a number of studies, it may be possible to modulate brain ion and water content by using naturally occurring hormones as well as specific ion transport inhibitors.


## ACKNOWLEDGEMENTS

This work was supported by National Institute of Health grants NS-23870 and HL-18575.


## REFERENCES


Atarashi K., Matsuoka H., Takagi M., Yamada K., Hirata Y., and Hayakawa H., 1988, Effects of intracerebroventricular infusion of aldosterone on blood pressure and sodium and potassium concentrations in cerebral spinal fluid in rats., *Clin. Exper. Therapy Pract.* A10:317-322.

Betz A. L., 1992, An overview of the multiple functions of the blood-brain barrier., *NIDA Res. Monogr.* 120:54-72.

Betz A. L., Firth J. A., and Goldstein G. W., 1980, Polarity of the blood-brain barrier: distribution of enzymes between the luminal and antiluminal membranes of brain capillary endothelial cells, *Brain Res* 192:17-28.

Betz A. L., Keep R. F., Beer M. E., and Ren X.-D., 1994, Blood-brain barrier permeability and brain concentration of sodium, potassium, and chloride during focal ischemia., *J. Cereb. Blood Flow Metab.* 14:29-37.

Bradbury M. W. B. and Kleeman C. R., 1967, Stability of the potassium content of cerebrospinal fluid and brain, *Am J Physiol* 213:519-528.

Bradbury M. W. B. and Stulcova B., 1970, Efflux mechanism contributing to the stability of the potassium concentration in cerebrospinal fluid., *J. Physiol.* 208:415-430.

Crook R. B., Farber M. B., and Prusiner S. B., 1984, Hormones and neurotransmitters control cyclic AMP metabolism in choroid plexus epithelial cells., *J. Neurochem.* 42:340-350.

Keep R. F., Xiang J., and Betz A. L., 1993, Potassium transport at the blood-brain and blood-CSF barriers., *Adv. Exp. Med. Biol.* 331:43-54.

Kimelberg H. K. and Norenberg M. D., 1989, Astrocytes., *Sci. Amer.* 260:66-76.

Klarr S. A., Keep R. F., and Betz A. L., 1993, Chronic central administration of potassium prevents blood pressure increase in DOCA-salt hypertensive rats., *Hypertension* 22:447.

Nilsson C., Lindvall-Axelsson M., and Owman C., 1992, Neuroendocrine regulatory mechanisms in the choroid plexus-cerebrospinal fluid system., *Brain Res. Rev.* 17:109-138.

Ohno K., Pettigrew K. D., and Rapoport S. I., 1978, Lower limits of cerebrovascular permeability to nonelectrolytes in the conscious rat, *Am J Physiol* 235:H299-H307.

Schielke G. P., Moises H. C., and Betz A. L., 1991, Blood to brain sodium transport and interstitial fluid potassium concentration during early focal ischemia in the rat, *J Cereb Blood Flow Metab* 11:466-471.

Sykova E., 1983, Extracellular K+ accumulation in the central nervous system., *Prog. Biophys. Mol. Biol.* 42:135-189.

Walz W. and Hertz L., 1983, Functional interactions between neurons and astrocytes. II. Potassium homeostasis at the cellular level., *Prog. Neurobiol.* 20:133-183.

# TRANSPORT OF THE NUCLEOSIDE, THYMIDINE, IN THE CENTRAL NERVOUS SYSTEM: THE BLOOD-CEREBROSPINAL FLUID AND BLOOD-BRAIN BARRIERS

Sarah A. Williams, Hugh Davson and Malcolm B. Segal

Sherrington School of Physiology
United Medical and Dental Schools of Guy's and St. Thomas's Hospitals
St. Thomas' Hospital
Lambeth Palace Road
London, SE1 7EH

The exchanges between blood, on the one hand, and the cerebrospinal fluid (CSF) and brain-tissue on the other, are governed by so-called "barriers", which may be assessed quantitatively by measuring rate of uptake of a given solute from blood into the CSF - the *blood-CSF barrier* - and into the brain tissue - the *blood-brain barrier*. From these uptakes a simple unidirectional transfer coefficient, $K_{in}$, can usually be derived. The two barriers are not independent since the uptake into one component of the central nervous system may include a contribution from, or to, the other because of the free diffusional communication between brain-tissue and CSF across the ependymal and pia-glial membranes. Typically, this interchange is seen with relatively lipid-soluble substances, such as the thioureas or ethyl alcohol; simultaneous measurement of uptake into brain and CSF indicates that the brain comes much more rapidly into equilibrium with blood than does the CSF. This is due, essentially, to the different geometries of the systems. As Figure 1 illustrates, direct penetration from blood into CSF takes place almost exclusively within the ventricles across the choroidal epithelium; once outside the ventricles, the fluid is no longer exposed directly to the blood, so that further gains from this can only be achieved by diffusion from the brain tissue. The volume of the subarachnoid fluid is several times greater than that of the ventricular fluid, and this geometrical arrangement means that the final equilibration with blood is delayed by the requirement that CSF should flow from ventricles to subarachnoid spaces. If there were no additions to the CSF after if had left the ventricles, the maximum value of $K_{in}$ would be given by the turnover-constant of the CSF; on average this represents a renewal-rate of some 0.045% per minute to give a constant of $4.5x10^{-3}min^{-1}$. In practice we find, with lipid-soluble solutes such as the thioureas or ethyl alcohol, much larger values of $K_{in}$ (Figure 2); thus with ethyl alcohol the rate of equilibration in rabbit has a half-life of one or two minutes, giving a $K_{in}$ of 7 to $3.5x10^{-1}min^{-1}$.

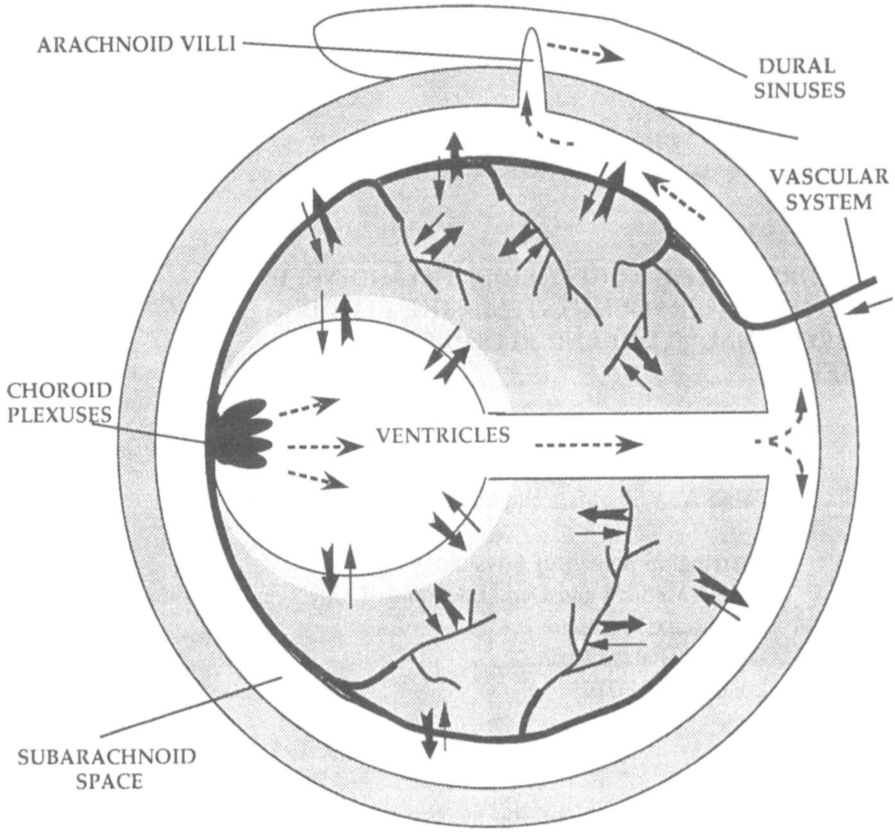


**Figure 1.** The relations between cerebrospinal fluid and brain tissue. When a high concentration of a substance is established in the plasma diffusion into CSF will occur rapidly across the choroid plexuses (thick arrows). Initially the concentration in the ventricles will be high in respect to that in the adjacent brain tissue, so the influx here will be large compared with brain adjacent to the subarachnoid fluid. Later, as the concentration in brain rises due to diffusion across the blood-brain barrier (thick arrows), the situation will be reversed, so far as brain tissue adjacent of the subarachnoid spaces is concerned, and CSF will now act as a "sink" to the brain.

When penetration of ethyl alcohol into brain was measured under the same conditions it was found that equilibration was virtually instantaneous, so that with this rapidly permeating substance uptake into the CSF must have been dominated by the rapid uptake into brain and subsequent diffusion into the ventricular and subarachnoid fluids. Figure 3 illustrates simultaneous penetration of thiourea and ethylthiourea into CSF and brain of the rabbit; here, too, the concentration in the brain tissue is always higher than in the CSF; in the graph the uptake into brain is measured as the amount in unit weight of tissue; correction for water-content of about 75% would raise the concentration in brain higher, so that there is no doubt that, during the approach to a steady state, there is always a gradient from tissue to CSF, i.e. the CSF is acting as a "sink" to the brain rather than as a "source". As lipid-solubility decreases, the situations become less clear-cut; thus with $^{24}$Na the rates of equilibration between blood, on the one hand, and CSF and brain on the other, are about equal if we assume that the uptake into brain is confined to the extracellular, or "sodium"-space; this is illustrated by Figure 4; thus exchanges between CSF and brain are not likely to be significant. With less rapidly equilibrating solutes, such as sucrose or sulphate, it is found that brain and CSF fail to come to equilibrium with the

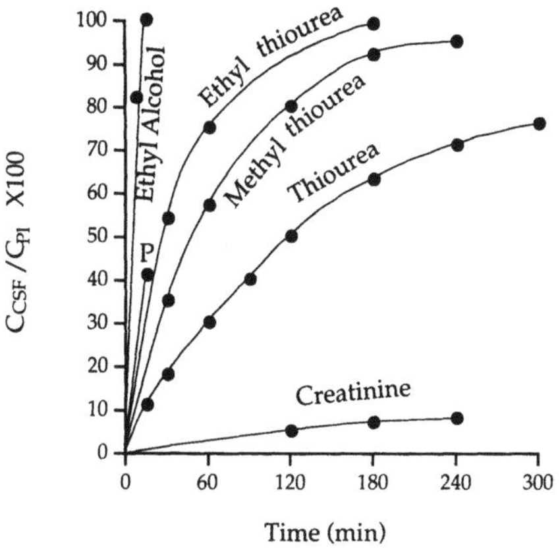


**Figure 2.** Penetration of different substances of increasing lipid-solubility into rabbit's cerebrospinal fluid. P = propyl thiourea. From Davson (1967).

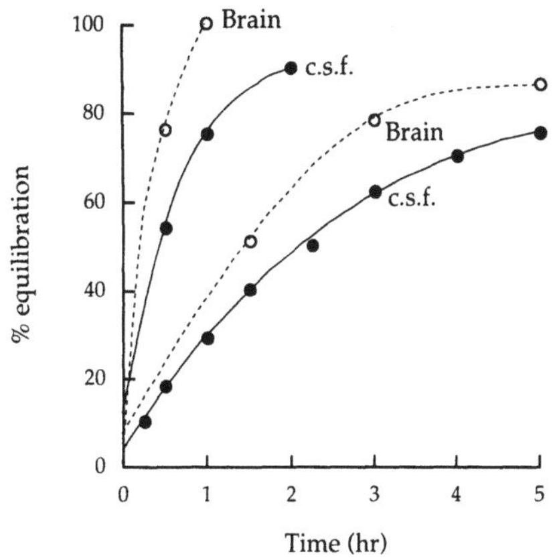


**Figure 3.** A comparison of the rate of entry of ethyl thiourea (upper curves) and thiourea (lower curves) from the blood into the brain (open circles) and into the cerebrospinal fluid (closed circles). From Davson et al. (1963).

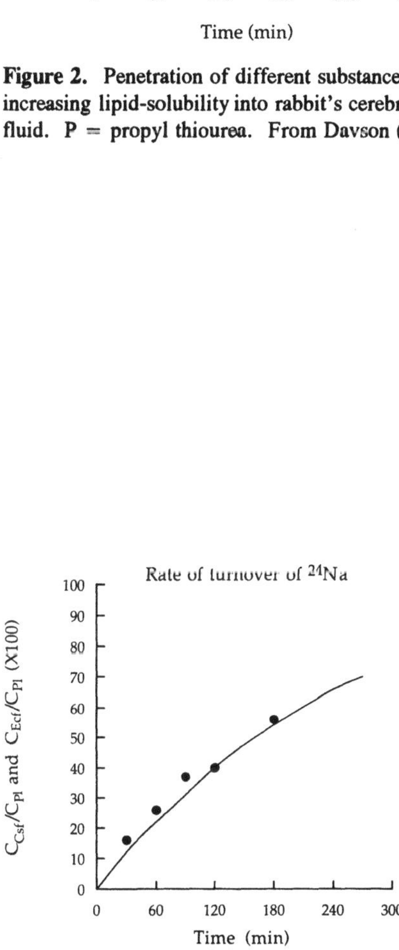


**Figure 4.** Penetration of $^{24}Na$ into the cerebrospinal fluid and brain tissue. Points represent the concentrations in the extracellular fluid of the nervous tissue, whilst the smooth curve represents the concentrations in the cerebrospinal fluid. From Davson (1955).

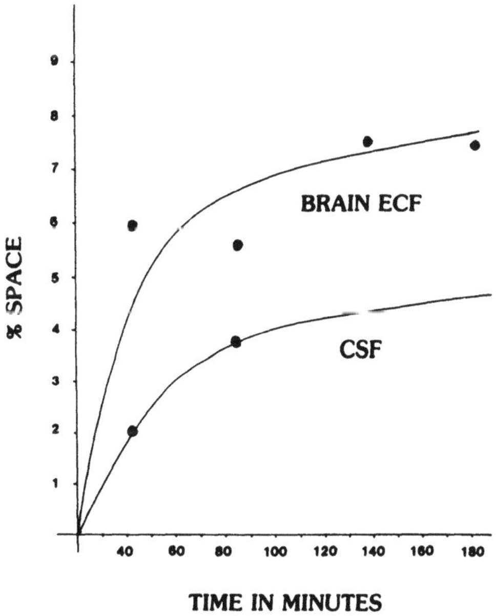


**Figure 5.** The entry from the blood of $^{35}SO_4^{2-}$ into brain extracellular fluid and cerebrospinal fluid; brain space has been increased by 100/15 to allow for distribution in extracellular fluid only. Note that the level in brain extracellular fluid is still far below that in plasma and is above the level in CSF. After Hollingsworth and Davson (1973).

plasma; instead, a steady-state is reached with the concentrations in CSF or brain extracellular fluid considerably less than in the plasma. Figure 5 illustrates the situation with sulphate. However, during the approach to the steady state, the concentration of the solute in the CSF is always less than that in the brain extracellular fluid, suggesting that the CSF acts as a "sink" for the brain rather than a source. The fact that the steady-state concentrations in CSF and brain extracellular fluid at the steady state are less than in plasma results from the condition that CSF flows and is drained away into the blood through open channels before it has a chance to come into full equilibrium with the plasma (Davson, 1956; 1967); the low steady-state level in brain extracellular fluid could result from a "sink-action" of the CSF, its concentration being always less than that in the extracellular fluid (Davson et al. 1961).

However, with the demonstration that the extracellular fluid also flows, but with a much slower turnover constant, at least a part of the low steady-state concentration in brain extracellular space can be accounted for by this flow rather than an action of the CSF (Abbott, 1992).

### CSF as a Source

Spector (1977; 1986) has argued that, with certain solutes, the supply to brain from the blood is largely, or even exclusively, through the CSF by virtue of the ability of the choroid plexuses to accumulate the solute, e.g. ascorbic acid, to give high concentrations in the freshly secreted CSF from which the vitamin would diffuse into the adjacent nervous tissue, a notion reminiscent of the Stern-Gautier hypothesis (Stern and Gautier, 1921, 1922, 1923).

**Table 1.** Hypothetical routes of entry of some metabolites according to Spector (1986).

| *Micronutrients* | *Blood-CSF barrier* | *Blood-brain barrier* |
|---|---|---|
| Ascorbic acid | Predominant | Minor |
| Folates | Predominant | Minor |
| Pantothenic acid | Some | Some |
| Riboflavin | Some | Unknown |
| Thiamin | Some | Some |
| Pyridoxine | Some | Unknown |
| Inositol | Some | Unknown |
| Thymidine | Predominant | Minor |
| Deoxycytidine | Predominant | Minor |
| Uridine | Some | Some |

### Ascorbic Acid

Hammerstöm (1966), using radioactive autography, observed that, following administration of labelled ascorbic acid, there were high concentrations initially in the choroid plexuses and adjacent nervous tissue; and uptake by brain seemed to result from a spread from the CSF. Spector (1977) showed that isolated choroid plexuses could accumulate ascorbic acid to give tissue/medium ratios as high as 10, and he concluded that the brain received this vitamin from blood by way of a primary accumulation in CSF. Later, the concept was extended to some other vitamins and to nucleosides, i.e. the base-sugar complexes that are constituents of the nucleotides (Table 1). To the extent that these compounds, namely the nucleosides, are involved in brain synthesis of the nucleic acids, their transport in the central nervous system is of interest; and in the present work we shall describe some experiments designed to test Spector's hypothesis, i.e. to assess the possibility that a nucleoside, such as thymidine, could be delivered to the brain predominantly by way of the CSF.

## Ventriculo-cisternal Perfusion

It was decided to study the exchanges of thymidine between blood and CSF in more detail in the intact animal, rather than employing the isolated plexus, since it is well established that mere accumulation of a solute by the choroid plexuses is no evidence for an active transport in the direction of blood to CSF; thus the very opposite is true of the anions, $I^-$, $CNS^-$ and p-aminohippurate. These ions are accumulated to a high degree by the isolated choroid plexuses (Welch, 1962 a, b; Becker, 1961), yet there is no doubt that these anions are actively removed from the CSF and driven towards the blood in the choroid plexuses; this process is not confined to the choroid plexuses, a similar active transport taking place from extracellular fluid to blood in the brain across the brain capillaries (Bito et al., 1966; Ahmed and Van Harreveld, 1969; Davson and Hollingsworth, 1973).

Figure 6 illustrates the result of perfusing an artificial CSF containing $^3$H-thymidine through the ventricles of the rabbit. Included in the perfusion fluid was the large-molecular weight Blue Dextran, which is not appreciably removed from the fluid, its concentration changing with time according as the perfusion fluid replaces the existing CSF; at the steady state its concentration in the emerging fluid, $C_{out}$ will not be equal to that of the entering fluid, $C_{in}$, because of the addition of newly secreted CSF, which would be free of the Blue Dextran.

It will be seen that the value of $C_{out}/C_{in}$ for $^3$H thymidine reaches a steady state after about 60 minutes, in which the value is about 0.4, indicating transport from CSF into the

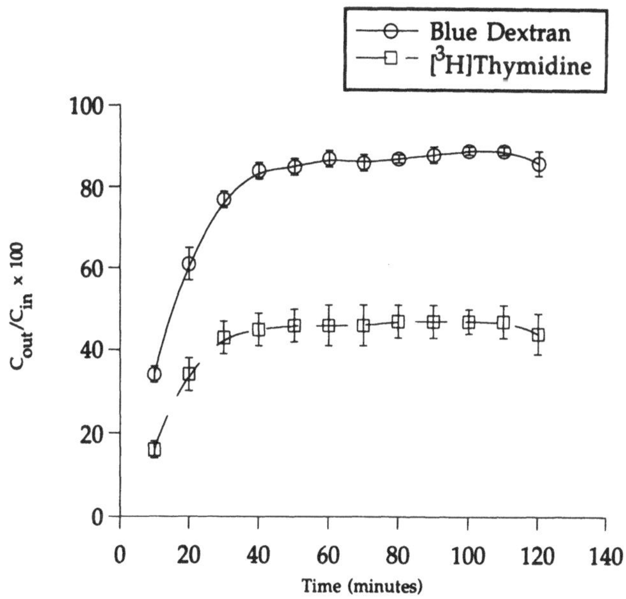


**Figure 6.** Ventriculo-cisternal perfusion in the rabbit by an artificial CSF containing Blue Dextran and $^3$H-thymidine. Ordinates: Ratio of emerging concentration, $C_{out}$ over entering concentration, $C_{in}$, x 100 Abscissae: Time after initiation of perfusion.

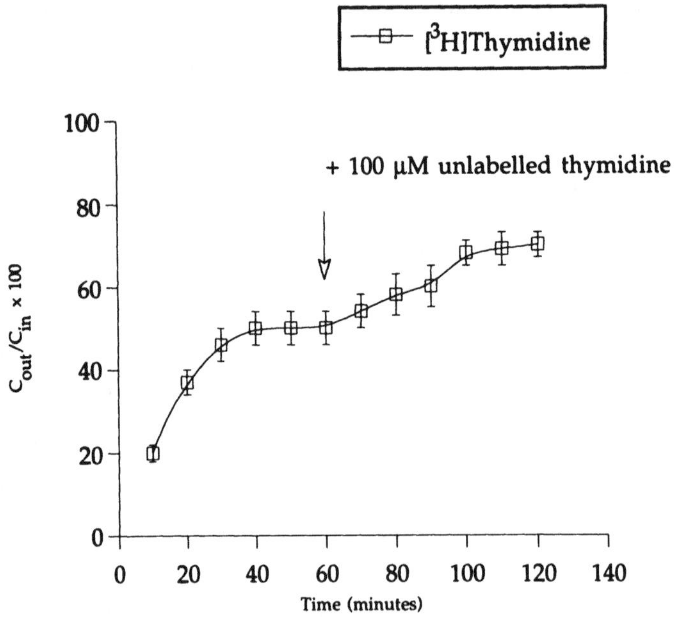


**Figure 7.** Effect of unlabelled thymidine on steady-state level of [$^3$H]-thymidine during ventriculo-cisternal perfusion in the rabbit. At the time marked by the arrow, the perfusion fluid was replaced with an identical one, but containing 100 μM unlabelled thymidine. Ordinates and Abscissae as in Figure 6.

blood by way of the choroid plexuses associated with transport into the brain; from the brain some diffuses into the blood and the remainder may be estimated by analysis of the brain tissue at the end of the experiment. From the steady-state levels of thymidine and Blue Dextran in the CSF a clearance can be calculated, which was on average 43.2 $\mu l\ min^{-1}$ for thymidine. From the measured amount remaining in the brain the separate clearances of thymidine into the blood and brain may be determined; these were 36.9 and 6.3 $\mu l\ min^{-1}$, indicating that most of the thymidine in the perfusion fluid had diffused into the blood, retention by the brain being relatively small. That the escape into the blood occurs

**Table 2.** The effect of 100 μM unlabelled thymidine on the steady-state ratios of [$^3$H]thymidine in the rabbit.

| [$^3$H]Thymidine | n | $R_{CSF}$ | Clearance ($\mu l.min^{-1}$ | $R_{BR}$ |
|---|---|---|---|---|
| alone | 7 | 45.1±3.6 | 45.7±4.7 | 11.3±0.8 |
| + 100 μM thymidine | 6 | 65.2±3.4 | 19.8±3.5 | 8.2±0.6 |
| P value | | $P<0.005$ | $P\leq0.001$ | $P<0.05$ |

Steady-state ratios, $R_{CSF}$, clearances and brain uptakes ($R_{BR}$) of labelled thymidine during ventriculo-cisternal perfusion under carrier-free conditions and in the presence of a 100 μM concentration of unlabelled thymidine. Limits are S.E.M.; n is the number of animals. Unpaired Student's *t*-test was used for the comparison of two means and the level of significance taken as $P<0.05$.

by a saturable process was shown by measuring the effects of added unlabelled thymidine, as illustrated by Figure 7. The changes in clearance and in the uptake by brain-tissue, $R_{Br}$, are shown in Table 2. The effect of the inhibitor of nucleoside transport, nitrobenzylthioinosine (NBMPR) is shown in Table 3; the inhibition is small but significant.

**Table 3.** Effect of 10 $\mu$M NBMPR on the clearance of [$^3$H]thymidine.

| [$^3$H]Thymidine | Clearance ($\mu$l.min$^{-1}$) | % Change |
|---|---|---|
| alone | 44.3±6.5 | - |
| + 10 $\mu$M NBMPR | 36.9±5.2 | 18.9±2.5 |
| P value | P<0.01 | - |

In these experiments the rabbit acted as its own control; thus ventriculo-cisternal perfusion was initiated with labelled thymidine and when an appropriate steady state had been reached, after 60-75 minutes, the perfusion fluid was switched to one containing the same radioactivity plus 10 $\mu$M of the nucleoside transport inhibitor, NBMPR. Limits are S.E. of 6 experiments. Paired Student's *t*-test was used for the comparison of two means and the level of significance taken as $P<0.05$.

## Comparison with Amino Acids

The behaviour of thymidine during ventriculocisternal perfusion is very similar to that observed with amino acids (Davson et al., 1982); this is illustrated by Figure 8 where values of $C_{out}/C_{in}$ for Blue Dextran, thymidine and alanine have been plotted against time; at the arrows, unlabelled solute was added to cause a rise in $C_{out}$ denoting a decreased clearance. In Table 4 the clearances of four different amino acids together with that of thymidine are shown, and the relative clearances to blood and brain indicated. In general, it may be said that the behaviour of thymidine during ventriculo-cisternal perfusion does not differ qualitatively from that of the amino acids; thus addition of unlabelled material suppresses clearance, and the relative contributions of clearance to blood and clearance to brain can be matched by an appropriate amino acid. Now it is well established that the transport of amino acids from blood to CSF and brain involves an active process that leads to a steady-state distribution between blood and CSF or blood and extracellular fluid, in

**Table 4.** Clearance of thymidine and various amino acids from CSF using ventriculo-cisternal perfusion in the rabbit.

| Solute | Total Clearance ($\mu$lmin$^{-1}$) | Clearance to Blood ($\mu$lmin$^{-1}$) | Clearance to Brain ($\mu$lmin$^{-1}$) |
|---|---|---|---|
| Thymidine | 43.2 | 36.9 | 6.3 |
| Alanine | 33.5 | 26.7 | 6.8 |
| Glycine | 60.4 | 28.9 | 31.5 |
| Arginine | 52.1 | 44.2 | 7.9 |
| Aspartic acid | 67.7 | 61.0 | 6.74 |

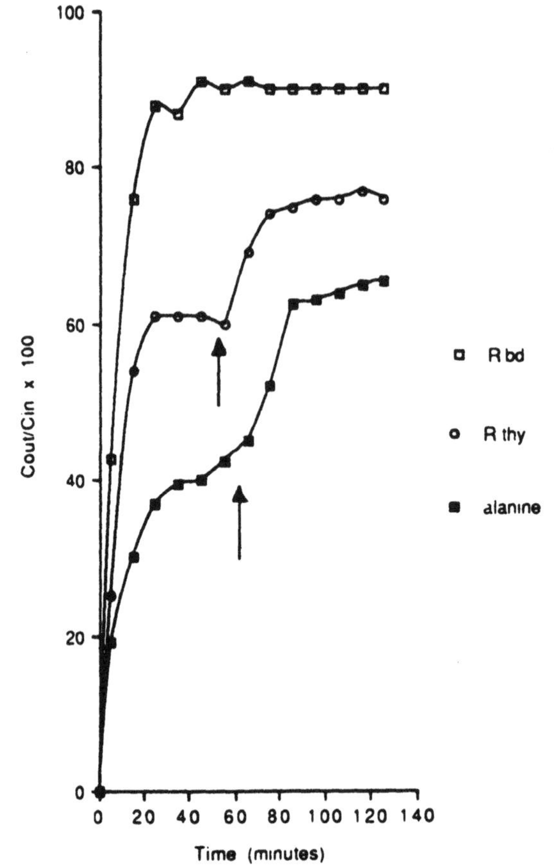


**Figure 8.** The ventriculocisternal perfusion of [$^3$H]-thymidine (open circles), alanine (filled squares) and Blue Dextran (open squares) in the rabbit.

which the concentration in blood plasma is several times *higher* than in the CSF or extracellular fluid. Thus, where metabolic alterations in the amino acids can be excluded, as with AIBA or cycloleucine, the steady-state concentrations in CSF are low compared with those in plasma (Davson et al., 1986). If the choroid plexuses were actively transporting thymidine from blood to CSF, then one might expect that addition of unlabelled thymidine to the perfusion fluid would increase the loss of labelled thymidine rather than to decrease the loss, as actually found. However, a more decisive experiment should involve a simultaneous direct measurement of uptake of thymidine from blood by CSF and brain. Employing labelled thymidine as a measure of thymidine transport introduces difficulties, to the extent that metabolic events while the labelled material is in the blood will confuse the issue, so that transport of the label need not necessarily indicate transport of the originally labelled material. However, the perfused isolated head preparation in the guinea-pig, described by Zloković et al. (1986), permits a study of transport across the blood-brain and blood-CSF barriers over a period of as long as 30 minutes without serious deterioration in the biochemistry of the brain and with no danger of loss of label while the material is in the "blood".

## Perfused Guinea-pig Head

The preparation described by Zloković et al. (1986) was modified to permit bilateral perfusion; in this situation the blood supplies to the choroid plexuses of both sides of the brain are adequately maintained. Included in the perfusion fluid containing the labelled thymidine was $^{14}C$-labelled mannitol; the penetration of this inert molecule is slow, so that any increases in rate would indicate a partial breakdown of the blood-brain barrier. Perfusions were terminated at different periods after initiation; the CSF was rapidly withdrawn from the cisterna magna, and the brain removed for analysis. Figure 9 illustrates penetration of blood into CSF and brain of the guinea-pig; penetration of mannitol is slow and comparable with that in earlier studies, e.g. (Zloković et al., 1988).

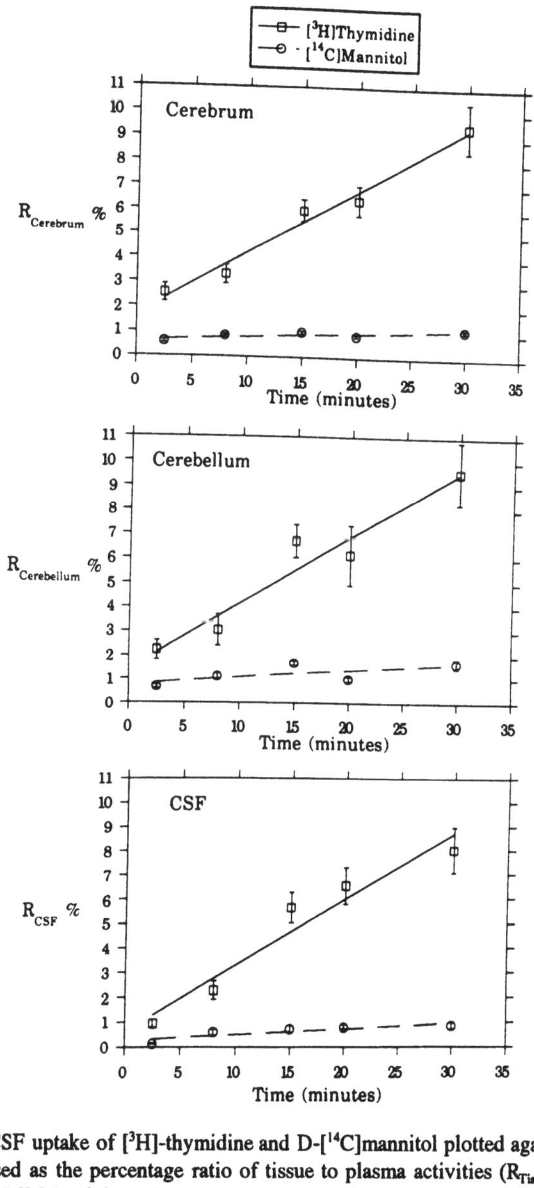


**Figure 9.** Brain and CSF uptake of [$^3$H]-thymidine and D-[$^{14}$C]mannitol plotted against time in the guinea-pig. Uptake is expressed as the percentage ratio of tissue to plasma activities ($R_{Tissue}$; ml.g$^{-1}$). Each point represents the mean ± S.E.M. of 5-9 animals. $K_{in}$ and $V_i$ values were determined as the slope and ordinate intercept of the computed regression lines. From Williams (1994).

Uptake into the brain has been expressed by the amount in unit weight of tissue; if, as is likely, uptake is mainly into the extracellular space, then the ordinate values for brain should be raised above those for CSF by a factor of about 5, in which case it becomes abundantly clear that the CSF could not possible be acting as a source for brain, but rather the reverse. A specific calculation makes this point clear. The rate of flow of CSF in the guinea-pig is some 3.5 $\mu l min^{-1}$ (Davson et al., 1987). If the newly secreted CSF contained $^3H$-thymidine, during intravascular perfusion of the brain, at the same concentration as that in the plasma; from the known weight of the brain (on average in our experiments 3.7±0.1 g), it is possible to estimate the maximum value of $R_{Br}$, at the end of 15 minutes of perfusion, this maximum being given by the assumption that the circulating CSF has given up all its $^3H$-thymidine to the brain on its passage through the ventricles. The estimated value of $R_{Br}$ on this basis is 0.0013 compared with some 0.05 actually found. Thus, only if the choroid plexuses were secreting a CSF at a concentration some 38 times that in the plasma, and if the CSF relinquished all its secreted $^3H$-thymidine to the brain, would it be possible to obtain the measured $R_{Br}$. That all the secreted $^3H$-thymidine is not relinquished to the brain is clear from the measured values of $R_{CSF}$.

**Table 5.** Calculated $K_{in}$ and $V_i$ for [$^{14}$C]mannitol and [$^3$H]thymidine into both brain and CSF determined using the *in situ* brain perfusion technique in the guinea-pig. (Williams, 1994).

| BRAIN | $K_{in}$ ($\times 10^{-3}$ ml.min$^{-1}$,g$^{-1}$) | $V_i$ (ml.100 g$^{-1}$) |
|---|---|---|
| [$^{14}$C]Mannitol | 0.18±0.05 | 0.62±0.10 |
| [$^3$H]Thymidine | 2.56±0.21 | 1.68±0.44 |
| CEREBROSPINAL FLUID | $K_{in}$ ($\times 10^{-3}$ ml.min$^{-1}$,g$^{-1}$) | $V_i$ (ml.100 g$^{-1}$) |
| [$^{14}$C]Mannitol | 0.23±0.08 | 0.26±0.16 |
| [$^3$H]Thymidine | 2.70±0.39 | 0.61±0.84 |

Values ± S.E.M. for 5 determinants.

## Transport Coefficients

In Table 5 values of $K_{in}$ and $V_i$, the rapidly equilibrating space, have been calculated. Quite clearly penetration of thymidine into the brain is significantly greater than that of mannitol, so that claims as to the impermeability of the blood-brain barrier to this nucleoside, based on single-pass experiments, are unsound, and the technique of isolated head perfusion, which permits studies over periods of time some 80 to 100 times longer than those employed in single-pass studies, permits studies of slowly penetrating, but nevertheless physiologically interesting substances such as vitamins, nucleotide intermediates and neuropeptides.

## Carrier Mediation

Figure 10 illustrates the effects of 10 mM unlabelled thymidine on the uptake of $^3$H-thymidine by cerebrum, cerebellum and CSF. Clearly the transport process exhibits self-saturation. The effect was not specific to thymidine, as similar experiments using unlabelled uridine and deoxycytidine were also effective. Unlabelled guanosine had no effect, however.

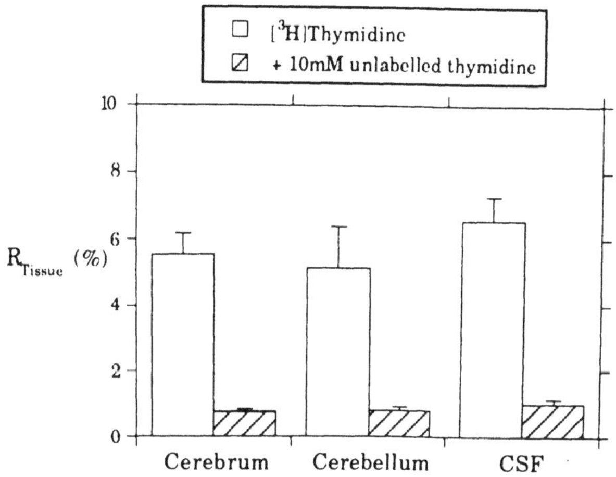


**Figure 10.** The effect of 10 mM unlabelled thymidine on the uptake into the guinea-pig brain of [$^3$H]thymidine. Uptake is expressed as the percentage ratio of tissue to plasma activities ($R_{Tissue}$; ml.g$^{-1}$) and D-[$^{14}$C]mannitol/ vascular space has been subtracted from the brain values. Values are the mean ± S.E.M. for 4-7 animals. From Williams (1994).

## The Spector Hypothesis

So far as the kinetics of exchange of thymidine between blood and brain are concerned, it would appear that Spector's hypothesis is incorrect, the transport to brain being by way of a carrier-mediated facilitated diffusion across the brain capillaries. As with other slowly penetrating substances, it would seem that the concentration-gradients between CSF and brain extracellular fluid during the approach to the steady state, after a change in plasma concentration, are such as to favour diffusion from extracellular fluid to the CSF; thus, as with other solutes studied, the CSF is favourably placed to act as a "sink" to the brain tissue. The hypothesis was initially proposed with respect to the kinetics of exchange of the vitamin, ascorbic acid. It was proposed by Hammerström (1966) on the basis of radioautographic evidence, as indicated earlier. It was supported by the studies of Spector and Lorenzo (1973) who demonstrated *in vitro* accumulation of the vitamin, but, as we have argued earlier, *in vitro* accumulation is not necessarily in favour of active transport from blood to CSF. The study of Ridge and Fairhurst (1976) demonstrated an *in vivo* accumulation of ascorbic acid in human CSF, so that values of $R_{CSF}$ as high as 6 may be observed at an optimal plasma concentration of about 5mM. It must be emphasized, however, that whenever a clear demonstration of active transport across the choroidal epithelium has been demonstrated, as, for example, with $K^+$, $Na^+$. $I^-$, etc., a similar active transport has been demonstrated to take place across the brain capillary endothelium. Thus, demonstration of an *in vivo* accumulation of solute in the CSF does not necessarily lead to the conclusion that the CSF acts as a prime or even significant source for the brain tissue. The initially high concentrations of labelled ascorbic acid in the brain tissue in the neighbourhood of the ventricles, described by Hammerström (1966), are convincing evidence that, in the initial stages of the approach to equilibrium, or the steady state, after a rise in plasma activity has been established, the CSF is, indeed, acting as a source to the brain; the geometry of the cerebrospinal system, however, is such that this tendency to act as a source is only temporary; once the CSF has moved out of the ventricles, further build-up in concentration must be slow and governed by the rate of renewal of the CSF as whole unless solute enters from the adjacent brain tissue. As we have indicated earlier, the brain tissue is far more satisfactorily situated with respect to the blood than is the subarachnoid CSF, so that, on virtually geometrical considerations only, we may expect the concentration relations between CSF and brain tissue to be reversed. This was demonstrated a long time ago by the studies of Davson and Pollay (1963) and Davson and Segal (1970) when they maintained a high concentration of $^{24}Na$ or $^{22}Na$ in the plasma of the rabbit during ventriculocisternal perfusion. Under these conditions uptake into the CSF perfusing the ventricles was governed by an initial rapid process, attributable to secretion from the choroid plexuses, followed by a much slower approach to a steady state due to diffusion from the brain tissue. Thus, whereas, on average, the CSF and brain, in the intact animal, come into equilibrium with the plasma $^{24}Na$ at about the same rates, there is no doubt that, in the earliest stage, the CSF within the ventricles is well ahead of the adjacent brain tissue. Thus, in more recent studies of the exchanges of $Na^+$ and $Cl^-$ between blood and brain, care has been taken to sample only parts of the brain that are remote from the ventricles, in this way making measurements of the blood-brain barrier uncomplicated by exchanges between CSF and brain. We may conclude, therefore, that the kinetics of exchange of ascorbic acid between blood and brain, and blood and CSF, are worthy of further study before subscribing to the view that the brain receives the vitamin exclusively from the CSF.

## REFERENCES


Abbott, N.J., 1992, The apparent 'sink-action' of the CSF - are the observations due to secretion and flow of interstitial fluid? *J. Physiol.* 446:496P.

Ahmed, N., and van Harreveld, H., 1969, The iodide space in rabbit brain. *J. Physiol.* 204:31.

Becker, B., 1961, Cerebrospinal fluid iodide. *Amer. J. Physiol.* 201:1149.

Bito, L.Z., Bradbury, M.W.B., and Davson, H., 1966, Factors affecting the distribution of iodide and bromide in the central nervous system. *J. Physiol.* 185:323.

Davson, H., 1955, A comparative study of the aqueous humour and cerebrospinal fluid in the rabbit. *J. Physiol.* 129:111.

Davson, H., 1956, "Physiology of the Ocular and Cerebrospinal Fluids." Churchill, London.

Davson, H., 1967, "Physiology of the Cerebrospinal Fluid." Churchill, London.

Davson, H., Begley, D.J., Chain, D.G., Briggs, F.O., and Shepherd, M.T., 1986, Steady-state distribution of cycloleucine and α-aminoisobutyric acid between plasma and cerebrospinal fluid. *Exp. Neurol.* 91:163.

Davson, H., and Hollingsworth, J.G., 1973, Active transport of $^{131}I$ across the blood-brain barrier. *J. Physiol.* 233:327.

Davson, H., Hollingsworth, J.G., Carey, M.B., and Fenstermacher, J.D., 1982, Ventriculo-cisternal perfusion of twelve amino acids in the rabbit. *J. Neurobiol.* 13:293.

Davson, H., Kleeman, C.R., and Levin, E., 1963, The blood-brain barrier. *In* "Drugs and Membranes", C.Adrian M. Hogben, ed., (Proc. 1st Int. Congr. Pharmacol., Stockholm). Pergamnon, Oxford, pp 71.

Davson, H., and Pollay, M., 1963, The turnover of $^{24}Na$ in the cerebrospinal fluid and its bearing on the blood-brain barrier. *J. Physiol.* 167:247.

Davson, H., and Segal, M.B., 1970, The effects of some inhibitors and accelerators of sodium transport on the turnover of $^{22}Na$ in the cerebrospinal fluid. *J. Physiol.* 209:131.

Davson, H., Welch, K., and Segal, M.B., 1987, "Physiology and Pathophysiology of the Cerebrospinal Fluid." Churchill Livingstone, London.

Hammersrtröm, L., 1966, Autoradiographic studies on the distribution of $^{14}C$-labelled ascorbic acid and dehydroascorbic acid. *Acta Physiol. Scan.* 70:Suppl.289.

Hollingsworth, J.G., and Davson H., 1973, Transport of sulphate in the rabbit's brain. *J. Neurobiol.* 4:389.

Ridge, B.D., and Fairhurst, E., 1976, Ascorbic acid concentrations in human plasma and cerebrospinal fluid. *Proc. Nutr. Soc.* 35:57A.

Spector, R., 1986, Vitamin homeostasis in the central nervous system. *New Engl. J. Med.* 296:1393.
Spector, R., and Lorenzo, A.V., 1973, Ascorbic acid homeostasis in the central nervous system. *Amer. J. Physiol.* 225:757.
Stern, L., and Gautier, R., 1921, Rapports entre le liquide céphalorachidien et la circulation sanguine. *Arch. Int. Physiol.* 17:138.
Stern, L., and Gautier, R., 1922, Les rapports entre le liquide céphalorachindien et les éléments nerveux de l'axe cerebrospinal. *Arch. Int. Physiol.* 17:391.
Stern, L., and Gautier, R., 1923, Rapports entre le liquide céphalorachidien des espaces ventriculaires et celui des espaces sous-arachnoidiens. *Arch. Int. Physiol.* 20:403.
Welch, K., 1962a, Active transport of iodide by choroid plexus of the rabbit *in vitro*. *Amer. J. Physiol.* 202:757.
Welch, K., 1962b, Concentration of thiocyanate by the choroid plexus of the rabbit *in vitro*. *Proc. Soc. Exp. Biol. Med.* 109:953.
Williams, S.A., 1994, "Nucleoside Uptake at the Blood-Brain and Blood-Cerebrospinal Fluid Barriers." Ph.D. Thesis, University of London.
Zloković, B.V., Begley, D.J., Djuricic, B.M., and Mitrovic, D.M., 1986, Measurement of solute transport across the blood-brain barrier in the perfused guinea-pig brain: method and application of N-methyl-alpha-aminoisobutyric acid. *J. Neurochem.* 46:1444.
Zloković, B.V., Lipovac, B.N., Begley, D.J., Davson, H., and Rakic, L., 1988, Slow penetration of thyrotropin releasing hormone across the blood-brain barrier of *in situ* perfused guinea-pig brain. *J. Neurochem.* 51:252.

# NASAL CEREBROSPINAL FLUID DRAINAGE IN MAN

Pia Löwhagen,[1] Barbro B. Johansson,[2] and Claes Nordborg[1]

[1]Department of Pathology, Division of Neuropathology
University of Göteborg, Sweden
[2]Department of Neurology, University of Lund, Sweden

## INTRODUCTION

Intracranially-applied tracers are effectively drained via the nasal route and accumulate in the deep cervical lymph nodes of experimental animals[1-12]. Recently, Löwhagen et al.[13] described the nasal drainage routes in man and reported that red blood cells drain via the perineural route from the subarachnoid space to the nasal mucosa in patients with intracranial haemorrhagic lesions. The object of the present paper was to study whether the drainage of cerebrospinal fluid (CSF) to the nasal mucosa could be visualized in man also in the absence of intracranial pressure increase and in patients without a cerebral disorder. For this purpose, the presence of brain-specific antigen was mapped immunocytochemically along the presumed routes of transport in post-mortem material comprising controls without a neurological disorder and patients with various brain lesions.

## MATERIAL AND METHODS

Samples were collected during autopsy from 6 cases without neurological disease and from 17 cases with various cerebral disorders (cerebrovascular disease; n: 10, brain tumour; n: 2, neurodegenerative disorders; n: 5). The average post-mortem time was 2.3 days (range: 1-4 days)

The olfactory tracts and bulbs were removed bilaterally together with the adjacent parts of the basal frontal lobes. The lamina cribrosa and subjacent parts of the nasal mucosa, the nasal septum and the walls of the nasal cavities were then dissected out. All the identified deep cervical lymph nodes were removed bilaterally, as were a number of the femoral ones. The specimens were fixed in a 4% phosphate-buffered formaldehyde solution for approximately one week. Coronal sections were then cut through the olfactory tracts and bulbs. To avoid the need for decalcification, the lamina cribrosa and nasal cavity were carefully dissected free from bone and cut in several frontal sections. All the lymph nodes were divided in two. The specimens were dehydrated in graded alcohols and embedded in

*New Concepts of a Blood–Brain Barrier*
Edited by J. Greenwood *et al.*, Plenum Press, New York, 1995

paraffin. Five-$\mu$m-thick sections were stained using various methods. Two different monoclonal antibodies (Dako anti-human GFAP, 1:300; Tp-GFAP 3, 1:50, kindly provided by Professor V.P. Collins, Department of Pathology, Göteborg) were used for the visualization of brain-specific antigen. Antibodies bound in the tissue were visualized using the avidin-biotin-peroxidase method, using 3-amino-9-ethylcarbazole (AEC) as a chromogen. The specimens were counterstained with haematoxylin. The frontal brain tissue was used as an internal positive control. Negative controls, in which the primary antibody was omitted, were made for every case. All the negative controls were totally free from immunoreactivity.

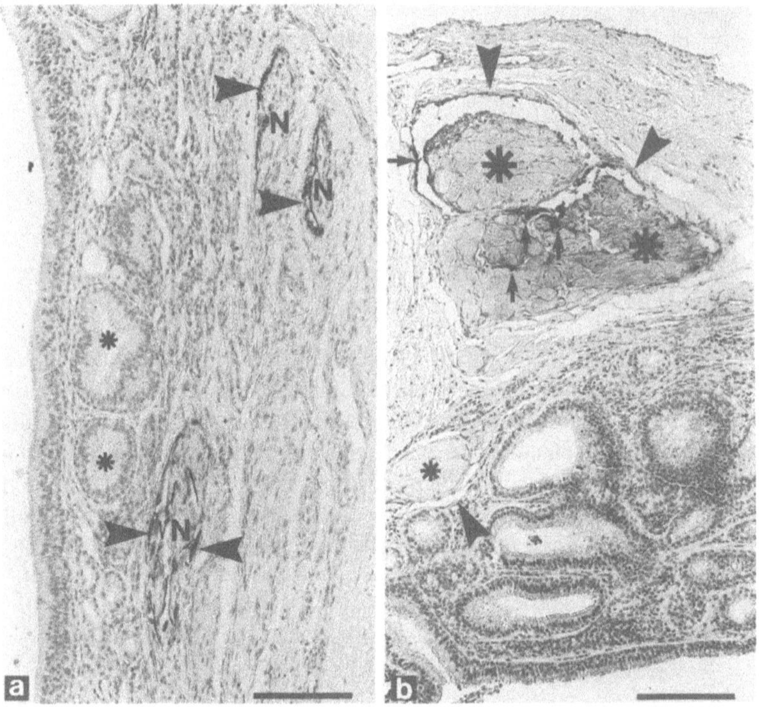


**Figure 1. a.** Immunoreactivity to GFAP in the perineural spaces of the nasal mucosal nerve branches (arrowheads). Nerve fascicles: N. Mucosal glands: *. (Anti-GFAP. Bar: 160 $\mu$m) **b.** Arachnoid villi in the nasal submucosa and mucosa with fibromeningotheliomatous cores (*) and surrounding spaces (arrowheads). Note the black immunoreactivity to GFAP (arrows). Nasal cavity: bottom. (Anti-GFAP. Bar: 200 $\mu$m)

## RESULTS

Staining with antibodies to GFAP revealed some positivity in the white matter and cortex of controls, as well as in cases with brain lesions. The specimens with the most pronounced positivity were found among the latter group. In all cases, the immunoreactivity was much greater in the olfactory tracts and bulbs than in the frontal lobes. A strong GFAP positivity was also seen in the perivascular, adventitial connective tissue of meningeal arteries, generally with an obvious spatial relationship to the strongly-positive olfactory tissue; the perivascular meningeal GFAP positivity was more widespread in three cases with brain damage.

The general morphology of the olfactory nerves and their passage to the nasal mucosa was in conformity with the descriptions by Erlich et al.[5] and Gomez et al.[6]. The olfactory nerve fascicles were surrounded by perineural spaces. In addition, leptomeningeal outpouchings with a fibromeningotheliomatous core were found penetrating the lamina cribrosa and ending underneath the nasal mucosa and occasionally even within it; these arachnoid villi did not bulge into large veins but were often situated close to smaller ones (Figure 1). They were sometimes found adjacent to nerve fascicles and fibromeningotheliomatous cores were also observed within the perineural spaces. Nasal arachnoid villi were found in 63% of the cases.

An immunoreactivity to GFAP was found in the perineural spaces surrounding olfactory nerve fascicles in all 6 control cases and in 15 of the 17 cases with brain disorder. A perineural GFAP positivity was observed in the nasal submucosa and mucosa in 5 of the 6 controls and in 11 of the patients with brain lesions (Figure 1). Positive staining was also occasionally encountered endoneurally and in some of the intranasal arachnoid villi (Figure 1). There was no apparent difference in the GFAP accumulation between patients with cerebrovascular disease, brain tumours, neurodegenerative disease or the controls. GFAP was not found in the mucosal stroma or in the nasal epithelium.

In the present study, immunoreactivity to GFAP was not detected in the deep cervical lymph nodes in any of the cases.

## DISCUSSION

The drainage of cerebrospinal fluid via the nasal route to the deep cervical lymph nodes constitutes a contact between the brain and the immune system which may be of pathogenetic significance in certain disorders of the central nervous system in man. The proximity to regional lymphatic tissue, such as lymph nodes, may be of importance for the generation of specific immunological responses in cerebral infections. A compartmentalization of the immune system with a local proliferation of virus-specific lymphocytes in the deep cervical lymph nodes has been demonstrated in experimental intracerebral infections and is known to precede the appearance of these cells in the blood and later in the brain and CSF[14]. The significance of the nasal CSF-drainage route in man is indicated by the fact that the deep cervical lymph nodes are involved in 20% of cases with known extracranial metastasis of brain tumours. At least 11 cases of metastasis exclusively to the deep cervical nodes are known[15-18].

Tracers applied experimentally in the ventricles or in the white or grey matter spread widely along perivascular basement membranes[3,12,19]. The present study may support the hypothesis that endogenous large molecules spread via this perivascular route into the subarachnoid space in man; an obvious periarterial immunoreactivity to GFAP was thus only seen in the immediate vicinity of the olfactory tissue, which showed the strongest GFAP positivity. The strong perivascular GFAP positivity in the olfactory groove might, on the other hand, also be caused by the greater transport and accumulation of CSF-borne substances in this region leading to a greater secondary uptake into the perivascular connective tissue. Red blood cells and iron pigment were concentrated here in cases with recent or older haemorrhagic brain lesions[13]. The accumulation of macromolecules and cells indicates that there is a transport route for CSF along the olfactory tracts and is in good agreement with the effective drainage of CSF from the olfactory region to the nasal mucosa.

The perineural extensions of the subarachnoid space have been shown to drain the CSF to the nasal mucosa in experimental tracer studies of rabbits[5,6]. Kida et al.[20] reported direct connections between the subarachnoid space and the nasal lymphatics in rats, thus indicating species differences when it comes to the anatomy of the drainage routes. In a recent study of autopsy material Löwhagen et al.[13] observed the spread of intracranially-applied Indian ink in perineural spaces down to the nasal mucosa in man. Furthermore, the tracer filled small cavities in the nasal mucosa and submucosa which had the appearance of arachnoid villi, thus indicating an additional route of drainage. Particles up to the size of a red blood cell were drained from the subarachnoid space to the nasal mucosa via the perineural spaces in cases with intracranial haemorrhagic lesions in man[13]. In the present study, brain-specific antigen was found in the perineural spaces and arachnoid villi in the nose not only in patients with different brain lesions but also in cases without neurological disease. The present observations indicate that endogenous substances are drained from the human CNS to the nose not only in brain disease but also in healthy subjects. The drainage of endogenous antigens by these routes was further indicated by the immunoreactivity to neurofilament in the olfactory perineural spaces and nasal arachnoid villi in man (Löwhagen et al., in preparation).

Astrocytes have been observed in olfactory nerve fascicles in the rat[21]. However, although some immunoreactivity to GFAP was occasionally encountered in the olfactory nerve fascicles, no GFAP-positive astrocyte-like cells were encountered in the present material. Instead, the immunoreactivity to GFAP was confined mainly to the surfaces of the spaces surrounding the nerve fascicles and to the surfaces of the meningeal spaces in the arachnoid villi. Since two different monoclonal antibodies directed at different epitopes produced identical results in terms of the localization of the GFAP immunoreactivity, it seems likely that the observed immunoreactivity reflects a real accumulation of brain-specific protein.

The centrifugal flow of endoneural interstitial fluid, which is known to occur in cranial and peripheral nerves, could theoretically explain why GFAP was found not only in spaces around olfactory nerves but occasionally also within their endoneurium[22]. Alternatively, GFAP might be taken up secondarily from the perineural space into the endoneurium; the olfactory nerves are loosely-arranged fibre bundles without any light-microscopically visible perineural sheath. The fact that numerous iron-containing macrophages were found endoneurally after a pure subarachnoidal haemorrhage indicates a secondary uptake from the perineural space in the nerve fascicles[13]. Likewise, the red blood cells, which were found endoneurally in a case of recent subarachnoidal haemorrhage, most probably emanated from the bleeding in the subarachnoid space, since there was no haematoma in the parenchyma from which they could have been transported in the endoneural interstitium.

The lymphatic drainage of the endogenous macromolecules which were found in the nasal mucosa is plausible, despite the fact that brain-specific antigen could not be demonstrated immunocytochemically in the deep cervical lymph nodes. The amount of antigen might be too small and the degradation rate of macromolecules within the lymph nodes too rapid to permit immunocytochemical detection. The demonstration of iron deposits in the deep cervical lymph nodes is not in itself an indication of drainage from an intracranial haemorrhage unless iron pigment or red corpuscles are shown in the perineural spaces which lead to the nasal mucosa. A recent study revealed that, in 2 of 5 cases with parenchymal or subarachnoidal bleedings, the lymph nodes were free from iron. Conversely, in 14 of 18 cases without an intracranial haemorrhage, there was some positivity in the Prussian blue staining[13]. Extracranial bleeding such as epistaxis thus appears to be a more common cause of iron deposition in the deep cervical lymph nodes.

*In conclusion*, the present study indicates that endogenous substances are transported by the perineural drainage route from the subarachnoid space to the nose in man also in the absence of brain disorders. Intranasal arachnoid villi might be an additional route of drainage. The fact that no GFAP was demonstrated in the deep cervical lymph nodes might be due to a rapid macromolecular break-down and to the passage into the blood in the nasal mucosa.

## Acknowledgements

The skilled technical assistance of Margareta Persson, Pia Bitto and Hans-Olof Ivarsson is gratefully acknowledged. This study was supported by the Swedish Medical Research Council (project number 14X-4968).


## REFERENCES


1. M.W.B. Bradbury, and D.F. Cole, The role of the lymphatic system in drainage of cerebrospinal fluid and aqueous humour, *J. Physiol.* 299:353 (1980).
2. M.W.B. Bradbury, H.F. Cserr, and R.J. Westrop, Drainage of cerebral interstitial fluid into deep cervical lymph of the rabbit, *Am. J. Physiol.* 240:329 (1981).
3. J.R. Casley-Smith, E. Földi-Börcsök, and M. Földi, The prelymphatic pathways of the brain as revealed by cervical lymphatic obstruction and the passage of particles, *Br. J. exp. Path.* 57:179 (1976).
4. H.F. Cserr, C.J. Harling-Berg, and P.M. Knopf, Drainage of brain extracellular fluid into blood and deep cervical lymph and its immunological significance, *Brain Pathology* 2:269 (1992).
5. S.S. Erlich, J.G. McComb, S. Hyman, and M.H. Weiss, Ultrastructural morphology of the olfactory pathway for cerebrospinal fluid drainage in the rabbit, *J. Neurosurg.* 64:466 (1986).
6. D.G. Gomez, J.D. Fenstermacher, R.P. Manzo, D. Johnson, and D.G. Potts, Cerebrospinal fluid absorption in the rabbit: olfactory pathways, *Acta Otolaryngol.* 100:429 (1985).
7. J.G. McComb, Recent research into the nature of cerebrospinal fluid formation and absorption, *J. Neurosurg.* 59:369 (1983).
8. M. Oehmichen, H. Grüninger, H. Wiethölter, and M. Gencic, Lymphatic efflux of intracerebrally injected cells, *Acta Neuropathol.* 45:61 (1979).
9. R.O. Weller, S. Kida, and E.-T. Zhang, Pathways of fluid drainage from the brain. Morphological aspects and immunological significance in rat and man, *Brain Pathology* 2:277 (1992).
10. H. Widner, B.-A. Jönsson, L. Hallstadius, K. Wingårdh, S.-E. Strand, and B.B. Johansson, Scintigraphic method to quantify the passage from brain parenchyma to the deep cervical lymph nodes in rats, *Eur. J. Nucl. Med.* 13:456 (1987).
11. H. Widner, G. Möller, and B.B. Johansson, Immune response in deep cervical lymph nodes and spleen in the mouse after antigen deposition in different intracerebral sites, *Scand. J. Immunol.* 28:563 (1988).
12. E.T. Zhang, H.K. Richards, S. Kida, and R.O. Weller, Directional and compartmentalized drainage of interstitial fluid and cerebrospinal fluid from the rat brain, *Acta Neuropathol.* 83:233 (1992).
13. P. Löwhagen, B.B. Johansson, and C. Nordborg, The nasal route of cerebrospinal fluid drainage in man. A light-microscopic study, *Neuropathol. Appl. Neurobiol.* 20: 543 (1994).
14. P.C. Doherty, J.E. Allan, F. Lynch, and R. Ceredig, Dissection of an inflammatory process induced by CD 8+ T cells, *Immunology Today* 11:55 (1990).
15. S. El-Gindi, M. Salama, M. El-Henawy, and S. Farag, Metastases of glioblastoma multiforme to cervical lymph nodes, *J. Neurosurg.* 38:631 (1973).
16. F.E. Glasauer, and H.P. Yuan, Intracranial tumors with extracranial metastases, *J. Neurosurg.* 20:474 (1963).
17. A. Ley, D. Campillo, and C. Oliveras, Extracranial metastasis of glioblastoma multiforme, *J. Neurosurg.* 18:313 (1961).
18. D.R. Smith, J.M. Hardman, and K.M.Earle, Metastasizing neuroectodermal tumors of the central nervous system, *J. Neurosurg.* 31:50 (1969).
19. M.L. Rennels, T.F. Gregory, O.R. Blaumanis, K. Fujimoto, and P.A. Grady, Evidence for a "paravascular" fluid circulation in the mammalian central nervous system, provided by the rapid distribution of tracer protein throughout the brain from the subarachnoid space, *BrainResearch* 326:47 (1985).

20. S. Kida, A. Pantazis, and R.O. Weller, CSF drains directly from the subarachnoid space into nasal lymphatics in the rat. Anatomy, histology and immunological significance, *Neuropathol. Appl. Neurobiol.* 19: 480 (1993).
21. P.C. Barber, and D. Dahl, Glial fibrillary acidic protein (GFAP)-like immunoreactivity in normal and transected rat olfactory nerve, *Exp. Brain Res.* 65: 681 (1987).
22. R.R. Myers RR, B.L. Rydevik, H.M. Heckman, and H.C. Powel, Proximodistal gradient in endoneurial fluid pressures, *Exp. Neurology.* 102:368 (1988).

# PATHOPHYSIOLOGY AND TREATMENT OF EARLY-ONSET HYDROCEPHALUS IN A RAT MODEL

Hazel C. Jones and Neil G. Harris

Department of Pharmacology and Therapeutics
University of Florida
Gainesville, FL 32610, USA

## INTRODUCTION

Congenital hydrocephalus occurs in humans at a rate of 0.5-1.5 per 1000 births (eg. Williamson et al., 1984) and a significant proportion of cases can be attributed to aqueduct stenosis (Hirsch et al., 1986). Fetal hydrocephalus can be diagnosed as early as the second trimester by high-resolution ultrasound (e.g. Pretorius et al., 1986). Infants born with a long-standing prenatal hydrocephalus have a high risk for early death or severe handicap despite postnatal treatment (Fernell et al., 1987; Oi et al., 1990) and for those with prenatal aqueduct stenosis the outcome of postnatal shunt treatment is often poor (Hanigan et al., 1991). Little is known about the pathophysiology of intrauterine hydrocephalus, the outcome of which may be related to the time of onset and the potential for perturbations in neuronal proliferation, migration and maturation, since these are processes which take place before birth in the human (Oi et al., 1989). The usual treatment for infantile hydrocephalus is the postnatal insertion of a shunt device to remove the excess cerebrospinal fluid and it has long been known that early decompression is required for optimal prognosis and development of, at least, average intelligence (Young et al., 1973; McCullough & Balzer-Martin, 1982). A recent survey showed that the outcome is improved if surgery is carried out within two months of birth but that IQ was below average in 40% of cases (Hirsch, 1994). Furthermore, shunt-treated children with hydrocephalus that have a normal IQ, frequently have residual deficits in visuospatial skills and in verbal problem-solving ability, yet the neuronal basis for these are not known (Zeiner et al., 1985).

The H-Tx rat first described by Kohn et al. (1981), has inherited congenital hydrocephalus due to aqueduct stenosis which occurs in late gestation (Jones & Bucknall, 1988). There are no other gross abnormalities in the brain or elsewhere and the condition may be likened to X-linked aqueduct stenosis which occurs in man (e.g. Bickers & Adams, 1949). Since the rat is an altricious species and is born at a very immature stage of brain development, the equivalent of human fetal hydrocephalus can be studied postnatally in the rat. Accumulated evidence suggests, that the rat at birth is at an equivalent stage of brain development to a human fetus around mid-gestation (Uylings et al., 1990). In the rat, the early postnatal period in particular, is a critical time for dendritic and axonal growth and synapse formation (Wise et al., 1979) whereas in the human much neuronal differentiation and myelination occurs before birth (Berry, 1982; Oi et al., 1989). The rat at 10 days after birth can be considered as equivalent to early postnatal in man.

Shunt treatment has been studied in several animal models of hydrocephalus. Hydrocephalus induced in adult animals, does not result in obvious pathological changes in the cortical grey matter although the periventricular white matter is affected and shunt treatment usually reduces the size of the ventricles, but the periventricular brain regions remain abnormal, suggesting that there may be permanent axonal damage (Rubin et al., 1976; Del Bigio & Bruni, 1988). Puppies with hydrocephalus induced at 1 month after birth, also had only small changes in cortical grey matter and when shunted at 2 months of age there was restoration of the cortical mantle, which occurred largely through gliosis, and myelin repair of the white matter (Yamada et al., 1992). Hydrocephalus in infant animals with an earlier onset, results in more extreme ventriculomegaly with damage to the cortical grey matter and if untreated, this would be expected to lead to severe neuronal deficits and a poor prognosis. However, the potential for recovery may be greater in infant animals when the normal maturational processes are ongoing, remaining plasticity can be anticipated, and there is reason to believe that alleviation with early shunt treatment will be effective despite this severity. The exact stage at which shunt treatment should be carried out to give maximum effect, has yet to be determined. In kittens with early postnatal kaolin-induced hydrocephalus and grey matter disruption, successful shunt placement resulted in an improvement in behaviour and near normal ventricles (McAllister et al., 1991). Nevertheless, light microscopy showed only a partial restoration of the cortex and its laminar structure with many but not all, of the pathological features of neurones restored (Hale et al., 1992). Golgi and electron microscopy studies, also showed some residual abnormalities after shunting (Kriebel et al., 1993) with only a partial restoration of synaptic density in kittens shunted 1 week after induction (McAllister et al., 1993b). However, glucose utilisation in the kitten model was abnormal only in cerebral white matter and this was restored by shunt treatment although myelination was delayed (Chumas et al., 1994).

The H-Tx rat has severe disturbances in the cortical grey matter with thinning and disruption of the laminar structure (Jones et al., 1991). This is already maximal in the posterior cortex at 10 days which is the age at which the first increase in CSF pressure is observed (Jones & Bucknall, 1987). Local cerebral blood flow and glucose utilisation are also reduced in the cortical grey matter of the H-Tx rat (Richards et al., 1989; Jones et al., 1993). Overall the grey matter effects are more severe in this model and this may be related to the prenatal onset of hydrocephalus, whereas results for the kitten model may be influenced by the timing of induction (4-10 day postnatal) representing a rapid-onset hydrocephalus rather than a long-standing prenatal type as occurs in the H-Tx rat.

## TREATMENT OF HYDROCEPHALUS IN THE H-TX RAT

In order to test the effect of treatment timing and type on the outcome of hydrocephalus in the H-Tx rat we have carried out ventriculosubcutaneous shunt operations at two ages, 3-6 days after birth and at 8-12 days. These ages were selected to correspond with the equivalent of prenatal and immediately postnatal ages, respectively, in the human. The technique used was similar to that described previously (Harris et al., 1994b). We have also carried out pharmacological treatment on a separate group of rats using daily doses of a carbonic anhydrase inhibitor, methazolamide. Carbonic anhydrase inhibitors reduce the rate of cerebrospinal fluid (CSF) secretion (Davson & Pollay, 1963; Vogh, 1980), reduce the CSF pressure (Knopp et al., 1957; Birzis et al., 1958) and have been used in the treatment of hydrocephalic infants (Huttenlocher, 1965; Shinnar et al., 1985). We used daily subcutaneous doses of methazolamide, 100mg/kg, starting at postnatal day 2 or 3. Ventricle decompression and cortical thickness was monitored in all groups by magnetic resonance imaging (MRI) carried out, in some rats before treatment and at regular intervals after treatment up to 21 days after birth, and compared to age-matched untreated hydrocephalic rats.

## Magnetic Resonance Imaging

Imaging was carried out using light halothane anesthesia in either a 2.0 Tesla or a 4.7 Tesla 30cm bore magnet using an inversion-recovery or spin-echo sequence, the latter sequence was adopted because it reduced the time required to acquire a full set of multislice images needed for quantitative analysis. Contiguous multislice coronal images at a preset thickness between 1.1 and 2.0 mm depending on the size of the rat, were analysed using Dispimage (University of London) to determine the volume of the lateral ventricles and the thickness of the cortical mantle, which was measured in four cortical regions; frontal, sensory motor, auditory and visual cortex (Harris et al.,1992). The ventricles in control rats were difficult to image because of their small size and instead, volumes were calculated from 1mm thick coronal slices of freshly fixed brain. From this we estimate the ventricles in control rats to be no larger than 25 $mm^3$ and this is consistent with values obtained previously with paraffin wax sections (Harris et al., 1994b). The ventricles of hydrocephalic rats increased linearly with age at mean rate of 52$\mu$l/day. Immediately after birth at 1-2 days the ventricles ranged from 45-90 $mm^3$, at 3-5 days the range was 49-290 $mm^3$ and by 3 weeks the volumes for individual rats ranged from 300-500 $mm^3$ in those with less severe hydrocephalus to 1800-1900 $mm^3$ in severely affected rats (Fig. 1).

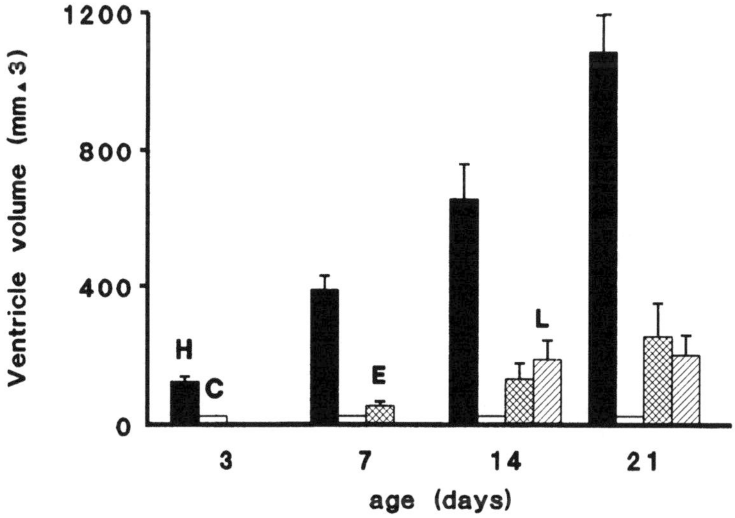


**Figure 1.** A plot of ventricle volumes measured from magnetic resonance images at four ages in H-Tx rats. Filled bars: H =unshunted hydrocephalic; open bars: C = control ventricles estimated from fixed brain slices; cross-hatched bars: E = rats shunted at 3-6 days after birth; diagonal bars: L = rats shunted at 8-12 days. Values are means +/- SEM, n=13-16.

## Ventricles and Cerebral Cortex in Treated Rats

Shunt-treated rats showed a rapid decrease in ventricle size within a few days of surgery (Fig 1.); this was evident in both early and late shunt groups and was observed in those rats that were imaged both before and after surgery. In rats not imaged before surgery a reduction in volume was inferred from a comparison with the mean values obtained for hydrocephalic rats at that age. The initial decrease in volume was greater for the late shunts, that is in rats with larger ventricles before surgery, than for the early shunt group which started with smaller ventricles, and by 14 days there was no significant difference between the two groups. The early shunt group showed a tendency to increase ventricle size after the initial decompression and this was particularly marked in two individuals which may have developed shunt obstruction. The late shunt group did not show this tendency but since the final analysis was carried out at 21 days, there was less time for an obstruction to occur in this group. At 21 days after birth, the mean volumes were still not significantly different; 255 +/-97 $mm^3$ for the early group and 201 +/-58 $mm^3$ for the late shunt group. The overall result for both groups was that at 21 days the ventricles were 20% of the untreated rats. Methazolamide treatment gave more variable results than shunt surgery (Fig. 2). In 3 pups it was not effective and the ventricles were not significantly reduced when compared to untreated hydrocephalic rats. In 5 pups the ventricles were intermediate between untreated and shunt-treated rats whereas in 7 pups it was effective and the ventricles were as small or smaller than shunted rats.

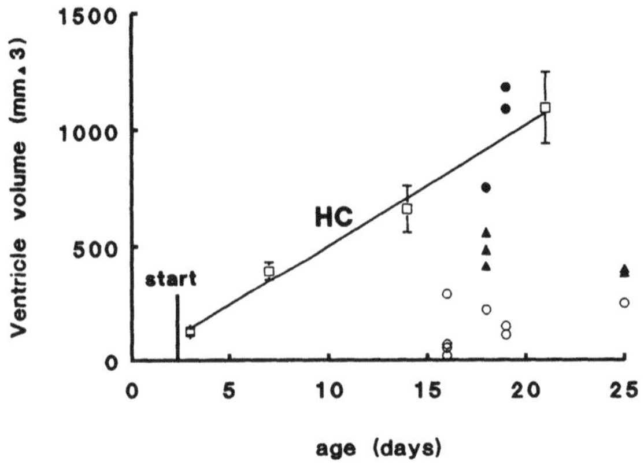


**Figure 2.** Ventricle volumes in hydrocephalic rats given daily doses of methazolamide from day 2 or 3. The solid line linked by open squares is the data from untreated rats as in Fig. 1. The ventricles in individual methazolamide-treated rats have been divided into those where treatment was as effective as shunts (n=7, open circles) those where treatment was not effective (n=3, closed circles) and those with intermediate ventricles (n=5, closed triangles).

The posterior cerebral cortex was reduced in thickness by around 50% in hydrocephalic rats imaged at 7 days, whereas shunt-treated rats had a cortical thickness approaching that of control rats with no significant differences between early and late shunt group at 14 or 21 days (Fig. 3). In methazolamide-treated rats, however, although treatment started at postnatal day 2 or 3, recovery of cortical thickness did not occur as rapidly as early shunt-treated rats in that the thickness of the mantle was close to untreated rats at 7 days but had increased to a value similar to shunted and control rats by 14 and 21 days.

## Comments

These results demonstrate that ventriculosubcutaneous shunts are effective in reducing lateral ventricle size in hydrocephalic H-Tx rat pups and that when examined at 21 days there was no difference between the ventricles of rats shunted soon after birth at 4 days compared to those treated at 11 days. However, the ventricles were not as small as in control rats and this is not an unusual observation (eg. Chumas et al., 1994). The carbonic anhydrase inhibitor, methazolamide was not effective in reducing ventricles size in all pups and there are two possible reasons for this: one is that drug doses given once per 24h may be insufficient for a maximal inhibitory effect, since plasma levels fall within 18h (Jones & Conroy unpublished observations). Another reason may be due to the fact that the inhibitor only reduces CSF secretion by 40 -60% (Vogh, 1980) and this may be insufficient to prevent further ventricular expansion in the presence of an obstruction. A previous study has shown that the resistance to drainage of CSF out of the lateral ventricles is about twice as high in hydrocephalic pups (Jones & Bucknall, 1987) and if the inhibition of secretion is not sufficient to counteract the high resistance, further expansion will take place.

We have also shown that shunt-treatment results in near normal cortical thickness soon after shunt placement, whereas the cortex takes longer to recover in inhibitor-treated rats. This delay in methazolamide-treated rats can be explained by the observation that the rat choroid plexus secretory processes may not be sufficiently mature in the first postnatal week to respond adequately to inhibitors (Johanson et al., 1992).

One question which arises from this study is, if the ventricles are still larger than control rats after treatment, does this affect the potential for full brain recovery? The answer must be that control-sized ventricles may not be necessary if the overall cranial volume remains slightly enlarged, so that a normal volume of brain tissue can be accommodated together with the extra CSF. In fact very small ventricles may adversely affect the continued function of the shunt through obstruction of the proximal end and producing the well-known 'slit-ventricle syndrome' which by itself can cause brain damage (Oi & Matsumoto, 1986).

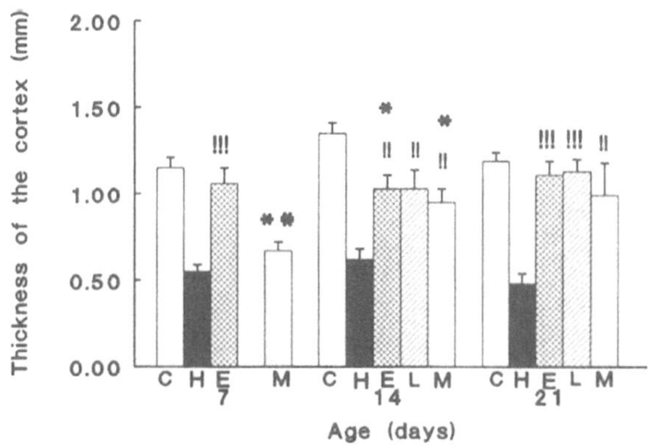


**Figure 3.** Thickness of the visual cortex from magnetic resonance images at 3 ages in control (C, open bars) hydrocephalic (H, filled bars), 3-6 day shunts (E, crosshatched bars), 8-12 day shunts (L, diagonal bars) and methazolamide-treated rats (M, dotted bars). Values are means +/-SEM. !! and !!! = significantly different from hydrocephalic (ANOVA), P< 0.01 and P<0.001 respectively; * and ** = significantly different from control, P< 0.05 and P< 0.01 respectively. The shunt-treated rats recovered a normal cortical thickness more rapidly than the methazolamide-treated rats.

## RECOVERY OF MICROSCOPICAL STRUCTURE AND FUNCTION AFTER SHUNT TREATMENT

To what extent does the cortical mantle, which is apparently restored in thickness when visualized macroscopically by NMR imaging, represent a normal tissue with full functioning potential? Follow up studies on hydrocephalic children indicate that residual effects lead to intelligence deficits despite shunt treatment before 12 months of age (Dennis et al., 1981). Experiments from another laboratory have shown that shunts at 7-9 days in H-Tx rats gives good recovery of ventricles and cortex and much better than late shunts at 4 weeks, when the outcome was low dendritic spine density and impaired learning ability (Suda et al., 1991; Miyazawa and Sato, 1991). From this and another report that cortical glucose utilisation is not restored after shunting at 4 weeks (Miyaoka et al., 1988), there does not seem to be much doubt that 4 weeks is too late for good recovery. This might be expected because hydrocephalus is rapidly progressive in H-Tx rats and death occurs between 4 and 7 weeks with a few deaths occurring between 3 and 4 weeks (Wada, 1988; Jones & Bucknall, unpublished observations). Furthermore, the rapid brain maturational processes are complete by 4 weeks in the rat. We have previously shown by histology that in H-Tx rats shunted at various ages ranging from 4 to 17 days and examined at 14 or 21 days, the posterior cortex has residual thinning and none of the rats had complete recovery of the cortical laminar structure (Harris et al., 1994b).

### Pyramidal Cell Structure

Pyramidal cells are the predominant neurone in the cerebral cortex and receive many different synaptic inputs. They are large cells with a characteristic architecture and an axon which forms the sole efferent pathway from the cortex. The branched basal dendrites radiate out individually from the soma and the single apical dendrite branches as it becomes more distant from the soma. In a study on infant rats with hydrocephalus induced with kaolin at day 1 and examined with Golgi staining at days 11-13, there were changes in the architecture of the pyramidal cell basal dendrites and changes in the spine density and morphology (McAllister et al., 1985). We have examined the structure of layer V pyramidal cells in the auditory cortex of H-Tx rats at 21 days using Golgi staining and computerised analysis of individual neurones (Harris et al., 1993). Cells from

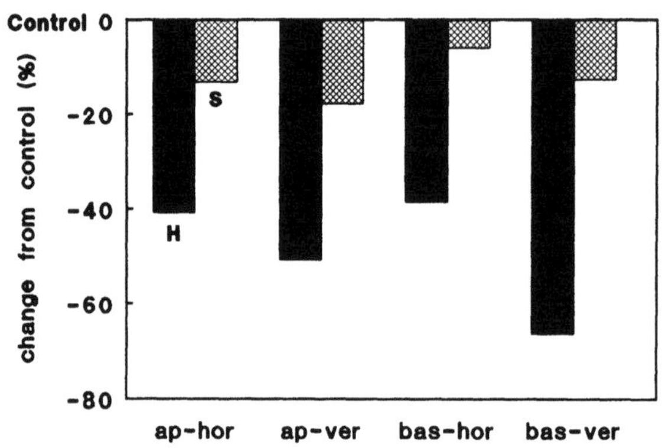


**Figure 4.** Percentage change from control rats in layer V pyramidal cells from the auditory cortex in hydrocephalic rats (H) and a group of early shunts (S); ap-hor =horizontal extent of apical dendrites; ap-ver = vertical extent; bas-hor = horizontal extent of basal dendrites; bas-ver = vertical extent. Values are means, n=3-4 rats and 8-15 cells.

hydrocephalic rats had large reductions in the horizontal and vertical extent and in the number of branches of both apical and basal dendrites (Fig.4). Also the angle of the shaft of the main apical dendrite which normally extends vertically to the pia, was much more variable in hydrocephalic rats. In rats which received shunt treatment at 3 days after birth there was considerable normalisation of pyramidal cell architecture and orientation but not up to levels measured for control rats.

Dendritic spines, an indicator of synaptic density, were reduced by 35% on the apical shaft in hydrocephalics and shunt treatment restored spine density to within 9% of controls (McAllister et al., 1993a) This suggests that even early shunt treatment may not be sufficient to normalise the cortical dendritic organisation in this model of hydrocephalus. However, we do not know whether a longer interval between shunt treatment and examination would have resulted in a more complete recovery. In another study on H-Tx rats, shunts were carried out at 7-9 days and layer II and III pyramidal cells from the frontoparietal cortex examined at 3 months of age were found to have normal dendritic spine density, total length and branch patterns (Suda et al.,1991; Suda et al., 1994b). However, we examined the deeper layer V pyramidal cells which are closer to the ventricle and from the more-severely thinned posterior cortex, thus making a direct comparison between the two studies difficult. Furthermore, Suda et al. (1994a) reported that there may be abnormal synaptic cytoskeletal proteins in the H-Tx rat which could affect shunt recovery. In kittens made hydrocephalic by kaolin injection at 10 days after birth, pyramidal cells from the visual cortex showed significant reductions in dendrite length, numbers of branches and spine density and shunt treatment (early and late) gave good recovery of dendritic architecture but left residual deficits in spine density (McAllister et al., 1993b) and in spine configuration (Kriebel at al., 1993)

A more detailed structural analysis on the effects of hydrocephalus and shunt treatment in the H-Tx rat has been obtained by electron microscopy of the deep layers (V and VI) of the frontal and visual cortex at 21 days of age (Boillat et al., 1993). The hydrocephalic rats had tissue which appeared to be oedematous due to the presence of swollen neurites and this was particularly marked in the more severely affected visual cortex. This is consistent with another report (Miyazawa & Sato, 1991) which described swollen dendrites and immature synapses in layer VI neuropil in hydrocephalic rats at 2 weeks of age. In our study, the pyramidal cells had watery cytoplasm with swollen mitochondria, expanded endoplasmic reticulum, dispersed ribosomes and multivesicular bodies. Rats which had received shunts at 4 days after birth had neuropil which was more compact and homogeneous although there were some residual swollen neurites with infrequent and apparently immature synapses. The pyramidal cells, however, had cytoplasm which appeared metabolically active, with many Golgi complexes, mitochondria rosettes of ribosomes and microtubules. Also present were centrioles which are suggestive for microtubule synthesis. Late shunts at 10 days did not show as much normalisation as the early shunts (Jones et al., 1994). Hence we can conclude that there is clear evidence of neuronal regeneration after shunt treatment but that normal synaptic connectivity may only be achieved after a longer time interval, or possibly may not be achieved at all. Electron microscopy of the cerebral cortex has been described in the kitten model for infantile hydrocephalus (Kriebel et al., 1993) in which hydrocephalus was induced by kaolin injection at 4 -10 days after birth and shunts were placed 10 days later. The deep cortical layers showed similar but more extreme changes to those in the H-Tx rat with an absence of synaptic connections. In shunted kittens, although the thickness of the cortical mantle was restored, the dendritic ultrastructure and the dendritic spines in particular, were still abnormal 15 days after treatment. It is uncertain whether a longer recovery period in the kitten would have resulted in a more complete normalisation of structure.

## FUNCTIONAL AND METABOLIC CHANGES IN HYDROCEPHALUS AND RECOVERY WITH TREATMENT

A potentially deleterious reduction in cerebral blood flow (CBF) has been recorded in human adult and infantile hydrocephalus with improvement after shunting (eg. Brooks et al., 1986, Huang & Chio, 1991, Goh et al., 1991). Hydrocephalic infants have episodes of raised CSF pressure which may cause ischemia and animal models of hydrocephalus also show low CBF (eg. Higashi et al., 1986). We have shown by autoradiography that hydrocephalic H-Tx rats have reductions in CBF which are not significant at 10 days but are significantly reduced by 21 and 30 days (Fig. 5). This was especially marked in the cerebral cortex where the low flow was suggestive of ischemia (Jones et al., 1993). Although the cause of low CBF in hydrocephalus is unknown, in the H-Tx rat the onset coincides with the onset of raised CSF pressure (Jones & Bucknall, 1987, Kaiser & Jones, 1991). It seems unlikely however, that changes in CBF are a primary cause of neuronal damage in the early-onset hydrocephalus in the H-Tx rat, since ventriculomegaly occurs before the abnormal blood flow or raised ICP are detected. Nevertheless, it might be important for the timing of treatment because delay until after increases in pressure and decreases in CBF have occurred may reduce the potential for full recovery. We have also studied the uptake of [$^{14}$C]deoxyglucose into brain as an indicator for glucose utilisation by quantitative autoradiography (Richards et al., 1989). We found that deoxyglucose uptake is significantly reduced by 30 days in hydrocephalic rats with the cortical regions most affected by the hydrocephalus showing the largest reduction of 33% (Fig. 5). Although some reductions were seen at 21 days they were not significant and overall, the effects were smaller than for CBF. We do not know if these effects are reversible by shunt treatment, but another study showed that shunts at 4 weeks in H-Tx rats do not restore glucose utilisation (Miyaoka et al., 1988).

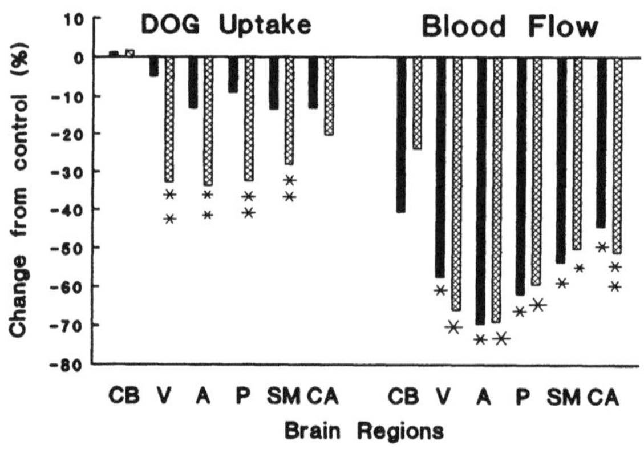


**Figure 5.** Percentage change from control rats in glucose utilisation using $^{14}$C-deoxyglucose (DOG) uptake and in local cerebral blood flow as measured with $^{14}$-C iodoantipyrine autoradiography in six brain regions at two postnatal ages, 21 days (filled bars) and 30 days (cross hatched bars). CB = cerebellar cortex; V =visual cortex; A =auditory cortex; P =parietal cortex; SM =sensory motor cortex; CA = caudate nucleus. * and ** = significantly different from control (ANOVA), $P< 0.05$ and $P< 0.01$ respectively. Reductions in cortical blood flow were close to 70%. (Data from Richards et al., 1989; Jones et al., 1993).

Increasing the CSF pressure in cats or dogs results in a decrease in cerebral perfusion pressure and an abrupt decrease in oxidative metabolism due to a reduced supply of $O_2$ and glucose (Zwetnow, 1970; Sutton et al., 1987). Hence it is likely that raised CSF pressure in hydrocephalus leads to disturbances in metabolism, although this has not been well documented. In hydrocephalic infants the CSF lactate/pyruvate ratio was elevated at low cerebral perfusion pressures and normalized after shunting (Raisis et al., 1975). In experimental hydrocephalus in animals, increases in brain lactate and decreases in phospholipids, cholesterol, ATP and cyclic AMP were recorded together with a 50% decrease in CBF (Higashi et al., 1986). NMR spectroscopy is a useful tool to assess metabolic status and neuronal viability non-invasively using phosphorus NMR for evaluation of oxidative phosphorylation potential and proton NMR for lactate and N-acetylaspartate (NAA). Hydrocephalic dogs were found by NMR spectroscopy to have a decrease in phosphocreatine /inorganic phosphorus ratio (PCr/Pi) but no change in ATP, and acute withdrawal of CSF did not reverse the situation, suggesting that raised pressure was not directly responsible for a decrease in phosphorylation potential (Tamaki et al., 1990). Hydrocephalic H-Tx rats at 20 days after birth in a study from another laboratory showed Pcr/Pi ratios which decreased progressively with the severity of the hydrocephalus and partly recovered after shunt treatment (Minamikawa, 1992). In contrast, we have carried out *in vivo* $^{31}P$ spectroscopy in H-Tx rats at 21 days and did not find any significant differences between normal and hydrocephalic rats (Jones & Harris, unpublished observations). We believe the absence of any changes in the hydrocephalic group may be due to technical problems encountered in obtaining comparable spectra with a non-localising pulse sequence since the geometry of the cerebral cortex changes in hydrocephalus.

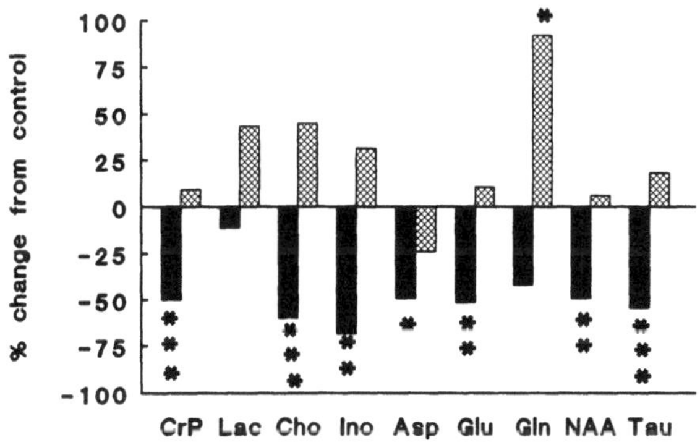


**Figure 6.** Percentage change from control rats (n=5) in cortical metabolites measured by *in vitro* NMR spectroscopy in hydrocephalic (n =4, filled bars) and rats with early shunts (n =3, cross-hatched bars). CrP =creatine and phosphocreatine; Lac = lactate; Cho = glycerophosphoryl-choline and phosphoryl-choline; Ino = inositol; Asp = aspartate; Glu =glutamate; Gln = glutamine; NAA = N-acetyl aspartate; Tau = taurine. *, ** and *** = significantly different from control rats (ANOVA), $P < 0.05$, 0.01 and 0.001 respectively.

NMR spectroscopy of tissue extracts has the advantage over *in vivo* spectroscopy in that insoluble compounds are removed in the sample preparation and extracts are examined in a high resolution spectrometer giving better peak separation. We have prepared perchloric acid extracts of cerebral cortex in H-Tx rats at 21 days using the funnel freezing technique (Ponten et al., 1973) and examined them by 1-dimensional $^{1}H$ NMR spectroscopy at 600MHz in a Varian Unity spectrometer (Harris et al., 1994a and unpublished observations). Many of the compounds resolved showed significant reductions in concentration in the cortex of hydrocephalic rats compared to controls (Fig. 6): creatine and phosphocreatine; glycerophosphorylcholine and choline, indicative of membrane synthesis; inositol, important for intracellular signalling; the excitatory amino acids, aspartate, glutamate and taurine, and N-acetyl aspartate, a neuronal marker. The inhibitory amino acids, glycine and GABA were not significantly affected and to our surprise lactate levels were normal. However, analysis of CSF samples from hydrocephalic rats has shown that lactate is elevated in the CSF (2.82 mmol/l compared to 1.9 mmol/kg in cerebral cortex) and the CSF lactate/pyruvate ratio is high (Harris & Jones unpublished observations). This indicates that there is impaired cerebral oxidative metabolism and that the CSF may be acting as sink for lactate produced in the brain. High CSF lactate is consistent with observations on CSF taken from human infants with hydrocephalus (Granholm & Siesjo, 1970; Raisis et al., 1975; Bejar et al., 1983). For substances which increase with age in rat brain, e.g. NAA, phosphorylcholine, aspartate and glutamate (Miyake & Kakimoto, 1981; Koller & Coyle, 1984; Parnavelas et al., 1988; Bates et al., 1989; Burri et al., 1990; Pettegrew et al., 1990), a reduced level in hydrocephalic rats could be interpreted as a failure of normal developmental processes. However for substances that decrease with age e.g. taurine (Bates et al., 1989; Burri et al., 1990) this argument no longer applies. In rats with early shunts none of these compounds were significantly reduced from control, instead most metabolites measured were similar to control rats, with the exception of glutamine which was significantly increased over control (Fig. 6).

In conclusion, the H-Tx rat is a model for early-onset fetal hydrocephalus in which the ventriculomegaly may affect maturation of the neuronal elements in the cortical and subcortical structures. Treatment at 3-5 days, equivalent to third trimester human fetus, and also at 10-12 days, equivalent to a newborn infant, gives good recovery macroscopically, when the brain is examined at 21 days. However, a more detailed examination suggests that cellular and subcellular organisation is not completely normal, although some metabolite levels have recovered. Electron microscopic evidence of regenerative processes suggests that given more time, complete normalisation may be achieved.

## ACKNOWLEDGEMENT

This work was supported by a Johnson & Johnson Focused Giving Award.

## REFERENCES

Bates, T.E., Williams, S.R., Gadian, D.G., Bell, J.D., Small, R.K. and Iles, R.A., 1989, $^1$H NMR study of cerebral development in the rat, *NMR Biomed.* 2:225.

Bejar, R., Saugstad, O.D., James, H. and Gluck, L., 1983, Increased hypoxanthine concentrations in cerebrospinal fluid of infants with hydrocephalus, *J. Pediatr.* 103:44.

Berry, M., 1982, The development of the human nervous system. *in*: "Brain and Behavioural Development," J.W.T. Dickerson and H. McGurk. ed. Surrey University Press, Glasgow.

Bickers, D.S. and Adams, R.D., 1949, Hereditary stenosis of the aqueduct of Sylvius as a cause of congenital hydrocephalus, *Brain.* 72:246.

Birzis, L., Carter, C.H. and Maren, T.H., 1958, The effect of acetazolamide on CSF pressure and electrolytes in hydrocephalus, *Neurology.* 8:522.

Boillat, C.A., Harris, N.G., Kaiser, G.L. and Jones, H.C., 1993, Electron microscopy of the cerebral cortex in control, hydrocephalic and shunted H-Tx rats, *Eur. J. Pediatr. Surg.* 3 suppl I:30.

Brooks, D.J., Beaney, R.P., Powell, M., Leenders, K., Crockard, H.A., Thomas, D.G.T., Marshall, J. and Jones, T., 1986, Studies on cerebral oxygen metabolism, blood flow, and blood volume, in patients with hydrocephalus before and after surgical decompression using positron emission tomography, *Brain.* 109:613.

Burri, R., Bigler, P., Straehl, P., Posse, S., Colombo, J.-P. and Herschkowitz, N., 1990, Brain development: $^{1}H$ magnetic resonance spectroscopy of rat brain extracts compared with chromatographic methods, *Neurochem. Res.* 15:1009.

Chumas, P.D., Drake, J.M., Del Bigio, M.R., Da Silva, M. and Tuor, U.I., 1994, Anaerobic glycolysis preceding white-matter destruction in experimental neonatal hydrocephalus, *J. Neurosurg.* 80:491.

Davson, H. and Pollay, M., 1963, Influence of various drugs on the transport of 131-I and PAH across the cerebrospinal fluid-blood barrier, *J. Physiol.* 167:239.

Del Bigio, M.R. and Bruni, J.E., 1988, Periventricular pathology in hydrocephalic rabbits before and after shunting, *Acta Neuropathol.* 77:186.

Dennis, M., Fitz, C.R., Netley, C.T., Sugar, J., Harwood-Nash, D.C.F., Hendrick, B., Hoffman, H.J. and Humphreys, R.P., 1981, The intelligence of hydrocephalic children, *Arch. Neurol.* 38:607.

Fernell, E., Uvebrant, P. and Wendt, L.von, 1987, Overt hydrocephalus at birth - origin and outcome, *Child's Nerv. Syst.* 3:350.

Goh, D., Minns, R.A., Pye, S.D. and Steers, A.J.W., 1991, Cerebral blood flow velocity changes after ventricular taps and ventriculoperitoneal shunting, *Child's Nerv. Syst.* 7:452.

Granholm, L. and Siesjo, B.K., 1970, Signs of tissue hypoxia in infantile hydrocephalus, *Dev. Med. Child Neurol.* 12 suppl 22:73.

Hale, P.M., McAllister, J.P., Katz, S.D., Wright, L.C., Lovely, T.J., Miller, D.W., Wolfson, B.J., Salotto, A.G. and Shroff, D.V., 1992, Improvement of cortical morphology in infantile hydrocephalic animals after ventriculoperitoneal shunt placement, *Neurosurgery.* 31:1085.

Hanigan, W.C., Morgan, A., Shaaban, A. and Bradle, P., 1991, Surgical treatment and long-term neurodevelopmental outcome for infants with idiopathic aqueductal stenosis, *Child's Nerv. Syst.* 7:386.

Harris, N.G., Inglis, B.A., Jones, H.C. and Briggs, R.W., 1994a, High-resolution $^{1}H$-NMR analysis of cerebral cortex from control and hydrocephalic H-Tx rats, Abstract, *Soc. Neurosci.* 20:1237.

Harris, N.G., Jones, H.C. and Patel, S., 1994b, Ventricle shunting in young H-Tx rats with inherited congenital hydrocephalus: a quantitative histological study of cortical grey matter, *Child's Nerv. Syst.* 10:293.

Harris, N.G., Jones, H.C. and Williams, S.C.R., 1992, MR imaging for measurements of ventricles and cerebral cortex in postnatal rats (H-Tx strain) with progressive inherited hydrocephalus, *Exp. Neurol.* 118:1.

Harris, N.G., McAllister, J.P. and Jones, H.C., 1993, The effect of untreated and shunt-treated hydrocephalus on cortical pyramidal neurone morphology in the H-Tx rat, *Eur. J. Pediatr. Surg.* 3 suppl I:31.

Higashi, K., Asahisa, H., Ueda, N., Kobayashi, K., Hara, K. and Noda, Y., 1986, Cerebral blood flow and metabolism in experimental hydrocephalus, *Neurol. Res.* 8:169.

Hirsch, J-F., 1994, Concensus: long-term outcome in hydrocephalus, *Child's Nerv. Syst.* 10:64.

Hirsch, J-F., Hirsch, E., Sainte Rose, C., Renier, D. and Pierre-Khan, A., 1986, Stenosis of the aqueduct of Sylvius: etiology and treatment, *J. Neurosurg. Sci.* 30:29.

Huang, C-C. and Chio, C-C., 1991, Duplex color ultrasound study of infantile progressive ventriculomegaly, *Child's Nerv. Syst.* 7:250.

Huttenlocher, P.R., 1965, Treatment of hydrocephalus with acetazolamide, *J. Pediatr.* 66:1023.

Johanson, C.E., Parandoosh, Z. and Dyas, M.L. 1992, Maturational differences in acetazolamide-altered pH and $HCO_3$ of choroid plexus, cerebrospinal fluid and brain, *Am. J. Physiol.* 262:R909.

Jones, H.C. and Bucknall, R.M., 1987, Changes in cerebrospinal fluid pressure and outflow from the lateral ventricles during development of congenital hydrocephalus in the H-Tx rat, *Exp. Neurol.* 98:573.

Jones, H.C. and Bucknall, R.M., 1988, Inherited prenatal hydrocephalus in the H-Tx rat: a morphological study, *Neuropath. Appl. Neurobiol.* 14:263.

Jones, H.C., Bucknall, R.M. and Harris, N.G., 1991, The cerebral cortex in congenital hydrocephalus in the H-Tx rat: a quantitative light microscopy study, *Acta Neuropathol.* 82:217.

Jones, H.C., Richards, H.K., Bucknall, R.M. and Pickard, J.D., 1993, Local cerebral blood flow in rats with congenital hydrocephalus, *J. Cereb. Blood Flow Metab.* 13:531.

Jones, H.C., Boillat, C.A., Harris, N.G., Kaiser, G.L., 1994, Electron microscopy of the cerebral cortex in young hydrocephalic and shunt-treated H-Tx rats, Abstract, *Society for Neuroscience Annual Meeting.*

Kaiser, G. and Jones, H.C., 1991, Cerebrospinal fluid pressure in 10-day old rats with congenital hydrocephalus, *Eur. J. Pediatr. Surg.* 1 suppl I:20.

Knopp, L.M., Atkinson, J.R. and Ward, A.A., 1957, Effect of diamox on cerebrospinal fluid pressure of cat and monkey, *Neurology.* 7:119.

Kohn, D.F., Chinookoswong, N. and Chou, S.M., 1981, A new model of congenital hydrocephalus in the rat, *Acta Neuropathol.* 54:211.
Koller, K.J. and Coyle, J.T., 1984, Ontogenesis of N-acetyl-aspartate and N-acetyl-aspartyl-glutamate in rat brain, *Dev. Brain Res.* 15:137.
Kriebel, R.M., Shah, A.B. and McAllister, J.P., 1993, The microstructure of cortical neuropil before and after decompression in experimental infantile hydrocephalus, *Exp. Neurol.* 119:89.
McAllister, J.P., Maughans, T.A., Shah, M.V. and Truex, R.C., 1985, Neuronal effects of experimentally induced hydrocephalus in newborn rats, *J. Neurosurg.* 63:776.
McAllister, J.P., Cohen, M.I., O'Mara, K.A. and Johnson, M.H., 1991, Progression of experimental infantile hydrocephalus and effects of ventriculoperitoneal shunts: an analysis correlating magnetic resonance imaging with gross morphology, *Neurosurgery.* 29:329.
McAllister, J.P., Harris, N.G., Boonswang, A., Kubit, J.D. and Jones, H.C., 1993a, Effects of surgical decompression on cortical dendrites in a model of congenital hydrocephalus, *American Association of Neurological Surgeons.*, San Diego.
McAllister, J.P., Morano, R.A., Amin, M. and Kriebel, R.M., Ireland, R.A., and Hale, P.M., 1993b, Synaptic and dendritic plasticity following early and late decompression for infantile hydrocephalus, *Eur. J. Pediatr. Surg.* 3 suppl I:39.
McCullough, D.C. and Balzer-Martin, L.A., 1982, Current prognosis in overt neonatal hydrocephalus, *J. Neurosurg.* 57:378.
Minamikawa, J., 1992, Experimental study on pathophysiology and treatment of congenital hydrocephalus evaluated by magnetic resonance imaging and magnetic resonance spectroscopy, *Arch. Jpn. Chir.* 61:35.
Miyake, M. and Kakimoto, Y., 1981, Developmental changes of N-acetyl-aspartic acid, N-acetyl-a-aspartylglutamic acid and b-citryl-L-glutamic acid in different brain regions and spinal cord of rat and guinea pig, *J. Neurochem.* 37:1064.
Miyaoka, M., Ito, M., Wada, M., Sato, K. and Ishii, S., 1988, Measurement of local cerebral glucose utilisation before and after V-P shunt in congenital hydrocephalus in rats, *Met. Brain Dis.* 3:125.
Miyazawa, T. and Sato, K., 1991, Learning disability and impairment of synaptogenesis in HTX rats with arrested shunt-dependant hydrocephalus, *Child's Nerv. Syst.* 7:121.
Oi, S., Ijichi, A. and Matsumoto, S., 1989, Immunohistochemical evaluation of neuronal maturation in untreated fetal hydrocephalus, *Neurol. Med. Chir.* 29:989.
Oi, S. and Matsumoto, S., 1986, Morphological findings of postshunt slit-ventricle in experimental hydrocephalus, *Child's Nerv. Syst.* 2:179.
Oi, S., Matsumoto, S., Katayama, K. and Mochizuki, M., 1990, Pathophysiology and postnatal outcome of fetal hydrocephalus, *Child's Nerv. Syst.* 6:338.
Parnavelas, J.G., Papadopoulos, G.C. and Cavanagh, M.E., 1988, Changes in neurotransmitters during development. *in*: "Cerebral Cortex," Vol 7, A. Peters and E.G. Jones. ed. Plenum, New York.
Pettegrew, J.W., Panchalingam, K., Withers, G., McKeag, D. and Strychor, S., 1990, Changes in brain energy and phospholipid metabolism during development and aging in the Fischer 344 rat, *J. Neuropath. Exp. Neurol.* 49:237.
Ponten, U., Ratcheson, R.A., Salford, L.G. and Siesjo, B.K., 1973, Optimal freezing conditions for cerebral metabolites in rats, *J. Neurochem.* 21:1127.
Pretorius, D.H., Drose, J.A. and Manco-Johnson, M.L., 1986, Fetal lateral ventricular ratio determination during the second trimester, *J. Ultrasound Med.* 5:121.
Raisis, J.E., Kindt, G.W., Miller, C.A. and McGillicuddy, J.E., 1975, Correlation of ICP with CSF lactate and lactate/pyruvate ratios in hydrocephalus. *in*: "Intracranial Pressure II," N. Lindberg, H. Ponten and M. Brock. ed. Springer-Verlag, Heidelberg.
Richards, H.K., Bucknall, R.M., Jones, H.C. and Pickard, J.D., 1989, The uptake of [$^{14}$C]deoxyglucose into brain of young rats with inherited hydrocephalus, *Exp. Neurol.* 103:194.
Rubin, R.C., Hochwald, G.M., Tiell, M., Epstein, F., Ghatak, N. and Wisniewski, H., 1976, Hydrocephalus; III Reconstruction of the cerebral cortical mantle following ventricular shunting, *Surg. Neurol.* 5:179.
Shinnar, S., Gammon, K., Bergman, E.W., Epstein, M. and Freeman, J.M., 1985, Management of hydrocephalus in infancy: use of acetazolamide and furosemide to avoid cerebrospinal fluid shunts, *J. Pediatr.* 107:31.
Suda, K., Sato, K., Miyazawa, T. and Arai, H., 1994a, Changes of synapse-related proteins (SVP-8 and drebrins) during development of brain in congenitally hydrocephalic HTX rats with and without early placement of ventriculoperitoneal shunt, *Pediatr. Neurosurg.* 20:50.

Suda, K., Sato, K., Takeda, N., Miyazawa, T. and Arai, H., 1994b, Early ventriculoperitoneal shunt-effects on learning ability and synaptogenesis of the brain in congenitally hydrocephalic H-Tx rats, *Child's Nerv. Syst.* 10:19.

Suda, K., Sato, K., Takeda, N., Wada, M., Miyazawa, T., Arai, H., Ito, M. and Miyaoka, M., 1991, Early Vs delayed ventriculoperitoneal shunt-effects on the impairment of the developing brain in congenitally hydrocephalic H-Tx rats. *in*: "Hydrocephalus Pathogenesis and Treatment," S. Matsumoto and N. Tamaki. ed. Springer Verlag, Tokyo.

Sutton, L.N., McLaughlin, A.C., Kemp, W., Schnall, M.D., Cho, B-K., Langfitt, T.W. and Chance, B., 1987, Effects of increased ICP on brain phosphocreatine and lactate determined by simultaneous $^{1}H$ and $^{31}P$ NMR spectroscopy. *J. Neurosurg.* 67:381.

Tamaki, N., Yasuda, M., Matsumoto, S., Yamamoto, T. and Iriguchi, N., 1990, Cerebral energy metabolism in experimental canine hydrocephalus, *Child's Nerv. Syst.* 6:172.

Uylings, H.B.M., Corbert, G.Van E., Parnavelas, J.G. and Kalsbeek, A., 1990, The prenatal and postnatal development of the rat cerebal cortex. *in*: "The Cerebral Cortex of the Rat," B. Kolb and R.C. Tees. ed. MIT, Cambridge,MA.

Vogh, B.P., 1980, The relation of choroid plexus carbonic anhydrase activity to cerebrospinal fluid formation: study of three inhibitors in cat with extrapolation to man, *J. Pharm. Exp. Ther.* 213:321.

Wada, M., 1988, Congenital hydrocephalus in HTX-rats: incidence, pathophysiology and developmental impairment, *Neurol. Med. Chir.* 28:955.

Williamson, R.A., Shauberger, C.W., Varner, M.W. and Aschenbrener, C.A., 1984, Heterogeneity of prenatal onset hydrocephalus: management and counseling implications, *Am. J. Med. Gen.* 17:497.

Wise, S.P., Fleshman, J.W. and Jones, E.G., 1979, Maturation of pyramidal cell form in relation to developing afferent and efferent connections of rat somatic sensory cortex, *Neuroscience.* 4:1275.

Yamada, H., Yokota, A., Furuta, A. and Horie, A., 1992, Reconstitution of shunted mantle in experimental hydrocephalus, *J. Neurosurg.* 76:856.

Young, H.F., Nulsen, F.E., Weiss, M.H. and Thomas, P., 1973, The relationship of intelligence and cerebral mantle in treated infantile hydrocephalus, *Pediatrics.* 52:38.

Zeiner, H.K., Prigatano, G.P., Pollay, M., Biscoe, C.B. and Smith, R.V., 1985, Ocular motility, visual activity and dysfunction of neuropsychological impairment in children with shunt uncomplicated hydrocephalus, *Child's Nerv. Syst.* 1:115.

Zwetnow, N.N., 1970, The influence of an increased intracranial pressure on lactate, pyruvate, bicarbonate, phosphocreatine, ATP, ADP and AMP concentration of the cerebral cortex of dogs, *Acta Physiol. Scand.* 79:158.

# THE FILTRATION CONCEPT OF THE BLOOD-CSF-BARRIER AS BASIS FOR THE DIFFERENTIATION OF CSF PROTEINS

K. Felgenhauer

Department for Neurology
Georg-August-Universität
Göttingen
Germany

## FILTRATION AND DIFFUSION AT THE BLOOD-CSF-BARRIER

The clear correlation between the molecular radius of plasma proteins and their decrease in concentration from blood to CSF was established twenty years ago (Felgenhauer, 1974) and was generally accepted as indicative of filtration and diffusion as the predominant transfer mode at the blood-CSF-barrier. The extension of this correlation however into the small molecular range down to chloride and urea appeared to be feasible, but was hard to accomplish because the serum peptides are quickly removed by the kidney and the CSF level of compounds with radii below 6 Angstroms is prone to modification by active transport, liphophilic passage and postfiltrational uptake.

Steady state equilibrium CSF/serum concentration ratios of numerous electrolytes, metabolites and amino acids depended - like proteins - upon hydrodynamic size (Felgenhauer et al., 1982), but the interpretation of this finding remained controversial. It is remarkable, that those components, whose CSF concentrations are most tightly controlled (Wright and Loo, 1994) have the smallest hydrodynamic size: chloride, potassium and sodium. On the other hand it should be stressed, that all values were obtained from lumbar CSF and therefore are not necessarily identical with those from the ventricles.

The scatter around the line for the small molecular components became smaller with increasing impairment of the blood-CSF-barrier and some normally "disobedient" CSF constituents approach diffusion determined levels such as glycine, alanine, glucose and glutamine. If the lines for the small and for the high molecular range are plotted together for increasing barrier impairments, the selectivity loss is much more evident in the small molecular range (see Figs. 1 and 3 of Felgenhauer et al., 1982).

Using a widely different experimental approach Nau et al. (1993) recently found comparable ratios. When, for instance, the osmotherapeutic agent glycerol was infused into patients with a ventricular drain and the concentration profiles in blood and CSF were compaired, the serum/CSF ratio of the area under the curve was not far from the earlier

*New Concepts of a Blood–Brain Barrier*
Edited by J. Greenwood *et al.*, Plenum Press, New York, 1995

steady state values. This was also true for other hydrophilic solutes like uric acid (Reiber et al., 1993) and methotrexate which supports the concept, that even aminoacids may reach the free CSF space predominantly by plain filtration such as serine, threonine, ornithine, methionine and lysine.

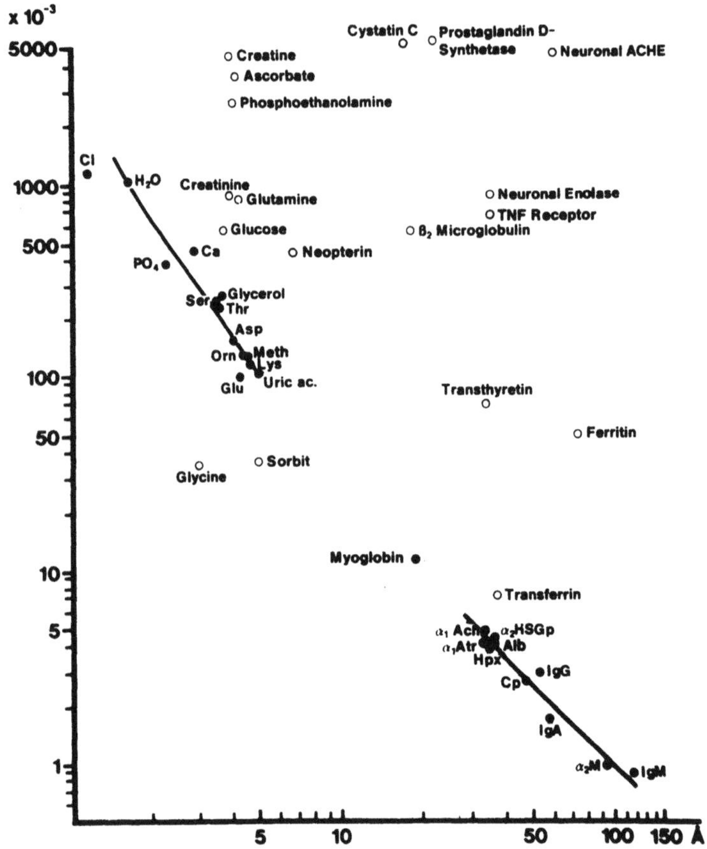


**Figure 1.** Correlation between hydrodynamic molecular radius (Å) and steady state CSF/serum concentration ratio for passively transferred, hydrophilic plasma compounds (full circles) and numerous "disobedient" CSF (e.g. sorbit) constituents (open circles). Unexpected high levels are indicative of active transport (e.g. glucose, ascorbate) or local synthesis (e.g. transthyretin in the choroid plexus, enolase in neurons). Unexpected low levels are indicative of secondary uptake (e.g. glycine) or metabolization (e.g. sorbit). The values have been collected over twenty years from many sources.

## FILTRATION AS UNIVERSAL TRANSFER MODE FOR SMALL AND LARGE HYDROPHILIC SERUM COMPONENTS

If the plot of hydrodynamic radius versus CSF/serum concentration ratio (Fig. 1) is considered, the evidence is rather strong, that both parts of the curve belong together. J. Gerhards (personal communication) has recently found one protein of appropriate size, that partially bridges the gap, namely myoglobin, thus supporting the concept of a general filtration line.

Uric acid with its extremely small day to day variation likewise follows the "law of CSF formation" and may be used as a suitable permeability marker for the small molecular range as the albumin for the high molecular range. With a normal intact barrier, the permeability of glucose, glutamine and neopterin have unexpected high levels.

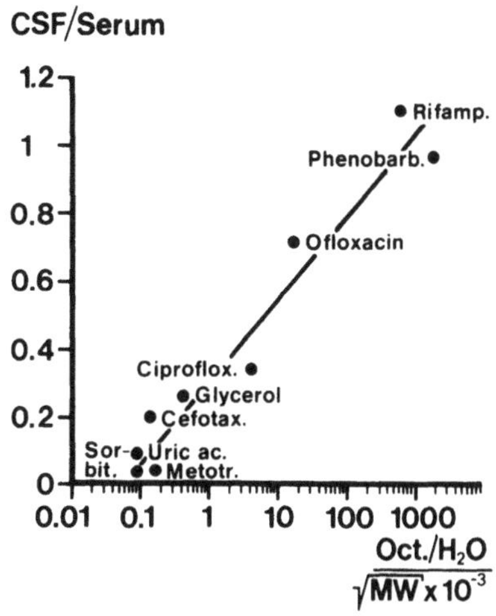


**Figure 2.** Overton's Law at the blood-CSF-barrier. CSF/serum ratios are the area under the curve ratios as obtained with ventricle drained intensive care patients (Nau et al., 1994). (Oct./$H_2O$ = Octanol/Water partition coefficient; MW = Molecular weight).

Lipophilicity is a greater factor in permitting the passage of molecules through the blood-CSF-barrier as compared to the blood-brain-barrier, where even the most lipophilic compounds are excluded if they are greater than 500 daltons (Unger et al., 1993). Overton's rule is also applicable to antibiotics such as cefotaxim and ciprofloxacin (Nau et al., 1994), both with octanol/water partition coefficients below 0.1 as well as to the very lipophilic rifampicin, having a mol. weight of 823, yet it crosses the blood-CSF-barrier easily (Fig.2).

If we accept the concept of a blood-CSF-barrier independent from the blood-brain-barrier, the most appropriate markers for a permeability curve would be

| | | |
|---|---|---|
| Water: | R = 1.8 Å | Q = 1000 x $10^3$ |
| Uric acid: | R = 5.2 Å | Q = 100 x $10^{-3}$ |
| Albumin: | R = 35.8 Å | Q = 4 x $10^{-3}$ |
| Coeruloplasmin: | R = 46.8 Å | Q = 2.7 x $10^{-3}$ |
| $\alpha_2$Macroglobulin: | R = 93.5 Å | Q = 0.9 x $10^{-3}$. |

In most diseases of the central nervous system the blood-CSF-barrier is impaired (Felgenhauer, 1992), which results in the increase of all serum components with the exception of chloride, a compound that is smaller than water and therefore its ratio decreases in barrier breakdown, e.g. in tuberculeous meningitis. Since by convention the ratio for water is always one, the slope of the curve indicates that selectivity of the filtration has decreased. Many years ago Bradbury found a decreased chloride and increased urea levels in tuberculeous meningitis (Fig.3). It is feasible to conclude that - at least when the barrier is open - both molecules reach the CSF space predominantly by diffusion. We can even calculate the hydrodynamic radius of water from his data (= 1.6 Å) and the selectivity loss due to the large permeability increase is evident.

## THE SIGNIFICANCE OF "DISOBEDIENCE" WITH THE BARRIER CURVE

Unexpected low CSF levels may be caused by postfiltrational metabolism in the striatum and the thalamus, e.g. of sorbitol (Wiesinger and Hamprecht, 1989) or by receptor mediated uptake in the lower brain stem, e.g. of glycine. Unexpected high

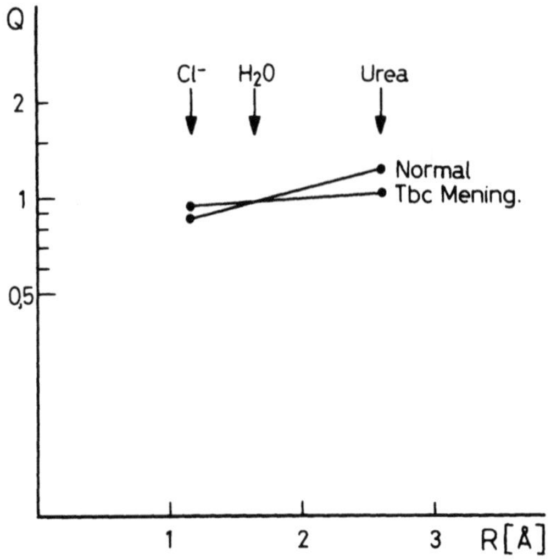


**Figure 3.** CSF chloride decreases in patients with a breakdown of the blood-CSF-barrier (e.g. meningitis tuberculosa), since its hydrodynamic size is smaller than that of water. Urea increases together with all other passively transferred serum components (Data from Prof. Bradbury).

levels of the molecule in CSF are indicative of active transport, e.g. ascorbate (Reiber et al., 1993) or release from local sources.

It is evident, that the major proteins of local origin, although less than 20% of the total (Fig. 4), originate from tissues adjacent to the free CSF space e.g. prostaglandin-d-synthetase (beta-trace) from the leptomeninges (Watanabe et al., 1995), as well as transthyretin (prealbumin) and cystatin C (gamma-trace) from the choroid plexus epithelium (Schreiber and Aldred, 1993).

There are two neuronal proteins that originate by more than 99% from the brain parenchyma: enolase and acetylcholinesterase (Mäder et al., 1991), the latter with a CSF/serum concentration ratio of five, which is indicative that the CNS is the major production site of the protein. There are proteins of uncertain origin like beta-2-microglobulin and ferritin (microglia? astrocytes?).

From the recent literature it seems reasonable to assume, at least for diagnostic purposes, that there is a CSF compartment which is separated from the circulating blood for some cytokines and their soluble receptors. This is certainly the case for the TNF-receptor, figure 5 (Puccioni-Sohler et al., 1994). Serum derived fractions of, lets say

| | | |
|---|---|---|
| Prostaglandin-D-Synthetase ($\beta$-trace) | 3.3 mg/dl | 8.9% |
| Transthyretin (Pre) | 1.6 " | 4.3% |
| Transferrin (tau) | 0.6 " | 1.6% |
| Cystatin C ($\gamma$-trace) | 0.6 " | 1.6% |
| $\beta_2$-Microglobulin | 0.1 " | 0.3% |
| | 6.2 mg/dl | 16.7% |

**Figure 4.** The major locally synthesized CSF proteins as inferred from the filtration curve of Fig. 1. The original protein names - generally overseen in present publications - are indicated in brackets: $\beta$-trace = beta-trace-protein; Pre = prealbumin; tau = tau-globulin; $\gamma$-trace = gamma-trace-protein.

less than 5% can be neglected in all those diseases, where the blood-CSF-barrier remains normal or is only slightly disturbed, e.g. in multiple sclerosis, chronic HIV encephalitis or tropical spastic paraparesis. Under such circumstances a lack of correlation between the concentration ratios of albumin and the cytokine is indicative of a prevailing intrathecal synthesis.

| | |
|---|---|
| Serum | 1.9 ng/ml |
| CSF | 1.1 ng/ml |
| CSF/Serum | $579 \times 10^{-3}$ (exp = $4 \times 10^{-3}$) |
| Local synthesis | > 99% |

**Figure 5.** The evaluation of the intrathecal TNF-receptor fraction in CSF as inferred from Fig.1. The hydrodynamic radius is estimated according to Felgenhauer (1974).

## THE CSF ANALYTICAL HALF BRAIN

The CSF-chemist should be aware, however, that he performs a partial brain analysis (Felgenhauer, 1993), since only about 50% of the central nervous system is accessible by lumbar puncture. In order to evaluate the total brain one must investigate the blood which carries away most of the neocortical proteins. This becomes evident, for instance, with the carcinoembryonic antigen from central nervous system carcinomas. Nine out of ten cerebral metastases release the tumour marker into the extracellular space, that is the CSF of meningeal infiltrates, but only 45% of the parenchymal metastases do so. These are predominantly localized in the periventricular and the paracisternal parts of the mediobasal telencephalon. Most upper-lateral tumours remain negative (Jacobi et al., 1986). This also holds true for lymphomas secreting monoclonal immunoglobulins ("paraproteins"). If the tumours are localized adjacent to the CSF space and accessible to lumbar puncture, paraproteins are detectable in the CSF, even if virtually absent in the circulating blood (Burkhardt et al., 1992). In contrast the CSF may contain only the serum derived fraction, if the tissue wall between tumour mass and ventricle edge is greater than 30 mm, which is apparently too wide for the tumour protein to reach the ventricle.

If a tissue width of that size is critical for proteins from the extracellular space to appear in the ventricular CSF ("brain-CSF-barrier") the CSF level of process markers should depend upon the distance from lesion edge to ventricle border. This is indeed the case with the intercellular adhesion molecule ICAM released by active multiple sclerosis lesions (Rieckmann et al., 1993). The CSF values in patients with acute lesions adjacent to the ventricle are much higher than in cases with distant lesions (Fig.6). The lesion of Fig.7 is just within the 30 mm area and its soluble proteins like cytokines, adhesion molecules and immunoglobulins should appear in the CSF.

Therefore CSF levels do not reflect the overall activity of an ongoing disease process nor the severity of the clinical symptoms as examplified by specific HTLVI antibody levels in tropical spastic paraparesis. If the antibody specificity index, that is the CSF/serum ratio of antibody per unit weight of total IgG, is plotted versus the duration and the severity of the disease, there is no correlation (Puccioni-Sohler et al., 1994).

In Creutzfeldt-Jakob disease the neuron specific enolase, in the CSF, is highest during the first half of the progressive disease, when the myoclonia starts and it is likely, that neurons from striatum and thalamus adjacent to the CSF are the main source (Felgenhauer et al., 1993).

| Distance (mm) | ICAM (ng/ml) | |
|---|---|---|
| 3 | 27.0 | (N = 3) |
| 8 | 13.0 | (N = 4) |
| 17 | 7.0 | (N = 7) |
| 28 | 1.5 | (N = 4) |

**Figure 6.** Soluble ICAM in the CSF and distance between acute MS lesions and wall of ventricles. Only acute MS lesions within a distance of 25-30 mm from the ventricles belong to the descending arm of the free CSF space, which can be accessed by lumbar puncture.

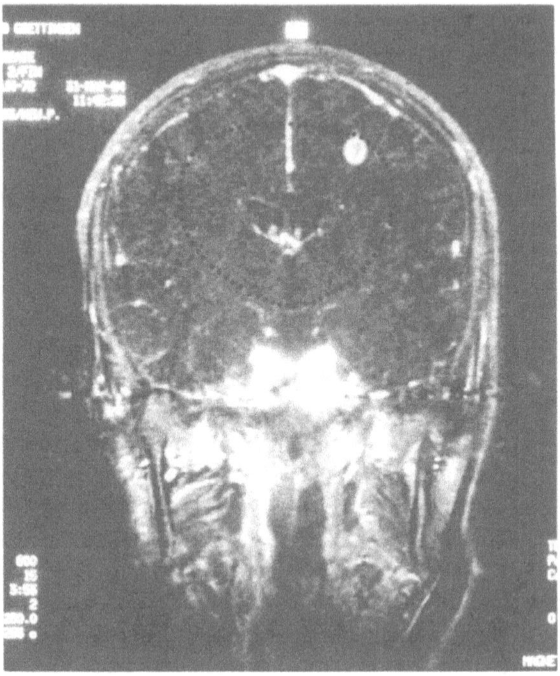

**Figure 7.** Acute, gadolinium enhanced MS lesion within a 30 mm distance from the ventricles. ICAM-1 = 3 ng/ml (compare Fig. 6).

## THE BLOOD CARRIES DIAGNOSTIC MARKERS FOR THE WHOLE BRAIN

In order to evaluate the total brain, including the telencephalic cortex remote from the ventricles it is necessary to study circulating blood. This becomes evident in the use of ICAM as marker for the overall inflammatory activity of multiple sclerosis. In active cases only the blood levels vary in parallel with the clinical signs and NMR-findings, whereas in stable MS, cases exhibit rather smooth and undramatic curves (Rieckmann et al., 1994).

The serum concentration of neuron specific enolase (NSE) may increase quickly after brain infarction and raise again several days later if a malignant brain oedema developes (Schaarschmidt et al. 1994). A clear correlation between the incipient NSE increase and the clinical severity cannot be expected since the symptoms depend upon the localisation and not the size of the infarction. In contrast the serum concentrations of NSE within the first three days after diffuse brain hypoxia are valid prognostic markers (Prange, 1994).

NSE values above 120 ng/ml are only found in patients with the worst Glasgow outcome scores 1 and 2 (death or severest defects).

The NSE increase in the CSF is much more sluggish as compared to the serum (Fig.8) and has no diagnostic significance, since the enolase comes mainly from the cortical infarction areas, and it needs 3 to 4 days to reach the CSF space. In the serum there is a first maximum due to neuronal destruction and opening of the brain-blood-barrier and a second maximum, if a malignant brain oedema develops.

## THE PROTEIN COMPOSITION OF THE EXTRACELLULAR SPACE

An intriguing question is the diffusional equilibrium between the cerebrospinal fluid and the brain extracellular fluid since their composition may differ and there are no barriers between these compartments. Although the blood-brain-barrier is possibly the most extensively studied barrier of the human body, the protein composition of the extracellular fluid is still unknown. Since brain capillaries have neither fenestrations nor pinocytotic vesicles, which are plentiful in the choroid plexus, one may assume that the level of serum proteins in the CSF is higher than that in brain extracellular fluid.

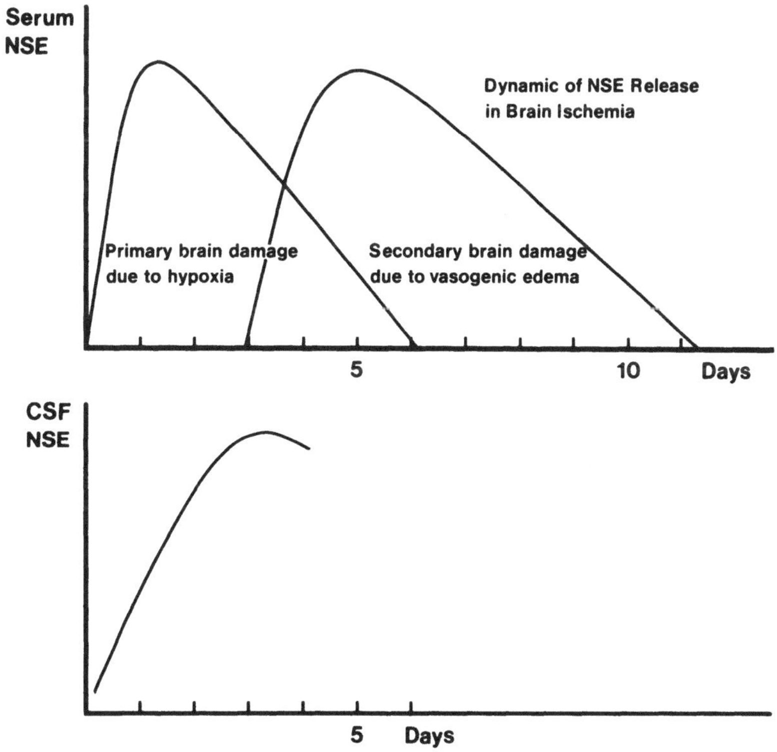


**Figure 8.** Prototypic behaviour of serum and CSF neuron specific enolase levels in brain infarctions (comment see text).

However the increase of serum derived proteins (e.g. albumin and IgG) along the neuraxis is compatible with the concept of an influx of these molecules from the brain parenchyma which would imply that the brain extracellular fluid should then contain higher albumin levels than CSF. This may be emphasised in the periventricular organs, which are regions without a blood-brain-barrier, but with a closed ependyma for hydrophilic compounds, e.g. adriamycin.

The extracellular fluid level of neuronal proteins like enolase should be much higher in the extracellular than in the cerebrospinal fluid. This would explain the immediate and limited increase in fluid serum NSE of up to 60 ng/ml in some patients after electroconvulsive shock therapy in view of the absence of neuronal damage (R. Thomas, personal communication).

## CONCLUSION

1. The blood-CSF-barrier is a typical filtration type barrier. The concentration ratio under steady state equilibrium condition depends upon the barrier permeability as well as upon the turnover rate of the free CSF.

2. The blood-brain barrier and the blood-CSF barrier are quite different.

3. The choroid plexus contains a macrofilter, seperating thrombocytes from the largest proteins and a leaky junction microfilter for size dependent molecular restriction. Although alterations in the composition of primary ventricular CSF occur on its passage down to the lumbar sac, its basic filtrational properties remain unchanged.

## REFERENCES


Burkhardt, D., Schipper, H.I., Kaboth, U., and Felgenhauer, K., 1992, IGA producing primary intracerebral lymphoma, *J. Neurol. Neurosurg. Psychiatry* 55:623.

Felgenhauer, K., 1974, Protein size and cerebrospinal fluid composition, *Klin. Wochenschr.* 52:1158.

Felgenhauer, K., 1974, Evaluation of molecular size by gel electrophoretic techniques, *Hoppe-Seyler's Z. Physiol. Chem.* 355:1281.

Felgenhauer, K., Liappis, N., and Nekic, M., 1982, Low molecular solutes and the blood cerebrospinal fluid barrier, *Klin. Wochenschr.* 60:1385.

Felgenhauer, K., 1992, Liquordiagnostik, *in*: Labor und Diagnose, L. Thomas, ed, Medizinische Verlagsgesellschaft, Marburg.

Felgenhauer, K., Weber, T., and Poser, S., 1993, Early clinical and laboratory diagnosis of Creutzfeldt-Jakob Disease, *Dev. Biol. Stand.* 80:125.

Felgenhauer, K., 1993, Barrier concepts and CSF analysis, *in*: CNS Barriers and Modern CSF Diagnostics, K. Felgenhauer, M. Holzgraefe, and H.W.Prange, eds., VCH Verlag, Weinheim.

Jacobi, C., Reiber, H., and Felgenhauer, K., 1986, The clinical relevance of locally produced carinoembryonic antigen in cerebrospinal fluid, *J. Neurol.* 233:358.

Mäder, M., Soerensen, K., Wiedmann, Th., Dickmann, U., and Felgenhauer, K., 1991, Neuronal acetylcholinesterase levels in cerebrospinal fluid and serum determined by a specific and sensitive immunoassay, *Eur. J. Clin. Chem. Clin. Biochem.* 29:51.

Nau, R., Zysk, G., Thiel, A., and Prange, H.W., 1993, Pharmacokinetic quantification of the exchange of drugs between blood and cerebrospinal fluid in man, *Eur. J. Clin. Pharmacol.* 45:469.

Nau, R., Sörgel, F., and Prange, H.W., 1994, Lipophilicity at pH 7.4 and molecular size govern the entry of the free serum fraction of drugs into the cerebrospinal fluid in humans with uninflamed meninges, *J. Neurol. Sci.* 122:61.

Prange, H.W., 1994, Pathophysiologie, Therapie und Prognose des hypoxisch-ischaemischen Hirnschadens, *Z. Kardiologie (Suppl.)* in press.

Puccioni-Sohler, M., Rieckmann, P., Kitze, B., Lange, P., Albrecht, M., and Felgenhauer, K., 1994, A soluble form of tumor necrosis factor receptor in cerebrospinal fluid (CSF) and serum of HTLV-1 associated myelopathy/tropical spastic paraparesis (HAM/TSP) and other neurological diseases, *J. Neurol.* in press.

Reiber, H., Ruff, M., and Uhr, M., 1993, Ascorbate concentration in human cerebrospinal fluid (CSF) and serum. Intrathecal accumulation and CSF flow rate, *Clin. Chim. Acta* 217:163.

Rieckmann, P., Nünke, K., Albrecht, M., Burchhardt, M., Wiltfang, J., Ulrich, M., and Felgenhauer, K., 1993, Soluble intercellular adhesion molecule-1 (sICAM-1) in cerebrospinal fluid: An indicator for the inflammatory impairment of the blood CSF barrier, *J. Neuroimmunol.* 47:133.
Rieckmann, P., Martin, S., Albrecht, M., Kitze, B., Weber, T., Tumani, H., Broocks, A., Lüer,W., Helwig, A., and Poser, S., 1994, Serial analysis of circulating adhesion molecules and TNF receptor in serum from patients with multiple sclerosis: cICAM-1 is an indicator for relapse, *Neurology* in press.
Schaarschmidt, H., Prange, H.W., and Reiber, H., 1994, Neuron-specific enolase concentrations in blood as a prognostic parameter in cerebrovascular diseases, *Stroke*, 24:558.
Schreiber, G., and Aldred, A.R., 1993, Origin and function of proteins in the cerebrospinal fluid, *in*: CNS Barriers and Modern CSF Diagnostics, K. Felgenhauer et al, eds, VCH-Verlag, Weinheim.
Unger, C., Krisch, B., and Eibl, H., 1993, Brain uptake of chemotherapeutic agents in the rat, *in*: CNS Barriers and Modern CSF Diagnostics, K. Felgenhauer et al, eds, VCH-Verlag, Weinheim.
Watanabe, K., Urade, Y., Mäder, M., Murphy, Ch., and Hajaishi, O., 1994, Identification of beta-trace as prostaglandin D synthetase, *submitted* for publication.
Wiesinger, H., and Hamprecht, B., 1989, Purification and characterization of sorbitol dehydrogenase from bovine brain, *J. Neurochem.* 52:342.
Wright, E.M., and Loo, D.F., 1994, The role of the choroid plexus in the blood-brain-barrier, *this volume.*

# BIOPHYSICS OF PROTEIN DIFFUSION FROM BLOOD INTO CSF: THE MODULATION BY CSF FLOW RATE

Hansotto Reiber

Neurochemisches Labor
University Göttingen
Robert-Koch-Str. 40
D - 37075 Göttingen, Germany

## INTRODUCTION

Many neurological diseases are accompanied by increased protein concentrations in cerebrospinal fluid (CSF), described as a blood-CSF barrier dysfunction. There is still widespread belief in a model based on "leakage", i.e. a morphological defect is necessary to explain the empirical data. In my contribution I will show with a mathematical treatment of a diffusion model that an increased molecular flux of proteins through a diffusion barrier is possible without any morphological changes if only the CSF flow rate is decreased.

The blood-CSF barrier (Davson, 1967) is a functional definition for the restricted transfer of molecules from blood into CSF, usually characterized by the molecular size-dependent CSF/serum concentration ratio. The CSF/serum quotient of albumin ($Q_{Alb}$), an exclusively blood-derived protein in CSF, is generally accepted as indicator of changes in the blood-CSF barrier function. We observe nonlinear changes in the ratios of individual protein quotients between molecules of different size, like immunoglobulins and albumin ($Q_{IgG} : Q_{Alb}$). This function is clinically relevant to discriminate a brain-derived fraction from a blood-derived fraction. In the absence of a physiologically relevant theory it is possible to fit all types of linear and nonlinear functions, in particular for a restricted range of empirical data. A nonlinear function is unambiguous if a sufficiently large range of concentrations in CSF albumin are investigated, i.e. sufficiently large blood-CSF barrier dysfunctions (Reiber, 1994). In contrast to the morphologically well defined blood-brain barrier (Bradbury, 1979), the blood-CSF barrier function includes several different influences like conditions of CSF production in the choroid plexus, CSF flow rate or diffusion controlled transfer of plasma proteins into CSF during its flow through the subarachnoid space and increased lumbar protein concentrations up to three times that of the ventricular concentration.

The involvement of CSF flow rate in experimental (cit. in Davson, 1967) or theoretical approaches (Rapoport, 1983) has a long tradition in CSF research. But, it has

*New Concepts of a Blood–Brain Barrier*
Edited by J. Greenwood *et al.*, Plenum Press, New York, 1995

been overlooked so far that the diffusion into a flowing solvent does not allow a simple linear approach to describe the concentration gradient between blood and CSF. For the first time I introduced a nonlinear approach by the application of Fick's second law of diffusion to describe the diffusion of proteins from blood into CSF (Reiber, 1993; Reiber, 1994). With my model of diffusion for the blood-CSF barrier function involving CSF flow rate, it was possible to show that a hyperbolic function is the correct mathematical solution (Reiber, 1994) to describe a changing concentration ratio of two different blood-derived molecules in CSF with increasing protein concentrations, i.e. decreasing CSF flow rate.

## THE PROTEIN FLUX/CSF FLOW MODEL

The diffusion-controlled transfer of proteins from blood to CSF and the CSF flow rate in the subarachnoidal space are the main physical parameters influencing the steady-state concentration of a serum derived protein in CSF. Fig. 1 represents an idealized model (Reiber, 1993) of a cut through the subarachnoidal space. Regarding the diffusion of proteins from blood into CSF, the following boundary conditions are suggested:

1. The protein concentration of a single protein at one side of the barrier (e.g. in the blood) remains constant, i.e. the reservoir of proteins is infinite.
2. Proteins diffuse through tissue, which does not need to be characterized with its very different structures like endothelial cell layer with tight junctions, fenestrated capillaries, intercellular fluid, ependymal cell layer with and without tight junctions.

This simplification is possible as the theory only compares the relation of proteins of different size, diffusing through the same tissue structures at the same time. It is reasonable to suggest that the *ratio of diffusion coefficients* is not changed in spite of different values of the single diffusion coefficient in different tissue structures.

3. Proteins diffusing from blood through the tissue enter the cerebrospinal fluid along its way in the subarachnoidal space. The theory does not need to state a laminar flow or CSF flow rate unique for the whole subarachnoidal space.

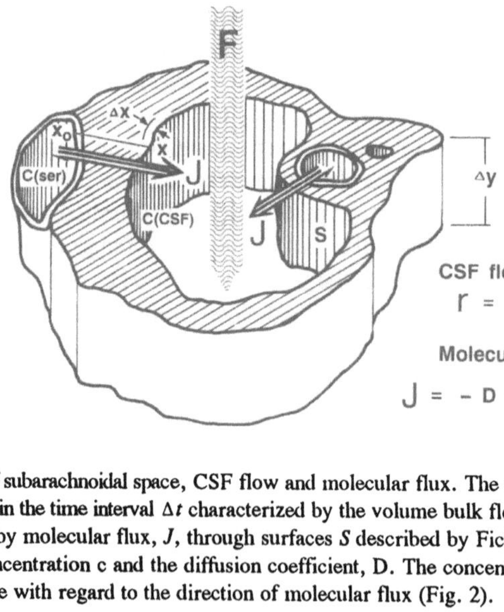


CSF flow rate

$$r = F/A$$

Molecular flux

$$J = -D\frac{dc}{dx} \sim f(1/r) \qquad (1)$$

**Figure 1.** Model of subarachnoidal space, CSF flow and molecular flux. The volume interval $\Delta\mathrm{Vol}=A\cdot\Delta y$ passes the cross section A in the time interval $\Delta t$ characterized by the volume bulk flow $F = \Delta Vol/\Delta t$. Proteins enter the CSF volume $\Delta Vol$ by molecular flux, $J$, through surfaces $S$ described by Fick's first law of diffusion in Eqn. (1) with the solute concentration c and the diffusion coefficient, D. The concentration gradient dc/dx is negative as its slope is negative with regard to the direction of molecular flux (Fig. 2).

For this model the following application of an implicit solution of the fundamental differential equation of diffusion (Fick's second law of diffusion) in Eqn. (2) was developed (Reiber, 1994). General solutions of the diffusion equation can be obtained for a variety of initial and boundary conditions provided that the diffusion coefficient is constant (Crank, 1975; Carslaw and Jaeger, 1959). By differentiation of Eqn. (3) (A is an arbitrary constant) we see that it is a basic solution of Eqn. (2).

$$\frac{\partial c}{\partial t} = D\frac{\partial^2 c}{\partial x^2} \tag{2}$$

$$c = \frac{A}{\sqrt{t}}e^{-x^2/4Dt} \tag{3}$$

The case relevant for the diffusion through the blood-CSF barrier into flowing CSF (Fig. 1) has different boundary conditions which can be described as diffusion in a semi-infinite media with constant concentration $c_o$ at one surface (i.e. the serum concentration $c_{ser} = 1$ at $x = 0$ in Fig. 2). This needs to be treated as an extended initial distribution (integration of Eqn. (3)) and a reflection at the boundary $x = 0$. According to Crank (1975, p. 14), we thus obtain for the concentration $c_{x,t}$ the function in Eqns. 4 and 4b. This function is shown in Fig. 2.

$$c_{x,t} = \frac{c_o}{2\sqrt{\pi Dt}}\int_x^{\infty} e^{-x^2/4Dt}\,dx \quad = \frac{c_o}{2\sqrt{\pi Dt}}\left(1 - \int_0^x e^{-x^2/4Dt}\,dx\right) \tag{4}$$

or with $z = x/2\sqrt{Dt}$ and $dx = 2\sqrt{Dt}\,dz$:

$$c_{x,t} = \frac{c_o}{\sqrt{\pi}}\int_z^{\infty} e^{-z^2}\,dz \tag{4b}$$

The Eqn. (4) can be re-written as a standard mathematical function, i.e. the error function, erf z, (Eqn. 4c) and the error function complement erfc z (Eqn. 4d). The error function enters into the solution of a diffusion problem as a consequence of summing the effect of a series of line sources, each yielding an exponential type of distribution (Carslaw and Jaeger, 1959, p. 482):

$$\operatorname{erf} z = \frac{2}{\sqrt{\pi}}\int_0^z e^{-z^2}\,dz \tag{4c}$$

$$\operatorname{erfc} z = \frac{2}{\sqrt{\pi}}\int_z^{\infty} e^{-z^2}\,dz = 1 - \operatorname{erf} z \tag{4d}$$

With Eqn. 4 we find then for the quotient Q (Crank, 1975, p. 14):

$$Q = \frac{c_{x,t}}{c_o} = \frac{1}{2}\text{erfc } z \qquad (5)$$

Approximations for erf z are calculated for large and for small values of z using solutions from trigonometric series (Crank, 1972, p. 21; and Carslaw and Jaeger, 1959). We examined values of large z as $c_{x,t} << c_o/2$ ($Q = 0.5$). Therefore the trigonometric series of the error functions complement in Eqn. (6) becomes relevant for calculation of concentration values from z according to Eqn. (5).

$$\text{erfc } z = \frac{1}{\sqrt{\pi}} e^{-z^2}\left(\frac{1}{z} - \frac{1}{2z^3} + \frac{1 \cdot 3}{2^2 z^5} - \frac{1 \cdot 3 \cdot 5}{2^3 z^7} + \ldots\right) \qquad (6)$$

Values of this function in Eqn. (6) are reported in tables (Carslaw and Jaeger, 1959, p. 482), extracts of which are given for values of z between 0.1 and 2.1 in Table 1.

In my recent paper (Reiber, 1994) I have shown with this erfc z for two molecules of different size (i.e. with different diffusion coefficients $D_B$ and $D_A$) that their concentration ratio (at $x_p$ in Fig. 2) in the flowing solvent with varying flow rate varies according to a hyperbolic function. The difficulty with such a proof was that the relevant ratio of $Q_B : Q_A$ at $x_p$ in Fig. 2 cannot be expressed in an explicit function from the general differential Equations 2 or 4. As a mathematical trick, by projection of curve A onto curve B (Fig. 2), I obtained the concentrations $Q_A$ and $Q_B$ on the same curve for different values of x (x' and

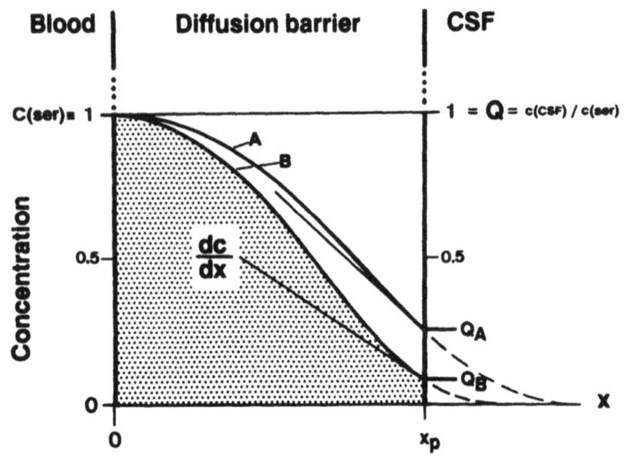


**Figure 2.** Protein concentration gradient between blood and CSF. The diffusion controlled protein transfer between blood and CSF is described by an idealized diffusion barrier (homogeneous and unique instead of multistructural, inhomogeneous). The concentration $c_i$ (in blood given as $c_{(ser)}$ and in CSF as $Q = c_{(CSF)}/c_{(ser)}$, i.e. normalized for $c_{(ser)} = 1$) along the effective diffusion pathlength, x ($x_p$, for a single patient), shows a sigmoidal curve, derived from Fick's second law of diffusion. At constant t (steady state), the local concentration gradient $dc_i/dx$ at $x_p$ determines the molecular flux, $J_i$ of molecules i into CSF according to Fick's first law of diffusion. The change of ($dc_i/dx$) along the idealized diffusion pathlength x, follows a Gaussian error curve with a maximum at $Q = 0.5$, i.e. half the serum concentration. The curves A and B can be interpreted as if they were from molecules of different size (A = albumin and B = IgG). The smaller molecule albumin with the larger diffusion coefficient $D_{Alb}$ reaches a higher tissue concentration (curve A) than IgG (curve B) and a larger local concentration gradient dc/dx at $x_p$. The function $Q_B = f(Q_A)$ can be shown to be a hyperbolic function., Eqn. (7).

$x_p$, respectively; Fig. 5 in Reiber, 1994). Then the position of any point x can be specified by the position ratio, a dimensionless parameter.

In fact, the only difference between the curve A and B is the mean displacement $\bar{x}_A = 2\sqrt{D_A t}$ and $\bar{x}_B = 2\sqrt{D_B t}$. The dimensionless position ratio is obtained by $x/\bar{x}$, i.e. the transfer of $Q_A$ (on curve A at $x_p$) to curve B gives the same value $Q_A$ at x' on curve B if $x' = x_p \cdot \bar{x}_B/\bar{x}_A = x_p (D_B/D_A)^{1/2}$.

With this transformation we obtain the concentration ratio $Q_B : Q_A$ for two different distances x' and $x_p$ on the same curve with the identical error function. $z_p$ represents the dimensionless effective diffusion pathlength corresponding to $x_p$ in Fig. 2 according to Eqn. (4b) with $z = x/2\sqrt{Dt}$. As a consequence of $Q_B : Q_A = \text{erfc } z_p : \text{erfc } z' = \text{erfc } z(D_B/D_A)^{1/2} : \text{erfc } z$, we get the general equation:

$$Q_B = \frac{\text{erfc } z\sqrt{D_B/D_A}}{\text{erfc } z} \bullet Q_A \qquad (7)$$

It can be shown by a mathematical fit that this Eqn. 7 represents a hyperbolic function corresponding to Eqn. (8). The values $Q_B$ and $Q_A$ calculated according Eqn. (5) with the tabulated values of erfc z (Carslaw and Jaeger 1959, p. 482) are shown in Table 1 for an arbitrary pair of molecules with the following ratio of diffusion coefficients: $D_B : D_A = 1:2.25$ corresponding to $z' : z_p = 1:1.5$ .

$$(y + c)^2/a^2 - x^2/b^2 = 1 \quad \text{or} \quad Q_B = a/b\sqrt{Q_A^2 + b^2} - c. \qquad (8)$$

$Q_B$(fit) in Table 1 shows the data calculated from $Q_A$ with Eqn. (8) and the values a(=c) and b calculated from two pairs of the extreme values $Q_B$ and $Q_A$ inserted into Eqn. (8). Covering three orders of magnitude the excellent concordance (Table 1) of $Q_B$(fit) from Eqn. (8) with the theoretical value $Q_B$ from Eqn. (7) compared for an arbitrary set of data demonstrates that Eqn. (7) represents a hyperbolic function. The shape of this hyperbolic curve is determined only by the ratio of the square roots of the both diffusion coefficients, as shown in Fig. 3.

Instead of this indirect way to show that Eqn. (7) represents a hyperbolic function it might be possible to find a direct transformation of Eqn. (7) into the form of Eqn. (8) using Laplace transforms (Carslaw and Jaeger, 1959). In general, this means that the concentration ratio of molecules of different size diffusing through the same matter into a flowing solvent is generally characterized by a hyperbolic function.

This new application of the differential equation of diffusion (2) for these biologically relevant boundary conditions could be confirmed by empirical data (Reiber, 1994).

## INFLUENCES OF CSF FLOW RATE

The pathological decrease of CSF flow rate, r, or volume bulk flow, F, has two consequences: 1. Increased protein concentration due to reduced volume exchange rate - a linear relation. 2. Initially an increasing molecular flux into CSF - a nonlinear function with inversion to a decreasing molecular flux above $Q = 0.5$.

**Table 1.** Concentration Ratios ($Q_B$:$Q_A$) of two diffusing molecules A and B at $z_p$ (effective diffusion pathlength).

| $Q_A$ *) | | $Q_B$ *) | | $Q_B$(fit) **) |
|---|---|---|---|---|
| z' | 1/2 erfc z | $z_p$ | 1/2 erfc z | |
| 0.1 | 0.4438 | 0.15 | 0.4160 | 0.4160 |
| 0.2 | 0.3886 | 0.3 | 0.3356 | 0.3318 |
| 0.3 | 0.3356 | 0.45 | 0.2622 | 0.2567 |
| 0.4 | 0.2858 | 0.6 | 0.1980 | 0.1923 |
| 0.5 | 0.2397 | 0.75 | 0.1444 | 0.1391 |
| 0.6 | 0.1980 | 0.9 | 0.1015 | 0.0971 |
| 0.8 | 0.1289 | 1.2 | 0.0448 | 0.0424 |
| 1.0 | 0.0786 | 1.5 | 0.0169 | 0.0160 |
| 1.2 | 0.0448 | 1.8 | 0.00545 | 0.00524 |
| 1.4 | 0.0238 | 2.1 | 0.00148 | 0.00149 |

*) Data for $Q_A$=0.5 erfc z'and $Q_B$=0.5 erfc Zp are calculated from the tables of Carslaw and Jaeger (1959, p.482), with the arbitrary ratio of diffusion coefficients $D_B$:$D_A$=1:2.25 or z':$z_p$=1:1.5 ($z'=z_p (D_B/D_A)^{1/2}$).
**) $Q_B$(fit) represents the data calculated from $Q_A$ with the hyperbolic function in Eqn.(8) for the fitted values a = c = 0.8736 and b = 0.4087.

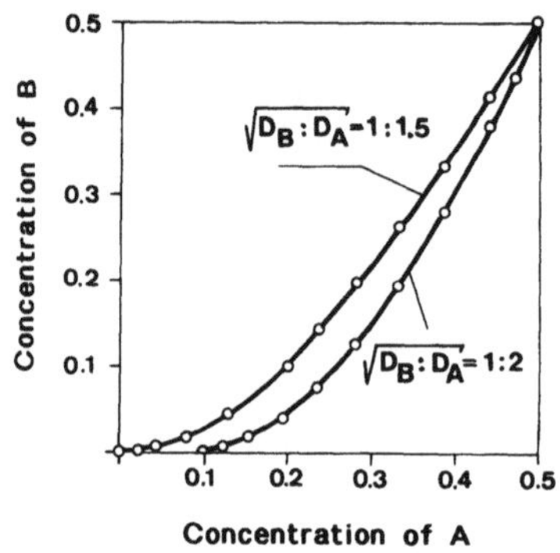


**Figure 3.** Concentration ratios of two pairs of diffusing molecules A and B at z (effective diffusion path-length). The concentrations $Q_A$ and $Q_B$ of substance A and B are calculated from tabulated values of error function complement erfc z according to Table 1. The two hyperbolic functions (Eqn. (7) in the text) are defined by arbitrary ratios of the diffusion coefficients $D_B/D_A$ = 1:2.25 and $D_B/D_A$ = 1:4.

With respect to the first consequence: This result is easily derived regarding the molecular flux $J_i$ in Fig. 1: If n molecules i enter $\Delta$Vol flowing through cross section A in the time interval $\Delta t$ we get for the local increase in CSF concentration of i : $\Delta c_{iCSF} = n_i / \Delta Vol$. With a reduced volume bulk flow F = $\Delta Vol/\Delta t$ or a reduced CSF flow rate r (flow rate or elimination rate of a single molecule, r = $\Delta y/\Delta t$, dimension cm/s)$\Delta c_{iCSF}$ is increased - inspite of a constant $n_i$ (a suggestion to be modified below). This linear relation can be expressed (Reiber, 1994) by Eqn. (9):

$$\Delta c_{iCSF} = \frac{J_i}{r} \cdot \frac{S}{A} = \frac{J_i}{r} \cdot \text{Const} = \frac{\text{molecular flux}}{\text{CSF flow rate}} \cdot \text{Const} \quad (9)$$

The value S/A like other structural conditions in the interval $\Delta y$ (Fig. 1) is a constant (Const) for all proteins compared in each single person, so that the idealized conditions in Fig. 1 are valid for a quantitative comparison of concentration ratios of different molecules. This linear form of Eqn. (9) is well established in the literature (Rapoport, 1983). But, what has been overlooked so far is the fact that the molecular flux, $J_i$, itself is a function of CSF flow rate. In Fig. 4 the consequences of CSF concentration after decreasing CSF flow rate are shown in a linear (b) and nonlinear (a) model of concentration gradients.

The increased $c_{CSF}$ or increased CSF/serum quotient, Q, due to reduced volume exchange (see above) must lead to an increased concentration $c_{x,t}$ at $x_p$ (Fig. 2). This reasonable assumption can be expressed with $c_{x,t} \gg c_{CSF}$. In Fig. 4 this is shown for a substance with a normal concentration quotient $Q_C$, subsequently increasing to $Q_B$ and $Q_A$. From the Fig. 4 we see that in the linear model the slope dc/dx , relevant for the molecular flux, decreases with increasing quotients. This is different for the nonlinear model (Fig. 4a): With increasing concentration quotient we get an increasing slope dc/dx, i.e. with increasing blood-CSF barrier dysfunction (= reduced CSF flow rate, Reiber, 1994) we get an increased molecular flux from blood into CSF. For empirical purposes (Reiber, 1994) this relation is expressed by $\Delta r \sim 1/\sqrt{\Delta Q_{Alb}}$ .

This dependence of $J_i$ from flow rate, r, is a consequence of dc/dt and with the differential equation of continuity (van Holde, 1971), dc/dt = -dJ/dx, we get dc/dt = D $d^2c/dx^2$ (Fick's second law of diffusion, Eqn. (2)).

In the case of the relevant nonlinear approach, dc/dx = f'(x) is the derivate of Eqn. 4 (Reiber, 1994). This has two consequences: With increasing $c_{CSF}$, due to decreasing CSF flow rate, the local concentration gradient dc/dx (at $x_p$, in Fig. 2) increases nonlinearly as long as $c_{CSF} < c_o/2$ or Q < 0.5 and it decreases only for the rare case of quotients Q > 0.5 ($500 \cdot 10^{-3}$).

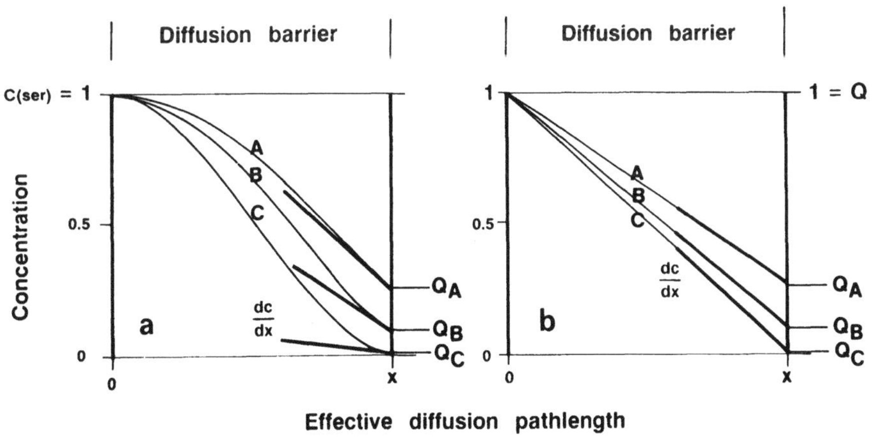


**Figure 4.** The curves A, B, C represent the concentration distribution of the same protein (e.g. IgM) at different times, before ($t_o$, curve C) and after ($t_1$,$t_2$, curve B and A) onset of a disease with decreasing CSF flow and subsequently increasing concentration of IgM in CSF. With a secondary increasing tissue concentration, $c_{IgM}$ (x, t), the gradient $(dc_{IgM}/dx)_{x,t}$ increases in the nonlinear model (a) with t (const x) and, as a consequence, the molecular flux, $J_{IgM}$, is increased as well. Vice versa, in the linear model (b) dc/dx and the molecular flux decrease with increasing CSF concentration $Q_C$ to $Q_A$.

This time- (and diffusion pathlength-) dependent change of the concentration at $x_p$ can be better understood if we regard the mean displacement - an important relation in diffusion. The mean displacement of diffusing particles describes the shift of diffusing molecules in relation to time and the diffusion constant: $\bar{x} = \sqrt{2Dt}$ where $\sqrt{Dt}$ represents a length (Jost, 1960, p. 25). $\bar{x}$ is comparable with the half width of the error curve. With increasing mean displacement (penetration depth), the maximal gradient dc/dx decreases (decreasing height, for dc/dx at $x_{0.5}$ (Fig. 6 in Reiber, 1994))

It is important to treat the system (Figs. 1, 2) as a semi-infinite system and not as a plane sheet (slab) with two constant concentrations at both surfaces, i.e. with two reflection planes, and dJ/dx = 0 for x = 0 and x = $x_p$. This condition dJ/dx = 0 represents a linear concentration gradient, $(c_o - c_{CSF})/2$, (Fig. 4b). This would have another interesting consequence. If the curves in Fig. 4 would represent different molecules (and constant flow), the smaller molecule A (albumin) with a smaller gradient dc/dx (according Fig. 4b) would show the smaller molecular flux into CSF compared to the larger molecule C (e.g. IgM). This is contradictory to the kinetics of proteins appearing in CSF after onset of neurological diseases (Reiber, 1994). Only in case of a nonlinear concentration gradient (Fig. 4a or Fig. 2) the smaller molecule A would have a larger molecular flux into CSF - a reasonable suggestion due to the theory of diffusion.

## CONCLUSIONS

The nonlinear treatment of blood/CSF concentration gradients of proteins explains increased molecular flux through the barrier with a decreased CSF flow rate without requiring any structural changes in the barrier. The consequences of a constant selectivity and constant permeability coefficients in blood-CSF barrier dysfunction for a new interpretation of observed data were described elsewhere (Reiber, 1994). The application of the model to the blood-CSF barrier function has led to the conclusion that a reduction of CSF flow rate is sufficient to explain quantitatively ($\Delta r \sim 1/\sqrt{\Delta Q_{Alb}}$) the increase of protein concentrations in CSF of patients with various neurological diseases (Reiber, 1994). My discovery of the hyperbolic function, initially introduced as an experimental fit (Reiber and Felgenhauer, 1987) could now be supported by the theory, which sufficiently explains many physiological phenomena (Reiber, 1994) and establishes the scientific base for evaluation of CSF data with CSF/serum quotient diagrams in clinical neurochemistry.

The molecular flux / CSF flow model can be applied to all calculations of concentration ratios in biological systems (also blood-brain barrier, membrane pore models, etc.) which have a constant concentration at one side and a consumption (chemical reaction) or elimination (solvent flow) at the other side of a diffusion barrier.

**Acknowledgement:**

I thank Prof. E. Thompson, London, for careful and competent proof reading.

## REFERENCES

Bradbury, M., 1979, "The Concept of a Blood-Brain Barrier," John Wiley & Sons, Chichester.
Carslaw, H.S., and Jaeger, J.C., 1959, "Conduction of Heat in Solids", Clarendon Press, 2nd ed., Oxford.
Crank, J., 1975, "The Mathematics of Diffusion", Clarendon Press, 2nd ed., Oxford.
Davson, H., 1967, "Physiology of the Cerebrospinal Fluid", Churchill, London.
Jost, W., 1960, "Diffusion in Solids, Liquids, Gases", Acad. Press, 3rd ed., N.Y.

Rapoport, S.I., 1983, "Passage of proteins from blood to cerebrospinal fluid", in: "Neurobiology of Cerebrospinal Fluid", J. H. Wood, ed., Vol. 2,:233-245, Plenum Press (1983).
Reiber, H., 1994, Flow rate of cerebrospinal fluid (CSF) - A concept common to normal blood-CSF barrier function and to dysfunction in neurological diseases, J. Neurol. Sci. 122:189-203.
Reiber, H., 1993, "Decreased flow of cerebrospinal fluid (CSF) as origin of the pathological increase of protein concentration in CSF", in: "CNS Barriers and Modern CSF Diagnostics", Felgenhauer, K., Holzgraefe, M., Prange, H. eds., Verlag Chemie, Weinheim, p. 305-317.
Reiber, H., and Felgenhauer, K., 1987, Protein transfer at the blood cerebrospinal fluid barrier and the quantitation of the humoral immune response within the central nervous system", Clin. Chim. Acta 163:319-328.
van Holde, K.E., 1971, "Physical Biochemistry", Prentice Hall Int. Englewood Cliffs, New Jersey.

# ISOLATED BRAIN MICROVESSELS AND CULTURED CEREBRAL ENDOTHELIAL CELLS IN BLOOD-BRAIN BARRIER RESEARCH: 20 YEARS ON

Ferenc Joó

Laboratory of Molecular Neurobiology,
Institute of Biophysics,
Biological Research Center,
P.O.B. 521,
6701-Szeged, Hungary

## INTRODUCTION

One of the most important functions of the endothelium is the regulation of the movement from the vascular to the extravascular space of water and solutes containing nutrients. The cerebral vasculature can be in general classified on the basis of topographical localization as the extra- and intraparenchymal vessels. Much of our knowledge on brain vessels has been collected from extraparenchymal vessels (for review see Edvinsson, 1982), but due to their importance in regulating the blood flow and penetration of substance circulating in the blood, there is an increased need to develop procedures for isolating intraparenchymal microvessels from brain tissue.

The brain of essentially all vertebrates contains capillary endothelial cells with unique characteristics, which can be best described as being a special mixture of endothelial and epithelial properties. Traditionally, studies of vascular permeability have employed intact animals or isolated-perfused organ preparations (Takasato et al., 1984). However, the endothelial cells *in situ* have remained inaccessible for direct biochemical studies.

A better understanding of the chemical composition of the intraparenchymal cerebral endothelial cells (CEC) and the molecular interactions regulating different endothelial functions has become available with the elaboration of a procedure, by means of which a fraction enriched with functionally and morphologically intact cerebral microvessels could be obtained (Joó and Karnushina, 1973). With the later recognition that most of the endothelial cells resist damage during isolation, remained viable, and could be maintained in tissue culture conditions (Panula et al., 1978), a new generation of model systems became available.

The results obtained from studying the freshly isolated cerebral microvessels have previously been reviewed in detail (Joó, 1985, 1993). Moreover, the information gained from the culture of cerebral endothelial cells has recently been updated (Joó, 1992).

Therefore, the primary aim of this present review is (i) to summarize the main features of CEC which are common with the endothelial cells of peripheral organs and (ii) to point at the unique properties of the CEC which are different from endothelial cells of other organs. Finally, the role of astrocytes will be shown to induce and maintain the "brain-like" features of cultured CEC.

## COMMON FEATURES OF CEC WITH ENDOTHELIAL CELLS OF OTHER ORGANS

It is generally accepted that endothelial cells *in vivo* respond in an active manner to humoral and physical stimuli to produce and secrete certain relaxing and contracting factors that contribute to vascular homeostasis and tone. Therefore, the CEC have several properties which are similar to, if not identical with, the endothelial cells of other organs (Table 1).

**Table 1.** Characteristics of cerebral endothelial cells. I.

| Endothelial features of cerebral endothelial cells | Reference |
|---|---|
| 1. Presence of general endothelial markers | |
| Factor VIII-related antigen | Phillips et al. (1979); Bowman et al. (1981) |
| Alkaline phosphatase | Karnushina et al. (1980a) |
| Binding of lectins | Diglio et al. (1981) |
| Uptake of acetylated-LDL | Krause et al. (1982); Tontsch and Bauer (1989) |
| γ-Glutamyl transpeptidase | Orlowski et al. (1974); Goetz et al. (1985); Meyer et al. 1990; Wolff et al. (1992) |
| 2. Active arachidonic acid metabolism | |
| Presence of cyclooxygenase | Gerritsen et al., 1980; Goehlert et al. (1981); Maurer et al.(1980); Gecse et al. (1981, 1982) |
| Presence of lipoxygenase | Moore et al. (1988a and b) |
| 3. Reactive second messenger systems | |
| Adenylate cyclase | Joó et al. (1975a, b); Palmer (1980); Baranczyk-Kuzma et al.(1992) |
| Guanylate cyclase | Karnushina et al. (1980b); Nakane et al.(1983); Homayoun et al. (1989) |
| Protein kinase C (PK C) | Markovac and Goldstein (1988); Catalán et al.(1989) |
| Calcium/calmodulin-dependent protein kinase II | Deli et al. (1993) |
| 4. Synthesis and secretion of substances | |
| Prostacyclin | Bicknell and Vallee, 1989; Marceau et al., 1989; Rosenthal and Jones, 1988 |
| Endothelium-derived relaxing | Durieu-Trautmann et al. (1993); Morin and Stanboli (1993) |
| and contracting factors | Yoshimoto et al. (1990), |
| Angiotensin-converting enzyme | Brecher et al. (1978); Gimbrone et al. (1979) |
| Growth factors | Estrada et al. (1990); Pákáski et al. 1990) |

## UNIQUE FEATURES OF CEC DIFFERENT FROM ENDOTHELIAL CELLS OF OTHER ORGANS

In addition to fulfilling the function of a common endothelial cell, the CEC possess unique properties, the majority of which, if not all, of which derive from their "epithelial" nature. Morphologically, the presence of tight interendothelial junctions, the paucity of pinocytotic vesicles in the endothelium and the absence of fenestrations are the hallmarks of this barrier (Reese and Karnovsky, 1967; Brightman and Reese, 1969). From physiological point of view, the CEC represent a very tight cellular barrier with high (1900 $\Omega/cm^2$) membrane resistance indicating a low ion permeability (Crone and Olesen, 1982).

**Table 2.** Characteristics of cerebral endothelial cells. II.

| Epithelial and other unique features of cerebral endothelial cells | Reference |
|---|---|
| 1. Structural characteristics | |
| Presence of tight junctions | Reese and Karnovsky (1967) |
| Low pinocytosis | Brightman and Reese (1969); Joó (1971) |
| High number of mitochondria | Oldendorf (1977) |
| 2. Functional aspects | |
| Barrier properties | |
| Impermeability | Ehrlich (1885) |
| High electrical resistance | Crone and Olesen (1982); Butt and Jones (1992) |
| Carrier functions | |
| Glucose transport | Goldstein et al. (1975); Goldstein et al. (1977); Dick and Harik, (1986) |
| Transport of amino acids | Ghandour et al. (1980); Sershen and Lajtha (1976); Hwang et al. (1983) |
| Choline transport | Shimon et al. (1988); Drewes and Singh (1988) |
| Polarity | Goldstein (1979); Betz and Goldstein (1978); Betz and Goldstein, 1980) |
| 3. Biochemical machinery | |
| Lectins | Fatehi et al. (1987) |
| Platelet-derived growth factor receptor | Smits et al. (1989) |
| Presence of histamine $H_2$-receptors | Joó et al. (1976); Karnushina et al. (1980a) |
| Transferrin receptors | Jefferies (1984) |
| Multidrug transport protein P170 | Thiebaut et al. (1989); Greenwood (1992) |
| Expression of brain specific antigens | Michalak et al. (1986); Risau et al. (1986); Miyatani et al. (1990) |

The addition of histamine (in $10^{-4}$ M concentration) to brain surface microvessels resulted in a 75% decrease in resistance (Butt and Jones, 1992). Cimetidine, a histamine $H_2$-receptor blocker inhibited completely the histamine-mediated increase in barrier permeability. The unique properties are summarized in detail in Table 2.

## INDUCTION BY ASTROCYTES AND NEURONES OF BLOOD-BRAIN BARRIER CHARESTERICS IN THE CEC

Among the nearest neighbours of the brain microvessels are the fine processes of astrocytes, in which the entire vascular cylinder is ensheathed. Grafted astrocytes to the anterior eye chamber were seen by Janzer and Raff (1987) to induce some of the BBB properties even in endothelial cells of peripheral origin. However, according to a careful analysis of Holash et al. (1993), neither the anterior eye chamber nor the chorio-allantoic membrane were suitable for BBB induction because of consequence of undesirable inflammatory responses. These authors concluded that, at present, there is no conclusive evidence for a significant role of mature astrocytes in initial expression of the BBB. On the other hand, co-culturing the CEC with primary glial cultures revealed marked changes of several endothelial parameters (Table 3) suggesting that several structural and functional characteristics of cerebral capillaries are indeed determined by the surrounding astrocytes. From the data summarized in Table 3, it can be concluded that, in general, the astrocytic environment can activate mainly those important processes in the cerebral endothelium, which are known to be important in the nourishment of brain. The increase, induced by astrocytes, of transendothelial electric resistance is most likely the result of reduction in the permeability to circulating macromolecules.

**Table 3.** Influence of astrocytes on the CEC

| Effects | Reference |
|---|---|
| Induction of $\gamma$-GTP | De Bault and Cancilla (1979); De Bault (1981) |
| Increase of neutral amino acid transport | Cancilla and De Bault (1983) |
| Influence on polarity of $\alpha$-$^3$H-methylaminoisobutyric acid transport | Beck et al. (1984) |
| Increase of $Na^+$, $K^+$-ATPase and non-specific alkaline phosphatase | Beck et al. (1986) |
| Induction of tight junctions | Arthur et al. (1987), Tao-Cheng et al. (1987) |
| Increase of $\gamma$-GTP by a product of astrocytes | Maxwell et al. (1987) |
| Induction of development of membrane specializations | Shivers et al. (1988); Tao-Cheng and Brightman (1988) |
| Stimulation of glucose uptake | Maxwell et al. (1989) |
| Increase of transmembrane resistance and $\gamma$-GTP | Dehouck et al. (1990) |
| Increase of transmembrane resistance | Kása et al. (1990, 1991) |
| Induction of formation of capillary-like structures | Laterra et al. (1990) |
| Increase of acetylcholinesterase activity | Pákáski and Kása (1992) |

Tontsch and Bauer (1991) reported recently on the induction of BBB-related enzyme activities in CEC by glioma cells and their plasma membranes and by plasma membranes from neuronal cells, while no change was observed in cultures of aortic endothelial cells. The results showed that conditioned media, neither from astrocytes, nor from C6 glioma cells affected $\gamma$-GTP activity, suggesting that *in vitro* induction of enzyme activity is not due to soluble factors. Interestingly, incubation with glial cells or their plasma membranes did not result in a significant increase of BBB-related enzyme activities. These data provide evidence that membrane-bound components, not only of glioma cells, but also of neurons, interact with CEC during cerebral ontogenesis and thus contribute to the establishment of metabolic and structural properties of the BBB.

## CONCLUDING REMARKS

As data accumulates, the CEC reveal their chemical composition, transport processes, and complex and subtle molecular interactions, by which the nourishment of the brain is accomplished. As we have seen, the tissue culture approach has proved its viability and has become a suitable model system for studying the BBB *in vitro*. In particular, it has led to a broadening of our knowledge of the molecular basis of permeability regulation in the CEC. Looking at further areas of possible application, it can be predicted that, in the future, this approach will be used more frequently in drug research, with the aim of developing new means for the improvement of penetration of therapeutic substances through the cerebral endothelium.

## FURTHER PERSPECTIVES

Apart from studying the basic molecular mechanisms underlying the regulation of transport processes in cerebral endothelial cells, the extension of studies using human capillaries to neurological problems such as multiple sclerosis or Alzheimer disease has paved the pathway for the application of the *in vitro* approach of BBB research to human neuropathology. There is no doubt that human brain banks offer an important reservoir of autopsy brain tissue of particular neurological diseases. As a possible example of this approach will be further research directed towards establishing tissue cultures of cerebral endothelial cells from frozen human brain material with diagnosed neurological diseases.

## ACKNOWLEDGEMENTS

I would like to thank the valuable contribution of every co-worker and collaborators who have helped the accomplishment of this task at each phase of research. Part of this review was prepared during a sabbatical leave spent in the Zentrum Anatomie, Georg-August Universität, Göttingen, F.R.G. and in Department of Biology, ÅBO Akademi University, BIOCITY, SF-20520 ÅBO, Finland within the framework of Exchange Programme of Scholars existing between The Hungarian Academy of Sciences and The Academy of Finland. The research was supported by Grants from the Hungarian Research Fund (T-14645), the Ministry of Public Welfare (ETT T-04 029/93) and the U.S.-Hungarian Joint Fund (JFNo. 392).

## REFERENCES

Arthur, F.E., Shivers, R.R., and Bowman, P.D., 1987, Astrocyte mediated induction of tight junctions in brain capillary endothelium: an efficient in vitro model, *Dev. Brain Res.* 36:155.

Baranczyk-Kuzma, A., Audus, K.L., Guillot, F.L., and Borchardt, R.T., 1992, Effects of selected vasoactive substances on adenylate cyclase activity in brain, isolated brain microvessels, and primary cultures of brain microvessel endothelial cells, *Neurochem. Res.* 17:209.

Beck D.W., Roberts R.L., and Olson J.J., 1986, Glial cells influence membrane-associated enzyme activity at the blood-brain barrier, *Brain Res.* 381:131.

Beck, D.W., Vinters, H.V., Hart, M.N., and Cancilla, P.A., 1984, Glial cells influence polarity of the blood-brain barrier, *J. Neuropathol. Exp. Neurol.* 43:219.

Betz, A.L., and Goldstein, J.W., 1978, Polarity of the blood brain barrier: neutral amino acid transport into isolated brain capillaries, *Science* 202:225.

Betz, A.L., and Goldstein, J.W., 1980, The basis for active transport at the blood-brain barrier, *in*: "The Cerebral Microvasculature," H.M. Eisenberg and R.L. Suddith, eds., Plenum Press, New York.

Bicknell, R., and Vallee, B.L., 1989, Angiogenin stimulates endothelial cell prostacyclin secretion by activation of phospholipase $A_2$, *Proc. Natl. Acad. Sci.* 86:1573.

Bowman, P.D., Betz, A.L., Ar, D., Wolinsky, J.S., Penny, J.B., Shivers, R.R., and Goldstein, G.W., 1981, Primary culture of capillary endothelium from rat brain, *In Vitro Cell. Develop. Biol.* 17:353.

Brecher, P., Tercyak, A., Gavras H., and Chobanian, A.V., 1978, Peptidyl dipeptidase in rabbit microvessels, *Biochim. Biophys. Acta*, 526:537.

Brightman, M.W., and Reese, T.J., 1969, Junctions between intimately apposed cell membranes in the vertebrate brain, *J. Cell Biol.* 40:648.

Butt, A.M., and Jones, H.C., 1992, Effect of histamine and antagonists on electrical resistance across the blood-brain barrier in rat brain-surface microvessels, *Brain Res.* 569:100.

Cancilla, P.A., and De Bault L.E., 1983, Neutral amino-acid transport properties of cerebral endothelial cells in vitro, *J. Neuropathol. Exp. Neurol.* 42:191.

Catalán R.E., Martínez A.M., Aragonés M.D., and Fernández I., 1989, Substance P stimulates translocation of protein kinase C in brain microvessels, *Biochem. Biophys. Res. Comm.* 164:595.

Crone, C. and Olesen, S-P., 1982, Electrical resistance of brain microvascular endothelium, *Brain Res.* 241:49.

De Bault, L.E., 1981, γ-glutamyl transpeptidase induction mediated by glial foot processes-to endothelium contact in co-culture, *Brain Res.* 220:432.

De Bault, L.E., and Cancilla, P.A., 1979, γ-glutamyl transpeptidase in isolated brain endothelial cells: Induction by glial cells in vitro, *Science*, 207:653.

Dehouck, M.-P., Méresse S., Delorme P., Fruchart J.C., and Cecchelli R., 1990, An easier, reproducible and mass production method to study the blood-brain barrier in vitro, *J. Neurochem.* 54:1798.

Deli, M., Joó, F., Krizbai, I., Lengyel, I., Nunzi, M., and Wolff, J.R., 1993, Calcium/calmodulin-stimulated protein kinase II is present in primary cultures of cerebral endothelial cells, *J. Neurochem.* 60:1960.

Dick, A.P.K., and Harik, S.I., 1986, Distribution of the glucose transporter in the mammalian brain, *J. Neurochem.* 46:1406.

Diglio C.A., Grammas P., Giacomelli F., and Wiener J., 1982, Primary culture of rat cerebral microvascular endothelial cells. Isolation, growth and characterization, *Lab. Invest.* 46:554.

Drewes, L.R., and Singh, A.K., 1988, Choline transport and metabolism in soman- or sarin-intoxicated brain, *J. Neurochem.* 50:868.

Durieu-Trautmann, O., Fédérici, C, N. Créminon, Foignant-Chaverot, N., Roux, F., Claire, M., Strosberg, A.D., and Couraud, P.O., 1993, Nitric oxide and endothelin secretion by brain microvessel endothelial cells: regulation by cyclic nucleotides, *J. Cell Physiol.* 155:104.

Edvinsson, L., 1982, Sympathetic control of cerebral circulation, *TINS*, 5:425.

Ehrlich, P., 1885, "Das Sauerstoff-bedürfnis des Organismus. Eine Farbenanalytische Studie." A Hirschwald, Berlin.

Estrada, C., Bready J., Berliner, J., Pardridge W.M,. and Cancilla P.A., 1990, Astrocyte growth stimulation by a soluble factor produced by cerebral endothelial cells, *J. Neuropathol. Exp. Neurol.* 49:539.

Fatehi, M. I., Gerhart, D.Z., Myers, T.G., and Drewes, L.R., 1987, Characterization of the blood-brain barrier: glycoconjugate receptors of 14 lectins in canine brain, cultured endothelial cells and blotted membrane proteins, *Brain Res.* 415:30.

Gecse, Á., Ottlecz, A., Mezei, Zs., Telegdy, G., Dux, E., and Karnushina, I., 1981, Capillaries as the source of prostacyclin and prostaglandin synthetases in brain, *in*: "Cardiovascular Physiology, Microcirculation and Capillary Exchange," A.G.B. Kovách, J. Hamar, and L. Szabó, eds., Vol. 7, Akadémiai Kiadó, Budapest.

Gecse, Á., Ottlecz, A., Mezei, Zs., Telegdy, G., Joó, F., Dux, E., and Karnushina, I., 1982, Prostacyclin and prostaglandin synthesis in isolated brain capillaries, *Prostaglandins*, 23:287.

Gerritsen, M.E., Parks, T.P., and Printz, M.P., 1980, Prostaglandin endoperoxidase metabolism by bovine cerebral microvessels, *Biochim. Biophys. Acta*, 619:196.

Ghandour M.S., Langley O.K., and Varga V., 1980, Immunohistological localization of γ-glutamyltranspeptidase in cerebellum at light and electron microscope levels, *Neurosci. Lett.* 20:125.

Gimbrone, M.A., Majeau, G.R., Atkinson, W.J., Sadler, W., and Cruise, S.A., 1979, Angiotensin-converting enzyme activity in isolated brain microvessels, *Life. Sci.* 25:1075.

Goehlert, U.G., Ng Ying Kin, N.M.K., and Wolfe, L.S., 1981, Biosynthesis of prostacyclin in rat cerebral microvessels and the choroid plexus, *J. Neurochem.* 36:1192.

Goetz, I.E., Warren, J., Estrada, C., Roberts, E., and Krause, D.N., 1985, Long-term serial cultivation of arterial and capillary endothelium from adult bovine brain, *In Vitro Cell. Develop. Biol.* 21:172.

Goldstein, G.W., 1979, Relation of potassium transport to oxidative metabolism in isolated brain capillaries, *J. Physiol. (Lond.)* 286:185.

Goldstein, G.W., Csejtey, J., and Diamond, I., 1977, Carrier mediated glucose transport in capillaries isolated from rat brain, *J. Neurochem.* 28:725.

Goldstein, G., Wolinsky, J., Csejtey, J., and Diamond, I., 1975, Isolation of metabolically active capillaries from rat brain, *J. Neurochem.* 25:715.

Greenwood, J., 1992, Characterization of a rat retinal endothelial cell culture and the expression of P-glycoprotein in brain and retinal endothelium in vitro, *J. Neuroimmunol.* 39:123.

Homayoun, P., Lust, W.D., and Harik, S.I., 1989, Effect of several vasoactive agents on guanylate cyclase activity in isolated rat brain microvessels, *Neurosci. Lett.* 107:273.

Holash, J.A., Noden, D.M., and Stewart, P.A., 1993, Re-evaluating the role of astrocytes in blood-brain barrier induction, *Exp. Dynamics,* 197:14.

Hwang, S. M., Miller, M., and Segal, S., 1983, Uptake of L-[$^{14}$C] proline by isolated rat brain capillaries, *J. Neurochem.* 40:317.

Janzer, R., and Raff, M.C., 1987, Astrocytes induce blood-brain barrier properties in endothelial cells, *Nature,* 325:253.

Jefferies, W.A., Brandon, M.R., Hunt, S.V., Williams, A.F., Gatter, K.C., and Mason, D.Y., 1984, Transferrin receptor on endothelium of brain capillaries, *Nature,* 312:162.

Joó, F., 1971, Increased production of coated vesicles in the brain capillaries during enhanced permeability of the blood brain barrier, *Brit. J. Exp. Path.* 52:646.

Joó, F., 1985, The blood-brain barrier in vitro: Ten years of research on microvessels isolated from the brain, *Neurochem. Int.* 7:1.

Joó, F., 1992, The cerebral microvessels in culture, an update, *J. Neurochem.* 58:1.

Joó, F., 1993, The blood-brain barrier in vitro: The second decade, *Neurochem. Int.* 23:499.

Joó, F., and Karnushina, I., 1973, A procedure for the isolation of capillaries from rat brain, *Cytobios,* 8, 41.

Joó, F., Rakonczay, Z., and Wollemann, M., 1975a, cAMP-mediated regulation of the permeability in the brain capillaries, *Experientia,* 31:582.

Joó, F., Szucs, A., and Csanda, E., 1976, Metiamide-treatment of brain oedema in animals exposed to $^{90}$Yttrium irradiation, *J. Pharm. Pharmacol.* 28:162.

Joó, F., Tóth, I., and Jancsó, G., 1975b, Brain adenylate cyclase: its common occurrence in the capillaries and astrocytes, *Naturwissenschaften,* 8:397.

Karnushina, I.L., Palacios, J.M., Barbin, G., Dux, E., Joó, F., and Schwartz, J.C., 1980a, Studies on a capillary-rich fraction isolated from brain: histaminic components and characterization of the histamine receptors linked to adenylate cyclase, *J. Neurochem.* 34:1201.

Karnushina, I., Tóth, I., Dux, E., and Joó, F., 1980b, Presence of the guanylate cyclase in brain capillaries: histochemical and biochemical evidences, *Brain Res.* 189:588.

Kása, P., Pákáski, M., and Joó F., 1991, Endothelial cells from human fetal brain microvessels may be cholinoceptive but do not synthesize acetylcholine, *J. Neurochem.* 56:2143.

Kása, P., Pákáski, M., Joó F., Sershen H., and Lajtha A., 1990, Human in vitro blood-brain barrier model system for studying the effects of drugs, *Eur. Soc. Neurochem., 8th General Meeting,* Leipzig.

Krause, D., Goetz I.E., Estrada C., and Roberts E., 1982, Culture of vascular endothelial cells from adult bovine brain, *Fed. Proc.* 41:1760.

Laterra J., Guerin C., and Goldstein G.W., 1990, Astrocytes induce neural microvascular endothelial cells to form capillary-like structures in vitro, *J. Cell. Physiol.* 144:204.

Marceau F., Tremblay B., Couture R., and Regoli D., 1989, Prostacyclin release induced by neurokinins in cultured human endothelial cells, *Can. J. Physiol. Pharmacol.* 67:159.

Markovac, J., and Goldstein G.W., 1988, Transforming growth factor beta activates protein kinase C in microvessels isolated from immature rat brain, *Biochem. Biophys. Res. Commun.* 150:575.

Maurer, P., Moskowitz, M. A., Levin, L., and Melamed, E., 1980, The synthesis of prostaglandins by bovine cerebral microvessels, *Prostaglandins Med.* 4:153.

Maxwell, K., Berliner, J.A., and Cancilla, P.A., 1987, Induction of γ-glutamyl transpeptidase in cultured cerebral endothelial cells by a product released by astrocytes, *Brain Res.* 410:309.

Maxwell, K., Berliner, J.A,. and Cancilla, P.A., 1989, Stimulation of glucose analogue uptake by cerebral microvessel endothelial cells by a product release by astrocytes, *J. Neuropathol. Exp. Neurol.* 48:69.

Meyer, J., Mischeck, U., Veyhl, M., Henzel, K., and Galla H.J., 1990, Blood-brain barrier characteristic enzymatic properties in cultured brain capillary endothelial cells, *Brain Res.* 514:305.

Michalak, T., White, F. P., Gard, A. L., and Dutton, G. R., 1986, A monoclonal antibody to the endothelium of rat brain microvessels, *Brain Res.* 379:320.

Miyatani, N., Kohriyama, T., Maeda, Y., and Yu, R. K., 1990, Sulfated glucuronyl paragloboside in rat brain microvessels, *J. Neurochem.* 55:577.

Morin, A.M., and Stanboli, A., 1993, Nitric oxide synthase in cultured endothelial cells of cerebrovascular origin: cytochemistry, *J. Neurosci. Res.* 36:272.

Moore, S.A., Prokuski, L.J., Figard, P.H., Spector, A.A., and Hart, M.N., 1988a, Murine cerebral microvascular endothelium incorporate and metabolize 12-hydroxyeicosatetraenoic acid, *J. Cell Physiol.* 137:75.

Moore, S.A., Spector, A.A., and Hart, M.N., 1988b, Eicosanoid metabolism in cerebromicro-vascular endothelium, *Am. J. Physiol.* 254:C37.

Nakane, M., Ichikawa, M., and Deguchi, T., 1983, Light and electron microscopic demonstration of guanylate cyclase in rat brain, *Brain Res.* 273:9.

Oldendorf, W. H., 1977, The blood-brain barrier, *Expl. Eye Res. Suppl.* 25:177.

Orlowski M., Sessa G., and Green J.P., 1974, γ-Glutamyl transpeptidase in brain capillaries: Possible site of a blood brain barrier for amino acids, *Science*, 184:66.

Panula, P., Joó, F., and Rechardt, L., 1978, Evidence for the presence of viable endothelial cells in cultures derived from dissociated rat brain, *Experientia*, 34:95.

Pákáski, M., and Kása, P., 1992, The cholinergic system in embryonic human brain capillaries and endothelial cell cultures, *Neurochem. Int.* 21:129.

Pákáski, M., Kása, P., Joó, F., and Wolff, J.R., 1990, Cerebral endothelial cell-derived laminin promotes the outgrowth of neurites in CNS neuronal cultures, *Int. J. Devl. Neurosci.* 8:193.

Palmer, G.C., 1980, Beta adrenergic receptors mediate adenylate cyclase responses in rat cerebral capillaries, *Neuropharmacology*, 19:17.

Phillips P., Kumar P., Kumar S., and Woghe M., 1979, Isolation and characterization of endothelial cells from rat and cow brain white matter, *J. Anat.* 129:261.

Reese, T.S., and Karnovsky, M.J., 1967, Fine structural localization of a blood-brain barrier to exogenous peroxidase, *J. Cell Biol.* 34:207.

Rosenthal, M.D., and Jones, J.E., 1988, Release of arachidonic acid from vascular endothelial cells: Fatty acyl specificity is observed with receptor-mediated agonists and with the calcium ionophore A23187 but not with melittin, *J. Cell Physiol.* 136:333.

Risau, W., Hallmann, R., and Albrecht, U., 1986, Differentiation-dependent expression of proteins in brain endothelium during development of the blood-brain barrier, *Dev. Biol.* 117:537.

Sershen, H., and Lajtha, A., 1976, Capillary transport of amino acids in the developing brain, *Expl. Neurol.* 53:465.

Shimon, M., Egozi, Y., Kloog, Y., Sokolovsky, M., and Cohen, S., 1988, Kinetics of choline uptake into isolated rat forebrain microvessels: evidence of endocrine modulation, *J. Neurochem.* 50:1719.

Shivers, R.R., Arthur, F.E., and Bowman, P.D., 1988, Induction of gap junctions and brain endothelium-like tight junctions in cultured bovine endothelial cells: local control of cell specialization, *J. Submicrosc. Cytol. Pathol.* 20:1.

Smits, A., Hermansson, M., Nistér, M., Karnushina, I., Heldin, C.H., Westermark, B., and Funa, K., 1989, Rat brain capillary endothelial cells express functional PDGF B-type receptors, *Growth Factors*, 2:1.

Takasoto, Y., Rapoport, S.I., and Smith, Q.R., 1984, An in situ brain perfusion technique to study cerebrovascular transport in the rat, *Am. J. Physiol.* 247:H484.

Tao-Cheng, J.H., and Brightman, M., 1988, Development of membrane interactions between brain endothelial cells and astrocytes, *Int. J. Devl. Neurosci.* 6:25.

Tao-Cheng, J.H., Nagy, Z., and Brightman, M.W., 1987, Tight junctions of brain endothelium in vitro are enhanced by astroglia, *J. Neurosci.* 7: 3293.

Thiebaut, F., Tsuruo, T., Hamada, H., Gottesman, M.M., Pastan, I., and Willingham, M.C., 1989, Immunohistochemical localization in normal tissues of different epitopes in the multidrug transport protein P170: Evidence for localization in brain capillaries and crossreactivity of one antibody with a muscle protein, *J. Histochem. Cytochem.* 37:159.

Tontsch, U., and Bauer H.C., 1989, Isolation, characterization and long-term cultivation of porcine and murine cerebral capillary endothelial cells, *Microvasc. Res.* 37:148.
Tontsch, U., and Bauer, H.C., 1991, Glial cells and neurons induce blood-brain barrier related enzymes in cultured cerebral endothelial cells, *Brain Res.* 539:247.
Wolff, J.E.A., Belloni-Olivi, L., Bressler, J.P., and Goldstein, G.W., 1992, γ-Glutamyl transpeptidase activity in brain microvessels exhibits regional heterogeneity, *J. Neurochem.* 58:909.
Yoshimoto, S., Ishizaki, Y., Kurihara, H., Sasaki, T., Yoshizumi, M., Yanagisawa, M., Yazaki, Y., Masaki, T., Takakura, K., and Murota, S.-I., 1990, Cerebral microvessel endothelium is producing endothelin, *Brain Res.* 508:283.

# STUDIES ON AN IMMORTALIZED BRAIN ENDOTHELIAL CELL LINE : CHARACTERIZATION, PERMEABILITY AND TRANSPORT

N. Joan Abbott [1], Pierre-Olivier Couraud [2], Françoise Roux [3] and David J. Begley [1]

[1] Physiology Group, Biomedical Sciences Division, King's College, Strand, London WC2R 2LS, UK;
[2] Laboratoire d'Immunopharmacologie Moléculaire, Institut Cochin de Génétique Moléculaire, 22 rue Méchain, 75014 Paris, France;
[3] U 26 INSERM, Hôpital Fernand Widal 200, rue du Faubourg Saint-Denis, 75475 Paris cedex 10, France.

## INTRODUCTION : NEED FOR *IN VITRO* PREPARATIONS

A great deal of information is now available concerning the structure and function of the blood-brain barrier (BBB) *in vivo*. However, for understanding of molecular, membrane and cellular mechanisms, it is necesary to work on preparations where it is possible to apply the modern techniques of cell biology and physiology at the single cell level. For this reason, several *in vitro* preparations of brain endothelium have been developed, with the long-term aim of establishing an isolated blood-brain barrier model.

The group in London has been interested in several aspects of brain endothelial cell function, from the manner in which the blood-brain barrier is established and maintained, to mechanisms controlling permeability and transport of ions, amino acids, proteins and drugs, to questions concerning modulation of the BBB, and whether modulation could be used for therapeutic purposes. We need to be able to measure transport across the endothelial membranes in the presence and absence of inhibitors, to study membrane ionic channels, to examine the conditions that lead to a 'tight' barrier, and to determine the intracellular events activated by potential BBB modulators. All these studies require conditions in which the composition of the bathing medium can be controlled. We chose to develop methods to study rat brain endothelial cells, primarily to permit parallel *in situ* and *in vitro* studies. This would also enable us to make use of the wealth of information on the *in situ* rat blood-brain barrier, a field in which Michael Bradbury has made many major contributions, and has published several authoritative reviews. As we also anticipated situations where we would wish to examine the endothelial cells from animals pretreated in certain ways (e.g. with toxins, hormones, diet), the rat had advantages over other sources of brain endothelial cells, chiefly bovine and porcine brain. This review deals with the development of the primary cultured rat brain endothelial cell preparation, and recent collaborative work with groups in Paris on an immortalized rat brain endothelial cell line, RBE4.

## Freshly-isolated brain microvessels

The earliest work on *in vitro* models of the brain endothelium used microvessel fragments isolated from the brain. Some procedures used mechanical dissociation followed by filtration to separate microvessels, but more successful cleaning of the vessels was achieved with enzymatic dissociation. Such preparations were useful for the identification of membrane receptors, and for studies of certain types of carrier-mediated transport (Pardridge, 1988). However, in these microvessels, the medium has access to both apical and basal membranes, so that neither polarized features of transport, nor transcellular movement, can be studied. The cells are also depleted of ATP, with consequent run-down of ionic gradients and disturbances of ion- and ATP-dependent transport. Recent electron microscopy of freshly- isolated microvessels has shown that although the tight junctional region is relatively intact, mitochondria are vacuolated, the luminal membrane is distorted by the presence of numerous microvillar structures, and many clusters of vesicular profiles opening to plasmalemmal invaginations are present (Lane et al., 1994). The abnormal morphology makes it hard to interpret the results of transport studies conducted on such freshly-isolated vessel fragments.

## Primary culture of rat brain microvascular endothelial cells

Although isolated brain microvessels are not themselves suitable for general experimental study, they provide a good starting material for primary cultures (reviewed by Joó, 1992). Hughes and Lantos (1986) developed a method for rat brain endothelial cells based on that of Bowman et al. (1981, 1982), which was further modified simultaneously by the London group (Abbott et al., 1992) and by the group of Françoise Roux in Paris (Roux et al., 1994). In the London method, grey matter from 2-month old Lewis rats is chopped into uniform 1-2 mm pieces, after careful removal of meninges and large surface vessels. Two stages of enzymic digestion in collagenase/dispase are used, the first to separate microvessels from other brain cells, the second to free pericytes trapped in the perivascular basement membrane. Finally, microvessel fragments are separated from single cell contaminants by centrifuging on a Percoll gradient. The culture medium was originally Hams F10 (80%) with 16% bovine plasma-derived serum (treated to remove PDGF), supplemented (4%) with a mixture containing glutamine, heparin, endothelial cell growth supplement (ECGS) and antibiotics, but recently Dulbecco's MEM substituted for Hams F10 has been found to give improved growth. For routine assays, cells are grown in multiwell plastic plates, coated with rat-tail collagen, but the cells can also be grown on porous filters and on glass treated to improve the binding of collagen (Nobles and Abbott, 1994; also submitted).

The Paris group has used a similar protocol, but found that the use of rat plasma-derived serum (RPDS) significantly delays the senescence of these primary cells. Vessel fragments are seeded on to an extracellular matrix (ECM) synthesized by cultured corneal endothelial cells, in order to stimulate the growth of the brain endothelial cells.

Contaminating cells surrounding endothelial cell colonies are removed under the microscope with a modified glass Pasteur pipette in order to produce pure cultures.

The primary cultured rat brain endothelial cells grow out from the vessel fragments as whorls of spindle-shaped cells, and reach confluence within 7-10 days. The cells show characteristic endothelial properties (expression of Factor VIII-related antigen, angiotensin converting enzyme, binding of lectin from *Ulex europeus*, tight junctions, expression of ZO-1), and preserve two key markers of brain endothelial phenotype, expression of the transferrin receptor and the drug efflux pump, MDR or P-glycoprotein (Abbott et al., 1992a; Greenwood, 1992). The cells have been used for analysis of lipid synthesis and its regulation be serum lipoproteins (Roux et al., 1989), for studies of membrane permeability and transport (ions, Abbott & Revest, 1990; the immunosuppressive drug cyclosporin, Begley et al., 1990; proteins, Ramlakhan, 1990), to characterize the receptors present capable of elevating intracellular calcium (Revest et al., 1991), and to demonstrate chemical toxicity (Greenwood et al., 1991; Romero et al., 1992).

A disadvantage of the primary culture is that it is difficult to eliminate all non-endothelial cell contaminants (pericytes, leptomeningeal cells, smooth muscle), which may form up to 5-10% of the population. This is not a major problem for single cell studies, where it is possible to identify endothelial cells based on their morphology, or for transport studies, where the results will reflect the activity of the majority endothelial cell type. However, when contaminating cells occupy space within the monolayer, they frequently cause a flaw or hole in the layer, because the endothelial cells appear to be unable to grow over them. This becomes a limitation when attempting to form a monolayer suitable for studies of transendothelial permeability and transport. The preparation of the primary culture is also labour intensive (around 1 full day to prepare a 60/96 well plate) and relatively expensive. For these reasons, several groups have been investigating the possibility of using immortalized cell lines as an alternative.

## THE RBE4 CELL LINE

### Immortalization

Primary cultures of rat brain endothelial cells prepared by the group of Françoise Roux have been successfuly immortalized by Pierre-Olivier Couraud. Second passage brain endothelial cells were transfected with the plasmid pE1A-neo, which carries the entire E1A region of adenovirus 2 (conferring immortalization without oncogenic transformation), and the neo gene for resistance to the aminoglycoside G418, geneticin (Roux et al., 1994). Cells were grown in medium containing G418, and clones selected after 2 weeks. The clone RBE4 showed typical endothelial morphology, and has proved to retain many further brain endothelial characteristics. Southern blot hybridization with the SmaI-SacI 764 bp E1A fragment indicated that the immortalizing sequence was stably integrated into the RBE4 genome (Roux et al., 1994). This section reviews the progess made in characterization and study of the properties of the RBE4 cell line, and assesses its suitability as an in vitro BBB model.

### Characterization

Parallel studies on the RBE4 cell line have been conducted in Paris (groups of F.

Roux and P.-O. Couraud) and in London (group of N.J. Abbott and D.J. Begley). Although there is general agreement concerning the endothelial nature of the cells, some differences have been reported between the groups, which may prove important in developing a consistent BBB model. Both groups use cells from the same 'parent clone' stock, between passage 30 and 60. Cells are grown in a culture medium containing alpha MEM/Ham's F10 (1:1), supplemented with 2 mM glutamine, 10% heat inactivated fetal calf serum, 1 ng/ml basic fibroblast growth factor, bFGF, and 300 μg/ml geneticin (G418). However, the London group routinely plates cells onto a substrate coated with laboratory prepared rat-tail collagen cross-linked in ammonia vapour, while the Paris groups have generally used purified type I collagen or fibronectin without treatment in ammonia vapour.

In the preconfluent state, RBE4 cells are generally spindle-shaped, but a multipolar morphology is also seen. At confluence, cells in the normal medium form a somewhat irregular monolayer, with some areas of 'overgrowth', where the cells continue to proliferate and bud from the monolayer. Cell proliferation is reduced, and growth to confluence is slower, in medium without bFGF, and on plates not coated with collagen (Roux et al., 1994). The endothelial nature of the cells is shown by the fact that in preconfluent cultures, all cells show binding of the lectin from *Bandeiraea* (= *Griffonia*) *simplicifolia*, and Factor VIII-related antigen (von Willebrand Factor) is expressed throughout the cultures, particularly in chains of cells in end-to end contact (Durieu-Trautmann et al., 1993; Roux et al., 1994).

Roux et al. (1993, 1994) report that RBE4 monolayers observed 2-3 days post-confluence are not perfectly homogeneous, tending to form 'islets' of cells of different morphology, giving rise to 'sprouts' growing above the level of the monolayer, which become organised into strands or tubular structures in high concentrations of bFGF. This process resembles the 'angiogenic response' induced by bFGF that has also been reported in primary endothelial cultures (Ingber and Folkman, 1989; Iruela-Arispe et al., 1991). The formation of the tubular structures in RBE4 cultures is also stimulated by glial factors (Roux et al., 1993, 1994).

The monolayers grown in London appear somewhat more uniform, without 'islets' or 'sprouts', but observations are generally made, and experiments done, around the time of confluence. Small differences in the cultures between the two groups may also arise from differences in the collagen coating used as substrate. It is possible that the pure type I collagen coating gives a thicker and less uniform extracellular matrix, conditions that appear to favour capillary tube formation and BBB phenotypic differentiation in primary cultures (Montesano et al., 1986; Laterra et al., 1990; Minakawa et al., 1991). There is agreement that addition of agents that elevate intracellular cyclic AMP reduces cell proliferation (Roux et al., 1994) and leads to a more regular spindle-shaped/cobblestone morphology without overgrowth (Rist et al., 1994).

The extent to which RBE4 cells retain BBB markers is of interest. Freshly-isolated brain capillaries contain high levels of γ-glutamyl transpeptidase (γ-GTP) and alkaline phosphatase (AP), but both enzymes are progressively lost in primary cultures of brain endothelial cells grown alone (De Bault and Cancilla, 1980; Meyer et al., 1990). Co-culture with astrocytes can cause re-induction of the enzymes. Roux et al. (1994) reported that AP and γGTP are expressed in the sprouts and strands that are produced by 'overgrowing' RBE4 monolayers a few days after confluence, but not generally in the cells of the monolayer. Elevation of cAMP caused an increase in the expression of AP, while culture of the cells in the presence of glial factors (C6 glioma and astrocyte membranes or conditioned medium) caused increased expression of both enzymes, γGTP and AP. These observations indicate that RBE4 cells retain the ability to express certain BBB markers, and that expression can be up- and down-regulated by factors in the substrate or medium, including factors produced by glial cells.

### Cytoskeletal differentiation

Studies on a variety of epithelial cell types have shown that certain cytoskeletal specializations are necessary to produce a tight monolayer, in particular, aggregation of filamentous actin at the cell margins in the region of the tight junctions, and expression of the tight-junction associated cytoskeletal protein, ZO-1. ZO-1 is expressed by primary cultures of pure endothelial cells along lines of cell:cell contact (Greenwood, 1992), but not at the free leading edge of cell outgrowth, consistent with a role in organizing tight junctions. However, as ZO-1 is also expressed in other cell types without tight junctions, it may not be a good diagnostic marker for tight junctional differentiation.

Studies on filamentous actin have proved instructive. Primary cultures of rat brain endothelial cells grown alone show a relatively homogeneous distribution of actin throughout the cytoplasm (Abbott et al., 1992a). Rubin et al. (1991) showed that in primary cultures of bovine brain endothelial cells, the tightness of the monolayer was correlated with the extent to which the cells showed a marginal distribution of filamentous actin, as determined by staining with fluorescent phalloidin. Pure endothelial cultures in normal medium showed diffuse cytoplasmic actin and low transendothelial resistance, while elevation of cyclic AMP (cAMP) and addition of astrocyte-conditioned medium (ACM) caused aggregation of marginal actin and a rise in resistance. In primary cultures of rat brain endothelial cells, elevation of cAMP and addition of ACM are again needed to cause the marginal distribution of actin (Rist et al., 1994). However, in the RBE4 cells, marginal actin appears at confluence, whether or not the cells have raised cAMP or added ACM (Rist et al., 1994). This suggests that the RBE4 cells retain a robust epithelioid phenotype, in which the cytoskeletal organization necessary for tight junction formation is triggered by cell:cell contact. In this respect, they may have advantages over cells in primary culture, where such organization appears to require additional triggering factors. These observations become relevant in the development of the RBE4 cells as an *in vitro* BBB model (see below).

### Permeability of the RBE4 monolayer

There are now several reports showing that primary cultures of brain endothelial cells can be grown on porous filters, and that the tightness of the monolayer can be increased, either by elevation of cAMP in the presence of ACM (Rubin et al., 1991; Wolburg et al., 1994), or by co-culture with astrocytes (Dehouck et al., 1990). RBE4 cells have been successfully grown on porous filter inserts (Romero et al., 1994), under conditions where astrocytes can be grown on the base of the culture wells below the filters. In studies using ICN Cellagen (collagen) filters, the permeability to [$^{14}$C]sucrose of the RBE4 monolayer was ~ 11.9 x $10^{-3}$ cm.min$^{-1}$, which dropped to 6.4 x $10^{-3}$ cm.min$^{-1}$ with elevation of cAMP, and 2.3 x $10^{-3}$ cm.min$^{-1}$ following co-culture with astrocytes, an approximately 5-fold reduction in permeability compared to untreated RBE4 cells alone. While these permeabilities are still greater than those of the BBB *in situ*, they approach the values reported for primary cultures of brain endothelial cells (Dehouck et al., 1990; Raub et al., 1992) which have been used for studies of transendothelial permeability. Further improvements in culture conditions are likely to improve the usefulness of the RBE4 cell line for studies of transendothelial permeability.

### Receptors and secretion

Brain endothelial cells have been shown to possess a number of receptor types, with activation leading to many different effector responses including secretion of substances to the luminal and abluminal sides of the cell, and modulation of tight junction permeability. Thus endothelium-dependent relaxation is mediated by the production and release of prostacyclin, and of endothelium-derived relaxing factor, identified as nitric oxide (NO) or a closely related compound. Endothelial cells also release a potent vasoconstrictor, endothelin (ET)-1. Durieu-Trautmann et al. (1993) have shown that RBE4 cells release NO when induced by cytokines IFNγ and TNFα, and the cells constitutively secrete ET-1. cAMP potentiates the cytokine-induced NO release, and modulates ET-1 secretion, with low concentrations inhibiting secretion, and higher concentrations stimulating it. The stimulation is blocked by the calcium-chelator EGTA and the calmodulin antagonist W7, but not by protein kinase C inhibitors, suggesting involvement of the calmodulin branch of the calcium second messenger system. In addition, ET-1 secretion is inhibited by a factor produced by primary cultures of astrocytes, tentatively identified as an iron-nitroso compound (Fédérici et al., in press). As both vascular smooth muscle and astrocytes may be targets for NO, and ET-1 is a powerful vasoconstrictor, these observations suggest that brain endothelial cells may play an important role in the direct and indirect regulation of cerebrovascular tone, and that RBE4 cells are a good model system for examining this activity.

Several vasoactive substances have been implicated in the modulation of endothelial tight junction permeability, including bradykinin, histamine, serotonin and ATP (reviewed in Abbott & Revest, 1991). Olesen (1989) has proposed that a rise in intracellular calcium may be responsible for some of the receptor-mediated increases in BBB permeability observed in pial vessels. This idea was tested directly in primary cultures of rat brain endothelial cells. Revest et al. (1991) used cells loaded with the fluorescent calcium-indicator fura-2, and showed that bradykinin, histamine and ATP (but not serotonin) caused elevation of intracellular calcium, by a mechanism that appeared to depend on release of calcium from intracellular stores. Studies on single intact frog mesenteric vessels have shown a good correlation between rises in intracellular calcium and increase in transendothelial permeability, consistent with a $Ca^{2+}$-dependent contractile event pulling open the tight junctions (He et al., 1990; Curry, 1992; He and Curry, 1993) .

In experiments on RBE4 cells, rises in intracellular calcium could be elicited by histamine, bradykinin and ATP, although not all cells were responsive to all three agents, and ATP gave the most consistent response. Further studies on the RBE4 ATP receptors showed that the order of effectiveness was ATP=UTP>ADP>AMP, consistent with the presence of a P2U purinoceptor (nucleotide-type receptor) (Nobles et al., 1994; 1995). All four agents are likely to be released in the region of cerebral blood vessels, from platelets on the blood side, nerve endings on the brain side, in normal physiology and in some pathological conditions. These studies show that brain endothelial cells possess receptors capable of generating changes in intracellular calcium for the modulation of their effector activities, including changes in endothelial permeability and cerebrovascular tone. They also confirm that for receptors as for some other properties, the behaviour of RBE4 cells is similar to that of brain endothelial cells in primary culture, and of the BBB *in situ*.

**P-glycoprotein (multidrug resistance protein, MDR, or drug efflux pump)**

P-glycoprotein (Pgp) is the term for a family of intrinsic membrane proteins that is able to transport a range of lipophilic molecules out of the cell, via an ATP-dependent process. Substrates include several cytotoxic compounds and lipophilic drugs including vinblastine, vincristine and cyclosporin, while chlorpromazine and verapamil appear to act as inhibitors. Pgp is expressed in certain epithelia at the boundary between the body and the external world, and at the luminal membrane of cerebral endothelial cells, but not in the endothelia of most other body tissues (Cordon-Cardo et al., 1989). This distribution implies that Pgp has an important role in transporting lipophilic agents out of the body and out of the nervous system, and hence plays an important protective role at the blood-brain barrier. This is confirmed by a recent study showing that a transgenic 'knock-out' mouse lacking the *mdr1a* gene controlling Pgp expression, is much less resistant to the central antihelminthic neurotoxin ivermectin than is the normal mouse (Schinkel et al., 1994)

Two antibodies against intracellular epitopes of Pgp, JSB-1 and C219, have been used to demonstrate expression of Pgp in primary cultured rat brain endothelium (Greenwood, 1992). Both JSB-1 and C219 give positive immunostaining in confluent RBE4 cells (Begley et al., 1994; Lechardeur et al., 1994). Immunoblots of the cultures using the C219 antibody confirmed staining of a 160-180 kDa band, corresponding to the Pgp band isolated from the cell membranes of KB/vbl multidrug resistant cells. These studies confirm the expression of Pgp in the RBE4 cell line.

Functional studies of Pgp activity in RBE4 cells have been conducted. The London group has shown that the uptake of [$^3$H] colchicine can be used as a measure of Pgp activity, with enhanced uptake (indicating inhibition of efflux) being seen in the presence of known inhibitors of Pgp, vincristine, chlorpromazine and verapamil (Begley et al., 1994). Interestingly, the agent AZT, used to treat patients with HIV, also caused increased [$^3$H]colchicine accumulation. The Paris group has used [$^3$H]vinblastine as substrate, and shown enhanced uptake and reduced efflux in the presence of verapamil and chlorpromazine. These studies confirm the presence of functional Pgp in the RBE4 cell line, and illustrate the usefulness of this model system for examining the role of Pgp in blood-brain barrier function. As no satisfactory QSAR (quantitative structure-activity relation) model yet exists to predict the nature of compounds that will be substrates or inhibitors for Pgp, a screening assay for examining new and existing drugs for interaction with Pgp is likely to be extremely useful, for studies to increase our basic understanding, and for pharmaceutical companies to use in drug assessment.

**Toxicology studies**

Certain neurotoxins exert their damaging effects on neurones as a result of primary damage to the blood-brain barrier, neuronal damage then being a secondary consequence of BBB breakdown (Romero et al., 1991; Abbott et al., 1992b). Cultured endothelial cells and astrocytes have proved useful in establishing the site and mode of action of a range of such toxins (Romero et al., 1992), and a number of assays have been devised to test for toxic damage to the plasmalemma, mitochondria, and specific enzyme systems. It was of interest to establish the relative sensitivity of primary cultured rat brain endothelial cells and RBE4 cells. For toxins and assays so far tested, the dose-response curves for primary endothelial cells and RBE4 cells are quite similar; thus using cumulative glucose consumption as a measure of cellular metabolism, the toxin dinitrobenzene at 0.5 mM caused an 80-90%

stimulation of consumption in both cell preparations (Romero, in preparation). This example illustrates the usefulness of the RBE4 cell line as a screening system for assessing the toxic effects of agents acting on the brain endothelium.

### The RBE4 cell line as an assay sytem for determining drug penetration across the blood-brain barrier

Pharmaceutical companies are interested in simple *in vitro* blood-brain barrier models that would enable them to test new drugs for penetration across the blood-brain barrier. Such models would be useful not only for improving permeability characteristics of drugs to be targeted to the brain, but also for confirming exclusion of agents designed to act peripherally. Culture models of brain endothelium are attractive candidates as model systems, since the cells can be grown on filters separating two small volume fluid compartments, and drug penetration across the layer simply determined, with many fewer problems associated with binding and metabolism than are encountered in *in situ* studies. Provided a sensitive assay for the drug exists (HPLC, fluorescence), its concentration on the 'trans' side of the filter can be determined, avoiding the need for radiolabelling of test compounds.

Primary culture assay systems for BBB permeability have been reported by several groups, including Pardridge et al. (1988) and Dehouck et al. (1992). As expected, the correlation between permeability determined *in vitro* and *in situ* improves with the tightness of the monolayer, but it is interesting that the rank order for penetration is well predicted even in the leakier cultures. As more is learned about the inductive processes that cause the brain endothelium to become tight, and to express the full BBB complement of polarized membrane molecules and transporters, it will become possible to make the culture models resemble ever more closely the *in vivo* condition. The RBE4 cell line is likely to contribute to and benefit from this improved understanding, and to make an *in vitro* permeability assay that is simple, reproducible and inexpensive to use.

### RBE4 cell implants in experimental gliomas

Once immortalized, the RBE4 cell line offers further opportunities for genetic manipulation. There has been a great deal of interest in the use of gene therapy to target therapeutic biological substances directly to the brain and to brain tumours. In a recent collaboration between P.-O. Couraud and J. Laterra and co-workers (Lal et al., 1994), RBE4 cells further modified to express the β-galactosidase reporter (RBEZ) were implanted into rat brain and flank together with glioma cells. The RBEZ cells survived in both implants, where they were found to be associated with microvascular profiles, and to express the Glut-1 glucose carrier characteristic of brain endothelial cells. These studies hold promise for the use of the RBE4 cell line for studies of gene therapy.

## CONCLUSION

This review has shown that RBE4 cells preserve many important properties of the *in situ* BBB, and that as for primary cultured brain endothelial cells, other BBB properties can be induced by factors from glial cells. The simplicity of the cell line model, and the absence of contaminating cells, holds promise for an *in vitro* BBB based on the RBE4 cells. The cells can already be used as an assay system for analysing the regulation of expression of blood-brain barrier-specific properties, for examining the P-glycoprotein multidrug resistance transporter, for studying the cellular mechanisms underlying receptor-mediated events, and for establishing the mode of action of toxic agents. With further improvements in the method for growing cells on filters, the RBE4 cell line is likely to provide a useful polarized cell layer for studying permeability and transport across the endothelium. Finally, the ability of genetically modified RBE4 cells to survive as implants in host animals opens up a range of applications in gene therapy of the nervous system.

## ACKNOWLEDGEMENTS

The London group is grateful to the MRC, King's College, The Wellcome Trust, Yamanouchi Pharmaceutical Co, and Merck, Sharp & Dohme for support. We have benefited for over 20 years from many excellent and helpful discussions with our close colleague Michael Bradbury, and we look forward to further joint projects on both *in situ* and *in vitro* preparations after his official retirement.

## REFERENCES

Abbott N.J. and Revest, P.A., 1990, Single-channel currents recorded from rat brain capillary endothelial cells in culture. *J. Physiol.* 423:2P.

Abbott, N.J. and Revest, P.A., 1991, Control of brain capillary permeability. *Cerebrovasc. Brain Metab. Rev.* 3:1.

Abbott, N.J., Hughes, C.C.W., Revest, P.A. and Greenwood, J., 1992a, Development and characterisation of a rat brain capillary endothelial culture : towards an *in vitro* blood-brain barrier, *J. Cell Sci.* 103:23.

Abbott, N.J., Revest, P.A. and Romero, I.A., 1992b, Astrocyte-endothelial interaction : physiology and pathology, *Neuropath. Appl. Neurobiol.* 18:424.

Begley, D.J., Squires, L.K., Zlokovic, B.V., Mitrovic, D. M., Hughes, C.C.W., Revest, P.A. and Greenwood, J., 1990, Permeability of the blood-brain barrier to the immunosuppressive cyclic peptide cyclosporin A. *J. Neurochem.* 55: 1222.

Begley, D.J., Chen, Z-D., Rollinson, C. and Abbott, N.J., 1994, Activity of P-glycoprotein (multidrug resistance protein) in cultured immortalized rat brain microvascular endothelial cells (RBE4). *J.Physiol.* 480:9P.

Bowman, P.D., Betz, A.L., Wolinsky, J.S., Penny, J.B., Shivers, R.R. and Goldstein, G.W., 1981, Primary culture of capillary endothelium from rat brain . *In Vitro* 17:353.

Bowman, P.D., Betz, A.L. and Goldstein, G.W., 1982, Primary culture of microvascular endothelial cells from bovine retina: selective growth using fibronectin coated substrate and plasma derived serum. *In Vitro* 18:626.

Cordon-Cardo, C., O'Brien, J.P., Casals, D., Rittman-Grauer, L., Beideler, J.L., Melamed, M.R. and Bertino, J.R. , 1989, Multidrug-resistance gene (P-glycoprotein) is expressed by endothelial cells at blood-brain barrier sites. *Proc. Natl. Acad. Sci. USA* 86: 695.

Curry, F.E., 1992, Modulation of venular microvessel permeability by calcium influx into endothelial cells. *FASEB J.* 6: 2456.

DeBault, L.E. and Cancilla, P.A., 1980, γ-glutamyl transpeptidase in isolated brain endothelial cells: induction by glial cells *in vitro*. *Science* 207:653.

Dehouck, M-P., Meresse, S., Delorme, P., Fruchart, J-C and Cecchelli, R., 1990, An easier, reproducible, and mass-production method to study the blood-brain barrier *in vitro*. *J. Neurochem.* 54: 1798.

Dehouck, M-P, Jolliet-Riant, P., Bree, F., Fruchart, J-C., Cecchelli R. and J-P Tillement, 1992, Drug transfer across the blood-brain barrier: correlation between *in vitro* and *in vivo* methods. *J. Neurochem.* 58:1790.

Durieu-Trautmann, O., Fédérici, C., Créminon, C., Foignant-Chaverot, N., Roux, F., Claire, N., Strosberg, A.D. and Couraud, P.O., 1993, Nitric oxide and endothelin secretion by brain microvessel endothelial cells: regulation by cyclic nucleotides. *J. cell. Physiol.* 155:104.

Fédérici, C., Camoin, L., Creminon, C., Chaverot, N., Strosberg, A.D. and Couraud, P.O. Cultured astrocytes release a factor that decreases endothelin-1 secretion by brain microvessel endothelial cells. *J. Neurochem.* (in press)

Greenwood, J., 1992, Characterization of a rat retinal endothelial cell culture and the expression of P-glycoprotein in brain and retinal endothelium *in vitro*. *J. Neuroimmunol.* 39:123.

Greenwood, J., Adu, J., Davey, A.J., Abbott, N.J. and Bradbury, M.W.B., 1991, The effect of bile salts on the permeability and ultrastructure of the perfused, energy-depleted, rat blood-brain barrier. *J. Cereb. Blood Flow Metab.* 11:644.

He, P. and Curry, F.E., 1993, Albumin modulation of capillary permeability: role of endothelial cell $[Ca^{2+}]_i$. *Am. J. Physiol.* 265: H74.

He, P., Pagakis, S.N. and Curry, F.E., 1990, Measurement of cytoplasmic calcium in single microvessels with increased permeability. *Am. J. Physiol.* 258: H1366.

Hughes,C.C.W. and Lantos, P.L., 1986, Brain capillary endothelial cells in vitro lack surface IgG Fc receptors. *Neurosci. Lett.* 68: 100.

Ingber, D.E. and Folkman, J., 1989, Mechanical switching between growth and differentiation during fibroblast growth factor- stimulated angiogenesis *in vitro*: Role of extracellular matrix. *J. Cell Biol.* 109: 317.

Iruela-Arispe, M.L., Diglio, C.A. and Sage, E.H., 1991, Modulation of extracellular matrix proteins by endothelial cells undergoing angiogenesis *in vitro*. *Arteriosclerosis and Thrombosis* 11: 805.

Joó, F., 1992, The cerebral microvessels in culture, an update. *J.Neurochem.* 58:1.

Lal, B., Indurati, R.R., Couraud, P-O., Goldstein, G.W. and Laterra, J., 1994, Endothelial cell implantation and survival within experimental gliomas. *Proc. Natl. Acad. Sci. USA* 91: 9695.

Lane, N.J., Revest, P.A., Whytock, S. and Abbott, N.J., 1995, Fine-structural investigation of rat brain microvascular endothelial cells : tight junctions and vesicular structures in freshly isolated and cultured preparations. *J. Neurocytol.* 24: 347-360.

Laterra, J., Guerin, C. and Goldstein, G.W., 1990, Astrocytes induce neural microvascular endothelial cells to form capillary-like stuctures *in vitro*. *J. Cell Physiol.* 144: 204.

Lechardeur, D., Bardoul, M., Roux, F. and Sherman, D., 1994, Functional expression of the P-glycoprotein MDR in brain capillaries endothelial cell primary cultures and in an immortalized brain endothelial cell line (RBE4). *J. Physiol.* 480:135P.

Meyer, J., Mischeck, U., Veyhl, M., Henxel, K. and Galla, H.J., 1990, Blood-brain barrier characteristic enzymatic properties in cultured brain capillary endothelial cells. *Brain Res.* 514: 305.

Minakawa,T., Bready, J., Berliner, J., Fisher, M. and Cancilla, P.A., 1991, *In vitro* interaction of astrocytes and pericytes with capillary-like structures of brain microvessel endothelium. *Lab. Invest.* 65: 32.

Montesano, R., Vassalli, .D., Baird, A., Guillemin, R. and Orci, L., 1986, Basic fibroblast growth factor induces angiogenesis *in vitro*. *Proc. Natl. Acad. Sci. USA* 83:7297.

Nobles, M. and Abbott, N.J., 1994, Development of methods for growing brain endothelial cells on treated glass. *J.Physiol.* 480:4P.

Nobles, M., Revest, P.A., Couraud, P-O. and Abbott, N.J., 1994, Evidence for nucleotide receptors in immortalized rat brain capillary endothelial cells in culture. *J. Physiol.* 475:64P.

Nobles, M., Revest, P.A., Couraud, P-O. and Abbott, N.J. 1995, Characteristics of nucleotide receptors that cause elevation of cytoplasmic calcium in immortalised rat brain endothelial cells (RBE4) and in primary cultures. *Br. J. Pharmacology* 115: 1245-1252.

Olesen, S-P., 1989, An electrophysiological study of microvascular permeability and its modulation by chemical mediators. *Acta Physiol. Scand.* 136 (Suppl 579): 1.

Pardridge, W.M., 1988, Recent advances in blood-brain barrier transport. *Annu. Rev. Pharmacol. Toxicol.* 28: 25.

Ramlakhan, N., 1990, Albumin binding and endocytosis by cultured rat brain endothelium. *J. Physiol.* 423, 34P.

Raub, T.K., Kuentzel, S.L. and Sawada, G.A., 1992, Permeability of bovine brain microveseel endothelial cells *in vitro*: barrier tightening by a factor released from astroglioma cells. *Exp. Cell Res.* 199:330.

Revest,P.A., Abbott,N.J. and Gillespie, J.I., 1991, Receptor-mediated changes in intracellular [$Ca^{2+}$] in cultured rat brain capillary endothelial cells. *Brain Res.* 549:159.

Rist, R.J., Romero, I.A. and Abbott, N.J., 1994, The effects of a cAMP agonist and astrocyte-conditioned medium on the F-actin cytoskeleton in cultured primary and immortalized rat brain capillary endothelial cells. *J. Physiol.* 480:8P.

Romero, I.A., Chan, M.W.K. and Rist, R.J., 1994, The permeability of [$^{14}C$] sucrose through an *in vitro* blood-brain barrier model, the RBE4 cell line. *J.Physiol.* 489:8P.

Romero, I.A., Cavanagh, J.B., Nolan, C.C., Ray, D.E. and Seville, M.P., 1992, 1,3-dinitrobenzene, a neurotoxin acting at the blood-brain barrier. *J. Physiol.* 446:498P.

Romero, I., Brown, A.W., Cavanagh, J.B., Nolan, C.C., Ray, D.E. and Seville, M.P., 1991, Vascular factors in the neurotoxic damage caused by 1,3-dinitrobenzene in the rat. *Neuropath. Appl. Neurobiol.* 17: 495.

Roux, F., Durieu-Trautmann, O., Bourre, J-M, Strosberg, A.D. and Couraud, P-O., 1993, Immortalized rat brain microvessel endothelial cells : 1 - Expression of blood-brain barrier markers during angiogenesis *in*: "Frontiers in Cerebral Vascular Biology : Transport and its Regulation", LR Drewes and A.L. Betz, eds, p201, Plenum Press, New York.

Roux, F.S., Mokni, R., Hughes C.C., Clouet, P.M., Lefauconnier, J-M., Bourre, J-M., 1989, Lipid synthesis by rat brain microvessel endothelial cells in tissue culture. *J. Neuropath. Exp. Neurol.* 48: 437.

Roux, F., Durieu-Trautmann, O., Chaverot, N., Claire, M., Mailly, P., Bourre, J-M., Strosberg, A.D. and Couraud, P-O., 1994, Regulation of gamma-glutamyl transpeptidase and alkaline phosphatase activities in immortalized rat brain microvessel endothelial cells. *J. cell. Physiol.* 159:101.

Rubin, L.L., Hall, D.E., Porter, S., Barbu, K., Cannon, C., Horner, H.C., Janatpour, M., Liaw, C.W., Manning, K., Morales, J., Tanner, L.I.., Tomaseli, K.J. and Bard, F., 1991, A cell culture model of the blood-brain barrier. *J. Cell Biol.* 115: 1725.

Schinkel, A.H., Smit, J.J.M., van Tellingen, O., Beijnen, J.H., Wagenaar, E., van Deemter, L., Mol, C.A.A.M., van der Valk, M.A., Robanus-Maandag, E.C., te Riele, H.P.J., Berns, A.J.M., and Borst, P., 1994, Disruption of the mouse mdr1a P-glycoprotein gene leads to a deficiency in the blood-brain barrier and to increased sensitivity to drugs. *Cell* 77: 491.

Wolburg, H., Neuhaus, J., Kniesel, U., Krauss, B., Schmid, E-M, Ocalan, M., Farrell, C. and Risau, W., 1994, Modulation of tight junction structure in blood-brain barrier endothelial cells. *J. Cell Sci.* 107: 1347.

# DESIGNING DRUGS FOR OPTIMAL NERVOUS SYSTEM ACTIVITY

Nigel H. Greig[1], Arnold Brossi[2], Xue-Feng Pei[1], Donald K. Ingram[3] and Timothy T. Soncrant[1]

[1]Laboratory of Neurosciences, National Institute on Aging, National Institutes of Health, Bethesda, Maryland 20892
[2]Scientist Emeritus, National Institutes of Health, and Department of Chemistry, Georgetown University, Washington DC 20057
[3]Molecular Physiology and Genetics Section, Nathan W. Shock Laboratories, Gerontology Research Center, National Institute on Aging, Baltimore, MD 21224

## INTRODUCTION

The blood-brain barrier (BBB) continues to intrigue and frustrate neuroscientists. Perhaps like no other organ, the brain, together with its BBB challenges, the investigator at a number of levels. This is certainly true at the basic science level in elucidating the mechanisms by which it regulates the microenvironment of the central nervous system, allowing the control and transport of D-glucose, certain amino acids, ions and peptides by selective, saturable and often stereo-specific carrier systems. These systems support cerebral metabolism, neurotransmitter and protein synthesis, and maintain the extracellular environment of the brain to allow it to function under optimal conditions despite wide fluctuations in the environment of the rest of the body. The high metabolic component of the cerebral capillary endothelium, constituting the enzymatic element of the BBB, likewise interests the basic scientist in elucidating the mechanisms by which it separates and protects the brain from a variety of neurotransmitters, hormones and endogenous messengers that possess one role in the periphery and another in brain. Additionally, the manner in which the structural and enzymatic components of the BBB interact to so ably protect the brain from exogenous toxins continues to capture our attention. In as much as the BBB protects and regulates the microenvironment of the brain to allow it to function optimally under diverse conditions during health, it is frustrating from a clinical viewpoint that these same mechanisms thwart our ability to intervene during disease, impeding us in our corrective measures. Specifically, it limits our ability (i) to deliver chemotherapeutic agents, such as during the treatment of infections and cancer, and (ii) to selectively manipulate levels of neurotransmitters and neuropeptides and enzymes, such as in the treatment of Parkinson's

*New Concepts of a Blood–Brain Barrier*
Edited by J. Greenwood *et al.*, Plenum Press, New York, 1995

and Alzheimer's diseases.

An in-depth analysis and review of the function and working of the BBB at the structural and molecular levels is provided in Chapters 1 and 2 by Bradbury and Risau, respectively; for additional reviews see Bradbury (1992) and Broadwell and Banks (1993). This chapter describes specific techniques to overcome the restrictive role of the BBB by designing therapeutics with optimal characteristics to penetrate it, enter brain and treat neurologic diseases. Two examples are examined which highlight the manner in which drug design can be undertaken to individualize an agent to treat a specific disease (brain tumors and Alzheimer's disease). Clearly, the successful chemotherapeutic treatment of any disease, whether neurological or systemic, requires the development of a pharmacologically active, specific and non-toxic agent. As important, however, is its delivery to the disease target at a therapeutic concentration for sufficient time (C × T) to optimize its activity. Such transport characteristics should be considered as an integral part of initial drug design, rather than as an after-thought. All too often, an experimental therapeutic, whether a small molecular weight chemical, a peptide or a protein, will demonstrate remarkable activity in an *in vitro* model. However, when the same agent is administered to an animal model it is only minimally active, particularly when the disease target is sequestered behind the BBB, even when only variably intact.

To maximize the delivery of a therapeutic into the brain, and hence optimize its neurologic activity, it is critical to initially understand the factors that co-determine the concentration of a compound in brain following its systemic administration.

## FACTORS DETERMINING DRUG DELIVERY AND TIME-DEPENDENT CONCENTRATION IN BRAIN

We have previously reviewed the factors which, together, determine the amount of a drug achieved and maintained in brain after its systemic administration (Greig, 1987, 1992). These include:

(i) The time-dependent plasma concentration profile of the agent, which governs the amount of drug available to enter brain. The plasma concentration of a compound depends on its absorption, volume of distribution, metabolism and elimination.

(ii) The binding of the agent to plasma and brain constituents, and associated binding off-rates. The binding of a therapeutic to plasma proteins may be either restrictive or non-restrictive to brain uptake and increases with its lipophilicity. In the former case, only the free fraction of drug is available to enter brain, whereas in the latter, drug stripping from the plasma protein occurs during the transit of the agent through the brain circulation (as with diazepam (Jones et al., 1988)), as reviewed by Greig (1989a).

(iii) Drug permeability at the BBB. As the brain's blood vessels are lined by endothelial cells that are joined by continuous belts of tight intercellular junctions (Reese and Karnovsky, 1967), blocking paracellular diffusion, and are devoid of fenestrations, the BBB possesses permeability characteristics similar to those of a continuous cellular layer separating blood from brain (Bradbury, 1979). The solubility of an agent in the lipid component of the endothelial cell membrane therefore governs its permeability across the cell and its penetrability into the brain (Fenstermacher, 1992). There is thus a close relation between the permeability of an agent at the BBB and its octanol/water partition coefficient (Levin 1980), the latter being a measure of its physico-chemical character (Figure 1). The rate of transfer across the BBB of a lipophilic compound with an octanol/water partition of $>1$ ($\log P > 0$) is rapid and is primarily limited by the availability of the drug (i.e., the rate of blood flow). In contrast, polar, hydrophilic agents with an octanol/water partition coefficient of $<0.1$ ($\log P < -1$) are restricted from entering brain as a consequence of their low cerebrovascular permeability. Both factors, permeability and the rate of cerebral blood flow, determine the transfer rate of compounds whose partition coefficients lie between 1 and 0.1.

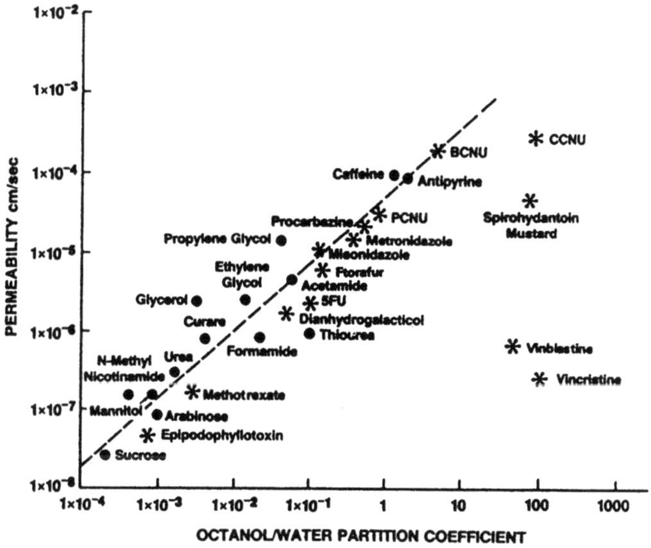


**Figure 1.** Relation between the permeability of a compound at the blood-brain barrier and its octanol/water partition coefficient. Asterisks denote pharmaceutical agents [the permeabilities of vincristine and vinblastine are lower than predicted from their octanol/water partition (Greig et al., 1990a)].

(iv) Facilitated transport systems. Exceptions to the proportionality between cerebrovascular permeability and octanol/water partition coefficient, as shown in Figure 1, occur for specific agents that share a transport mechanism at the BBB. Such systems may either augment drug entry into the brain (as is the case for melphalan (Greig et al., 1987) and NAM (Takada et al., 1992, see Smith, Chapter 29) at the BBB, and for AZT at the choroid plexus (Terasaki and Pardridge, 1988) or potentially limit brain uptake by expelling drug.

(v) Metabolic/degradative enzymes at the BBB may metabolize lipophilic drugs to more water-soluble ones prior to their brain entry; thereby reducing their cerebrovascular permeability (Hardebo and Owman, 1990).

(vi) Cerebral blood flow, which primarily affects lipophilic agents (Fenstermacher and Cowles, 1977) and is large for brain (approximately 20% of cardiac output), but can vary dramatically, up to 20-fold, between different brain regions (Fenstermacher et al., 1989).

The physicochemical characteristics of a compound, which in large part define its penetrability at the BBB, derive from its chemical structure. Conceptually, most drugs can be considered as organic molecules that are formed from carbon-containing backbones to which different functional groups are attached. The number and nature of the specific functional groups, the number of hydrocarbons, and their combined, three-dimensional structural arrangement provide the compound its pharmacological properties (i.e., ability to interact with a binding site to inhibit an enzyme or stimulate/antagonize a receptor), as well as its transport characteristics at biological barriers. Most interactions between drugs or peptides and their target involve the interaction of charged groups to initiate binding and an associated pharmacological action. Thus, agents with pharmacological activity are invariably either weak bases or acids. With regard to transport, agents that possess a "balanced" lipo/hydrophilicity, providing them solubility in the hydrophilic phases of plasma and extracellular fluid as well as in the lipophilic component of cell membranes that constitute the BBB and gastrointestinal tract, possess the requisite characteristics to enter the body, following oral administration, reach the brain via the blood stream (if not metabolized), cross the BBB and diffuse through the extracellular compartment to their target.

## PHARMACOLOGICAL PRINCIPLES RELATED TO DRUG DELIVERY, ACTION AND DRUG DESIGN

Prior to initiating attempts to augment drug delivery to brain, it is important to first assess whether or not a maximum pharmacological response has already been achieved (i.e., the position achieved on the dose/response curve). In the event that a maximal response has indeed been achieved, increasing brain drug delivery is of little value. In the event that one has not, it is critical to first assess which specific factor(s) limit the therapeutic potential of the agent. This information, together with knowledge of the physico-chemical characteristics of the "lead" compound will aid in assessing whether improved pharmacological activity might be better achieved by (i) enhancing drug delivery to brain (i.e., manipulating the agent's pharmacokinetics), and which technique might best achieve this, or (ii) enhancing the interaction between the drug and its target (i.e., manipulating its pharmacodynamics).

For example, drug ineffectiveness due to achievement of only a subtherapeutic concentration at its target in brain can potentially be overcome in one of several ways. (i) Administration of a higher initial dose, in the event that systemic adverse effects or toxicity would not preclude this, would result in correspondingly higher brain levels. (ii) Chemical modification of the drug could be undertaken to improve its transport to its target, and (iii) manipulation of the BBB could be undertaken to likewise improve access of a compound to its target in brain (Neuwelt, Chapter 30). In the event that an insufficient time-dependent concentration of drug ($C \times T$) at the target limits its therapeutic activity, then (i) administration of drug as a long-term infusion (Soncrant et al., 1993), (ii) its formulation as a slow-release depot (Knapp et al., 1990), or (iii) chemical modification to reduce its rate of metabolism and clearance, can be undertaken to augment $C \times T$. Finally, in the event that adverse actions/toxicity limit the therapeutic potential of a drug, then (i) co-administration of a selective antagonist to the adverse effect(s), such as the use of the $5HT_3$ antagonist ondanserton to reduce emesis during cancer chemotherapy (Calabresi and Schein, 1993), or (ii) chemical modification to provide receptor/enzyme subtype selectivity can be undertaken to maximize drug action.

It is important to remember that chemical modification of a drug, however minor, results in a new chemical entity. Despite its close relation to its parent compound, extensive preclinical pharmacology is required before approval for clinical studies can be obtained, and extensive clinical trials then are required prior to its approval for wide clinical use. The expense and time encountered in undertaking these tasks dictates that modification of a drug should be "specific" to the individual limitations of the compound to optimize its transport to and interaction with the disease target, rather than being "generic" in nature (Torrence et al., 1993), whereby resulting activity is improved instead of being maximized.

Characterization of the drug's target therefore is essential to allow the incorporation of design features to provide target-selective action. For example, knowledge of its location in brain is valuable. In the event that it is intracellular, then conferring a high lipophilicity ($\log P > 2.0$) on an experimental therapeutic will not only augment its brain uptake, but will drive the compound into the lipophilic environment of the intracellular compartment. Ideally, this will maximize drug levels at the target and sequester compound to reduce its back-diffusion out of brain. In contrast, for an extracellular localized target, such as a cell surface receptor or enzyme, a more marginal lipophilicity (log P 0-1.0) will not only allow BBB penetration, but also sufficient aqueous solubility to allow significant accumulation of drug in brain extracellular fluid.

Knowledge of a drug's mechanism of action is also important for optimal drug design. The continued presence of certain classes of drugs (i.e., receptor agonists) at their target is required to maintain activity. For such agents C × T is important for pharmacological action, and their pharmacokinetics and phamacodynamics are closely related. For other classes of drugs (i.e., enzyme inhibitors and receptor antagonists) that bind with high affinity or irreversibly to their target, their C × T is largely irrelevant to their pharmacological activity once initial binding has occurred. Finally, knowledge of the active form(s) of a drug, and of its metabolic route is critical to incorporate into drug design.

Two examples to highlight how chemical modification of two very different drugs can be undertaken to incorporate into them specific structure-activity and physico-chemical properties to provide them optimal activity for the treatment of specific neurologic diseases, brain tumors and Alzheimer's disease, are now described below.

## LIPOPHILIC CHLORAMBUCIL DERIVATIVES FOR BRAIN TUMOR TREATMENT

### Background

Malignant brain tumors remain virtually incurable despite numerous attempts to modify their dismal outcome. Primary malignant brain tumors account for 1.5% of all cancers, and number approximately 15,000 annually within the U.S. They occur predominantly and with increasing incidence in the elderly (Greig et al., 1990). Metastatic brain tumors occur in some 20% of all cancer patients (Posner, 1977), approximately 200,000 within the U.S. annually, also predominantly in the elderly. The median duration of survival for those with primary brain tumors is 9 months, whereas it is only 6 months for those with metastatic brain tumors (Greig, 1984).

Despite their limited clinical value, the nitrosoureas BCNU and CCNU remain the drugs of choice for primary brain tumor chemotherapy. In contrast, metastatic brain tumors are treated primarily with radiotherapy and corticosteroids alone, and, occasionally, water-soluble drugs, that are active against the primary extracerebral tumor, are added. The response rate of primary brain tumors to nitrosourea therapy is approximately 50%. However, heterogeneic subpopulations of tumor cell, which are either insensitive to or rapidly develop resistance mechanisms to nitrosoureas, soon multiply to kill the patient. Water-soluble drugs, administered to broaden the spectrum of anticancer activity, must contend with a partially intact BBB which dramatically restricts their access to their target, minimizing their activity and allowing the induction of resistance (Greig, 1987). The BBB within the central, often necrotic, region of brain tumors is generally only partially intact

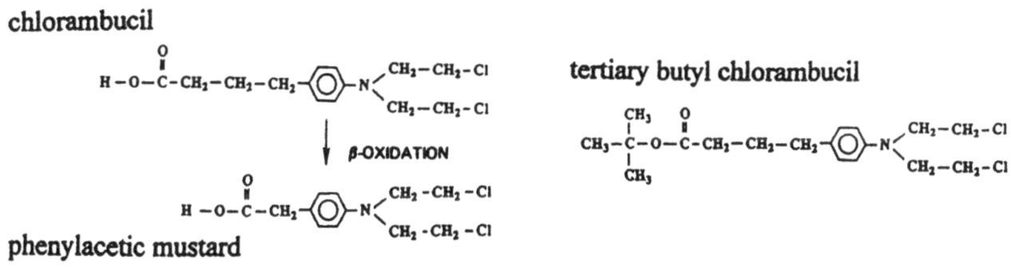


**Figure 2.** Chemical structures of chlorambucil, its active metabolite phenylacetic mustard, and of the lipophilic analogue tertiary butyl chlorambucil.

and has a variable blood flow (Tyson and Fenstermacher, 1989). In contrast it is invariably intact in the tumor periphery, which contains the majority of actively proliferating cells, as well as in the adjacent brain into which the tumor infiltrates (Greig, 1989b). No anticancer drugs have been specifically designed with optimal properties for access and activity throughout all regions of brain tumors, whether with an intact or a reduced BBB.

Anticancer drugs other than the nitrosoureas have proved active against brain tumors *in vitro*. Amongst these are the classical bifunctional nitrogen mustards, which includes chlorambucil (Figure 2). Bifunctional nitrogen mustards, in contrast to nitrosourea alkylating agents, have been shown to alkylate intracellularly by forming covalent links at the N(7) position of guanine, when surrounded by two other guanines, rather than at the O(6) position preferred by the nitrosoureas. Consequently, despite similarities between their mechanisms of action, resistance and insensitivity to nitrosourea treatment would not necessarily be conferred to a nitrogen mustard. For this reason, and as chlorambucil is relatively non-neurotoxic (if administered directly into brain) and has proven clinical activity in the management of breast cancer (Calabresis and Schein 1993), which frequently metastasizes to brain, it was chosen as a lead compound for drug modification.

**Chlorambucil Tertiary-Butyl Ester**

Extensive modification of the carboxylic acid group of chlorambucil (Figure 2) was undertaken to modify the agent's lipophilicity. This group has a pKa of 4.6 (Ehrsson et al., 1981), is highly charged a physiological pH, and, together with extensive binding to plasma proteins (99%), restricts chlorambucil's brain uptake (Greig et al., 1988). The pharmacological activity of chlorambucil derives from its nitrogen mustard group, which is distant from the carboxylic acid group and hence is only marginally affected by chemical masking of the latter. In initial studies (Greig et al., 1990d), aliphatic (straight chain) esters of up to 8 carbons length were synthesized. These increased the lipophilicity of the resulting derivatives from log P -0.66 for the parent compound, chlorambucil, to up to log P 7.55 for the octyl ester. However, short chain esters were rapidly cleaved in plasma to regenerate chlorambucil within seconds, whereas long chain esters proved difficult to solubilize for in vivo studies and became restrictively bound to plasma proteins (99.9%) when administered to animals. Hence all were restricted from entering brain and possessed low brain/plasma concentration ratios, similar to that of chlorambucil (0.02). Aromatic esters, including phenyl and prednisolone esters, proved equally ineffective in increasing chlorambucil's brain delivery as these also were rapidly metabolized to their parent compound (Greig et al., 1990d). A series of lipophilic chlorambucil amides were also synthesized, designed with lipophilicities ranging from log P 0.74, for a pyrimidine derivative, to log P 2.3 for a nicotinamide derivative. Although several of the chorambucil amides achieved significantly higher levels in brain, compared to equimolar chlorambucil administration, these were still low and their brain/plasma concentration ratios did not exceed 0.05.

As a consequence of these studies, increasing steric hindrance was introduced into esters of chlorambucil to provide the resulting compounds sufficient stability in plasma to allow them to reach the brain after administration. However, their lipophilicity was limited to prevent the restrictive plasma protein binding associated with the long chain aliphatic esters, but to allow a high BBB permeability and intracellular uptake. These studies culminated in the development of the tertiary-butyl ester of chlorambucil, TBC (Figure 2), as an experimental drug for brain tumor chemotherapy (Greig et al., 1990e).

Following administration to rats, TBC reached and maintained substantially higher levels in brain, compared to equimolar chlorambucil administration, and substantially lower plasma levels. Figures 3 and 4 illustrate the plasma (left panel) and brain (right panel) pharmacokinetics of equimolar TBC and chlorambucil, respectively. In each case, the concentration of all metabolites with anticancer activity was measured, and a total active concentration also was determined. As can be seen (Figure 3), TBC concentrations rapidly declined in plasma following its i.v. administration. This was a consequence of (i) its ester hydrolysis to release chlorambucil, as a result of high levels of unspecific esterases in plasma and liver, and (ii) its large distribution volume. Although TBC was highly bound to plasma protein, 99%, this was nonrestrictive and, due to its lipophilicity, log P 3.2, it readily entered brain and maintained levels that were >10-fold higher than concomitant ones in plasma. Its high lipophilicity and intracellular uptake thereby allowed its time-dependent sequestration in brain, with little back-flow despite minimal levels in plasma. TBC proved stable in brain and, as it possesses intrinsic anticancer activity, does not require breakdown to chlorambucil (Genka et al., 1993). Calculated from the total concentration of active drugs, TBC possessed a brain/plasma ratio of 0.85 and of >10 for TBC alone. These compare favorably to those of chlorambucil, 0.02, and BCNU, 0.3 (Oliverio, 1976).

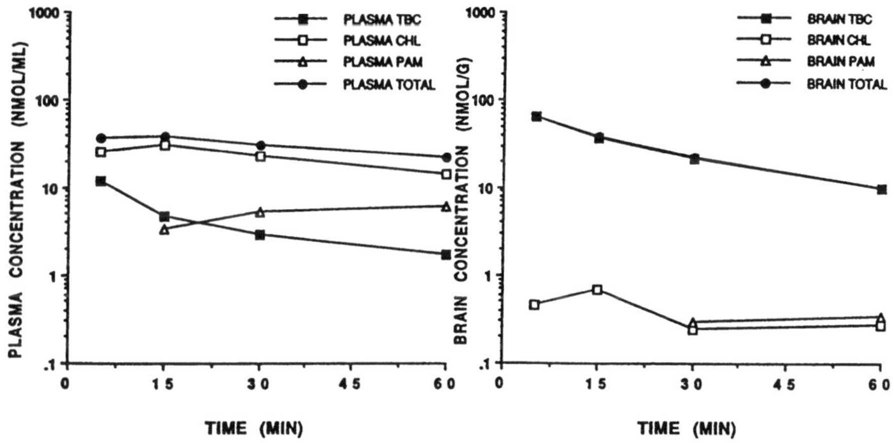


**Figure 3.** Time-dependent plasma (left) and brain (right) concentration of tertiary butyl chlorambucil (TBC) and its active metabolites, chlorambucil (CHL) and phenylacetic mustard (PAM), as well as total active drug levels (TOTAL) after i.v. administration of 13 mg/kg tertiary butyl chlorambucil to rats.

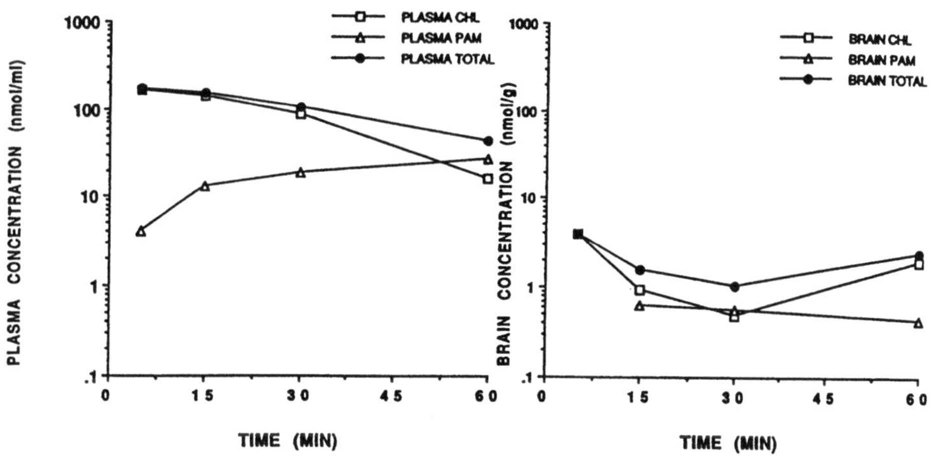


**Figure 4.** Time-dependent plasma (left) and brain (right) concentration of chlorambucil (CHL) and its active metabolite, phenylacetic mustard (PAM), as well as total active drug levels (TOTAL) after i.v. administration of 10 mg/kg chlorambucil to rats (equimolar to 13 mg/kg tertiary butyl chlorambucil).

Thus, as illustrated in Figure 5, the time-dependent concentration of active drug deriving from TBC was 14.5-fold greater than that achieved from equimolar chlorambucil (right panel), whereas plasma levels were 3.5-fold lower (left panel). As a consequence of its lower plasma concentration, TBC proved approximately 10-fold less toxic than chlorambucil, which allows the administration of higher doses to achieve correspondingly higher levels in brain. This improved delivery of anticancer drug to brain resulted in significant increases in the survival of rats with intracerebral metastatic brain tumors. Furthermore, TBC demonstrated high activity, sufficient to predict a clinical response, against 4 of 6 human primary malignant brain tumors *in vitro*, that were removed from patients that had failed BCNU chemotherapy (Ali-Osman et al., 1991).

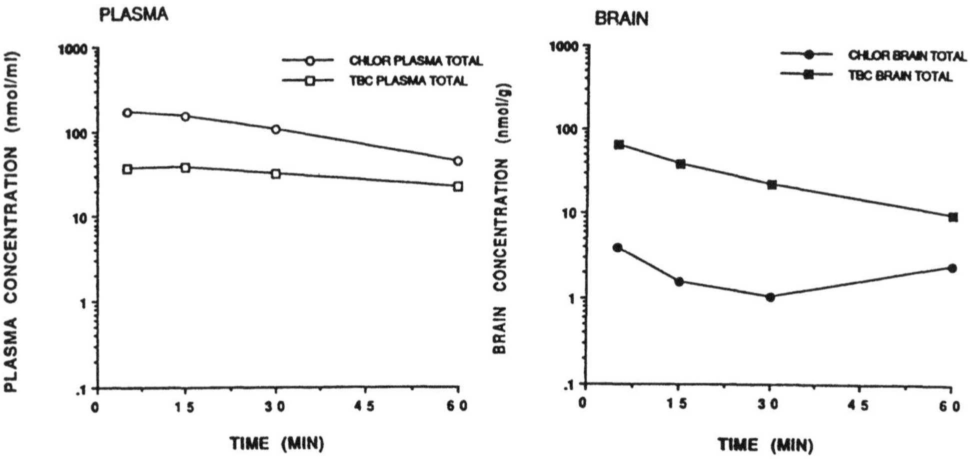


**Figure 5.** Total concentrations of active drug derived from equimolar i.v. administration of chlorambucil and tertiary butyl chlorambucil (10 and 13 mg/kg, respectively) in rat plasma (left) and brain (right).

In conclusion, chemical modification of the widely used water-soluble drug chlorambucil was undertaken to produce the agent TBC with specific features optimal for brain tumor therapy, allowing it to work on dual levels. First, the lipophilic compound TBC (i) readily crosses the BBB and enters brain, where (ii) it is maintained at high levels, achieving the required C × T for an anticancer alkylating agent, despite its rapid disappearance from plasma, (iii) it possesses high intrinsic anticancer alkylating activity, with (iv) an excellent toxicity profile, and (v) is minimally susceptible to cross-resistance with BCNU therapy. Secondly, TBC rapidly generates water-soluble chlorambucil in the circulation which can enter central tumor regions where localized areas of BBB breakdown occur (Greig, 1984, 1989b). Hence active drug reaches throughout the entire tumor and adjacent brain into which it invades, irrespective of the state of the BBB.

## DRUG DEVELOPMENT FOR ALZHEIMER'S DISEASE

Presently, there is considerable interest in the use of anticholinesterases in the treatment of Alzheimer's disease as the agent tacrine (Cognex) was recently approved as a prescription drug for wide clinical use. Chemical modification of cholinesterase inhibitors provides an example of how improved neuropharmacological activity can be achieved by optimizing the interaction between a drug and its target, in addition to improving its brain delivery.

## Background

Alzheimer's disease is the fourth leading cause of death in Western societies, afflicting some 4 million Americans and an additional 7 million persons in other developed countries. The median survival time from diagnosis is approximately 8 years. The disease is characterized by cognitive impairments and partial or total loss of memory. From a neurochemical perspective, it is characterized by a reduction in presynaptic markers of the cholinergic system and a dramatic loss of cortically-projecting cholinergic neurons, particularly to brain areas related to learning and memory (Whitehouse et al., 1982). Specifically, there is a dramatic loss of cholinergic neurons that innervate forebrain structures, and a decreased synthesis of the neurotransmitter acteylcholine, ACh, in the neocortex (Davies and Maloney 1976). Although multiple neurotransmitter deficiencies have also been linked to this disorder, these occur always in addition to, and not instead of, the cholinergic deficiency which occurs early during the disease when memory deficits predominate (Becker and Giacobini 1991). These findings, together with the well established role of cholinergic neurotransmission in memory storage and processing led to the hypothesis that cholinergic augmentation might improve cognition in Alzheimer's disease (Bartus et al., 1982).

The ability of the cholinesterase inhibitors physostigmine and tacrine to ameliorate cognitive impairments of Alzheimer's disease has been extensively studied (Becker et al., 1991; Knapp et al., 1994). It has been proposed that a substantial inhibition of brain acetylcholinesterase, AChE, the enzyme that metabolizes ACh, is probably required before beneficial clinical effects would be expected (Becker and Giacobini 1988). Small changes likely can be compensated for by homeostatic mechanisms. In studies involving physostigmine or tacrine in which cholinesterase levels were either directly measured or could be estimated, cognitive improvements were generally associated with levels of cholinesterase inhibition of 30% or more. Unfortunately, this level of inhibition coincides with the inevitable appearance of disabling, dose-limiting adverse effects for these agents (Becker et al., 1991). The cognitive response of patients improves with the higher the dose of physostigmine or tacrine administered, concomitant with higher inhibition of brain AChE, but the appearance of side effects precludes the use of both physostigmine and tacrine at adequate drug levels for optimal action. This problem, together with the short duration of action of these drugs, has long complicated the true evaluation of anticholinesterases for the treatment of Alzheimer's disease.

The described adverse effects are primarily peripheral. Therefore drug modification studies were undertaken around the structure of physostigmine to dissociate the beneficial actions of anticholinesterase therapy in brain from adverse peripheral ones by designing brain targeted and enzyme sub-type selective agents as experimental therapeutics. Physostigmine, unlike tacrine, is an excellent candidate for chemical modification as its is not associated with hepatic toxicity.

## Phenylcarbamate Analogues of Physostigmine: Phenserine

Extensive modification of physostigmine was undertaken in two regions, related to its binding to two distinct sub-sites on the the enzyme. (i) A variety of carbamate analogues were synthesized (R group in Figure 6) (Brzostowska et al., 1992). This region carbamylates, and the carbamoyl group is transferred to the esteratic sub-site of the enzyme and thereby inhibits its ability to hydrolyse ACh. (ii) Modification of the N(1)-position in the C ring of physostigmine (Figure 6) also was undertaken to develop entirely new generations of antichloinesterase agents (Yu et al., 1991, He et al., 1992). This region interacts with the anionic sub-site of the enzyme, involved in the attraction and orientation of the substrate for interaction with the esteratic sub-site (Quinn, 1987, Soreq and Zakut,

1993). The resulting compounds were then quanitatively assessed for their ability to inhibit human AChE and butyrylcholinesterase, BChE.

Whereas alkylcarbamates, such as the heptylcarbamate of physostigmine and physostigmine itself, proved potent in co-inhibiting AChE and BChE, with a slight preference for the latter, arylcarbamates, such as the phenylcarbamate phenserine, possessed a unique selectivity for AChE (Figure 6), and the chemistry and pharmacological properties of phenserine and its analogues have recently been reviewed (Greig et al., 1995a). Substitution onto the phenyl ring structure into the ortho- and para-positions, provided further selectivity for AChE and BChE inhibition, respectively. This unique property conveyed by the phenyl substitution arises from its defined 3-dimensional structure which can achieve a favored orientation (depending on its substitution) with respect to the rest of the molecule (the **A**,**B**,**C** rings) to allow it to differentiate between the highly similar binding sites of AChE and BChE (Flippen-Anderson et al., 1993). A wide variety of carbamates, such as the tertiary butyl, proved inactive against both enzyme sub-types.

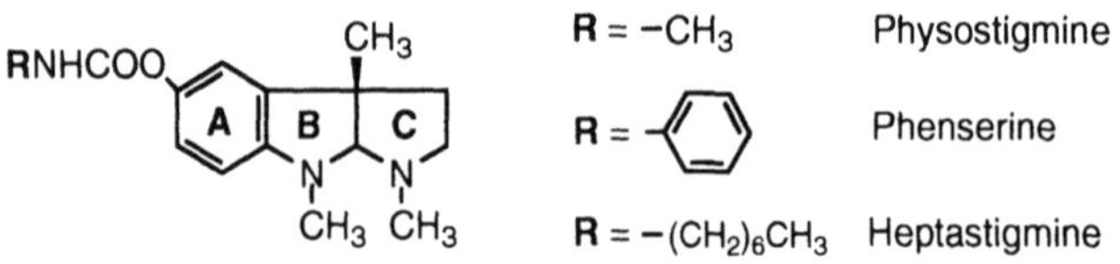


| | $IC_{50}$ Value (nM) | | |
|---|---|---|---|
| Compound | AChE | BChE | Selectivity |
| Physostigmine | 27.9 ± 2.4 | 16.0 ± 2.9 | 2-fold BChE |
| Phenserine | 22.2 ± 1.4 | 1552.0 ± 272.0 | 70-fold AChE |
| Heptastigmine | 21.7 ± 2.0 | 5.0 ± 0.1 | 4-fold BChE |
| Tacrine | 190.0 ± 40.0 | 47.0 ± 10.0 | 4-fold BChE |
| Galanthamine | 800.0 ± 60.0 | 7313.0 ± 827.0 | 9-fold AChE |

$IC_{50}$ values were determined in duplicate on a minimum of 4 different occasions

**Figure 6.** Chemical structures of physostigmine, heptastigmine and phenserine, and $IC_{50}$ values (concentration required to inhibit 50% of enzyme activity) against human erythrocyte AChE and plasma BChE, compared to other anticholinesterases of clinical interest.

AChE and BChE have multiple roles (Chatanet and Lockridge, 1989; Massoulie et al., 1993). Although those for BChE remain to be fully elucidated, it is found in brain in association with neuronal cells, potentially possessing a modulatory role for the dopaminergic system, as well as in association with glial cells and the BBB. Additionally, it is found in high amounts in liver, suggesting a role in lipid metabolism, in heart and in serum, where it is involved in the metabolism of several drugs. Clearly, the AChE-selectivity of alkylcarbamates is of potential pharmacological benefit to minimize BChE-induced effects of anticholinesterase therapy. Likewise, as AChE is a critical component of the cholinergic system throughout the body, affecting neuromuscular and cardiovascular regulation as well as cognition, its selective inhibition in brain for the treatment of Alzheimer's disease should minimize its deleterious peripheral actions.

Phenserine, as a consequence of its physico-chemical properties, has optimal properties for central nervous system activity. Specifically, substitution of the methylcarbamyl group of physostigmine with a phenylcarbamyl increases the lipophilicity of the compound from log P 0.1 to 0.9. As illustrated in Figure 7A, following administration of phenserine i.v. to rats, it achieves brain levels that are 10-fold higher than concomitant ones in plasma, and disappears from both rapidly with a half-life of approximately 10 minutes (Greig et al., 1995a; 1995b). In contrast, brain concentrations of physostigmine mirror those in plasma, with a brain/plasma ratio of 1.2 and an equally rapid disappearance half-life (Somani and Khalique, 1986). As shown in Figure 7B, however, unlike physostigmine whose duration of AChE inhibition is short, phenylcarbamylation of the enzyme with phenserine results in a more stable enzyme intermediate and a correspondingly longer duration of inhibition (>8 hours for a lower dose), as compared to both physostigmine and tacrine. As discussed earlier, the formation of a stable enzyme intermediate does not require the continued presence of drug for maintenance of enzyme inhibition, and hence a short C × T is optimal for pharmacological activity of this class of drug, to minimize total body exposure and adverse effects once binding has occurred. In contrast, the unselective alkylcarbamate heptastigmine has a much higher lipophilicity (log P 3.2), a high brain uptake and long duration of action, but a much slower disappearance half-life of 5 hours (Becker et al., 1991) due to its intracellular sequestration.

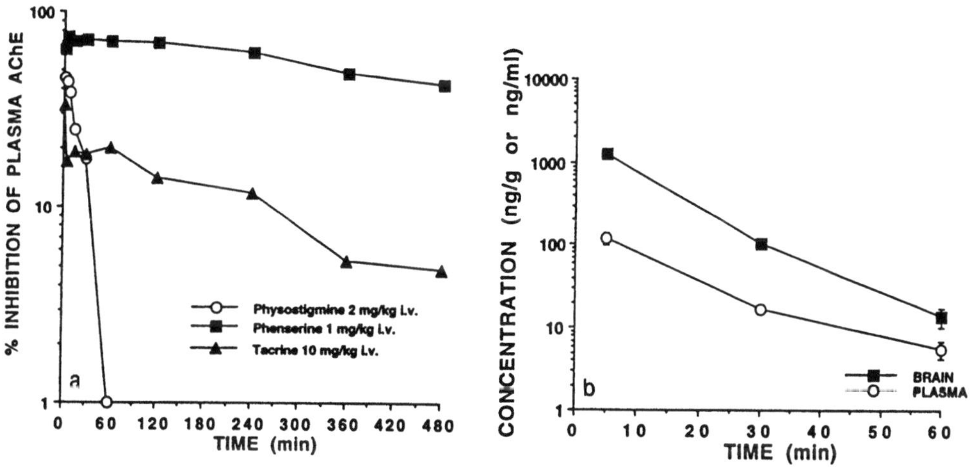


**Figure 7A.** (left panel) Time-dependent inhibition of AChE in plasma following i.v. administration of physostigmine, phenserine and tacrine (2, 1 and 10 mg/kg , respectively) to rats.
**Figure 7B.** (right panel) Time-dependent concentration of phenserine in brain and plasma following 1 mg/kg i.v. administration to rats.

In accord with its potency against AChE, high brain uptake and long duration of action, phenserine produced substantial and dose-dependent increases in brain extracellular concentrations of ACh (Greig et al., 1995b). Additionally, as a consequence of its targeting on two levels, (i) selectivity and optimization for the enzyme sub-type, and (ii) preferential brain uptake, phenserine demonstrated an unusually wide therapeutic window in behavioral tests as a cognition enhancer in rats (Iijima et al., 1993; Ikari et al., 1993) that has not been seen with other agents (Iijima et al., 1992; Greig et al., 1994a). Specifically, phenserine was studied in a widely used animal model that previously has been utilized to select out therapeutics of potential for Alzheimer's disease (Iversen et al., 1991). This model quantitates the ability of animals to learn a complex maze, where sequential correct choices are required to proceed through it (Spangler et al., 1986, 1989; Ingram et al., 1994). Phenserine dramatically improved memory acquisition in (i) young rats that were administered the cholinergic antagonist scopolamine, which induces a learning impairment as observed in Alzheimer's disease (Drachman and Leavitt, 1974), and in (ii) normal elderly rats (Ikari et al., 1995). The latter demonstrate an age-related decline in memory acquisition, and hence do not require cholinergic manipulation to produce a learning impairment. In both young and elderly animals, phenserine's effects on cognition were robust, occurred over a wide dose-range, and achieved without toxicity. Indeed, its toxicological assessment demonstrated that it possessed an excellent profile as a drug candidate for use in the elderly, proving >20-fold less toxic than physostigmine.

## SUMMARY

The therapeutic value of many promising drugs is diminished by the presence of the BBB, which possesses both structural and enzymatic components. In addition to a drug's permeability at the BBB, which is related to its octanol/water partition, several additional factors co-determine the amount that eventually enters and is maintained in brain. These include the agent's plasma and tissue binding, and whether these are restrictive or nonrestrictive, its time-dependent concentration in plasma, related to its distribution, metabolism and elimination, and, finally, to its susceptibilty to BBB enzymes, potential utilization of a BBB transport system and to cerebral blood flow.

For many therapeutics, simple alterations in dose, scheduling or route of administration will elevate brain levels above the threshold required for activity. For others, however, alternative strategies such as chemical modification are required. Few generic approaches to chemical modification have proved of real clinical value (Greig 1991). However, careful drug selection, together with the characterization of its neurological target, and the tailoring of chemical modifications to the specific compound and its mechanism of action have resulted in the rational design of a variety of selective and highly active neuropharmaceuticals.

## REFERENCES

Ali-Osman, F., Greig, N.H., John, V., and Lieberburg, I.M., 1991, Activity of tertiary butyl chlorambucil ester against 2-chloroethylnitrosourea-resistant human malignant glioma cell lines, *Proc. Am. Assoc. Cancer Res.* 32:318.

Bartus, R.T., Dean, R.L., and Beer, B., 1980, Memory deficts in aged cebus monkeys and facilitation with central cholinomimetics, *Neurobiol. Aging* 1:145.

Becker, R.E., and Giacobini, E., 1988, Mechanisms of cholinesterase inhibition in senile dementia of Alzheimer's type: clinical, pharmacological and therapeutic aspects, *Drug Dev. Res.* 12:163.

Becker, R.E., and Giacobini, E., 1991, "Cholinergic Basis of Alzheimer's Disease," Birkhauser, Boston.

Becker, R., Moriarty, P., and Unni, L., 1991, The second generation of cholinesterase inhibitors: clinical and pharmacological effects, *in:* "Cholinergic Basis of Alzheimer's Disease," R. Becker and E. Giacobini, eds., Birkhauser, Boston.

Bradbury, M.W.B., 1979, "Concept of a Blood-Brain Barrier," John Wiley, Chichester.

Bradbury, M.W.B., 1992, "Physiology and Pharmacology of the Blood-Brain Barrier, Handbook of Experimental Pharmacology Vol. 103", Springer Verlag, Berlin.

Brozostowska, M., He, X.S., Greig, N.H., Rapoport, S., and Brossi, A., 1992, Phenylcarbamates of (-)-eseroline, (-)-N1-noreseroline and (-)-physovenol: selective inhibitors of acetyl- and, or butyrylcholinesterase, *Med. Chem. Res.* 2:238.

Calabresi, P., and Schein, P.S., 1993, "Medical Oncology, Basic Principles and Clinical Management," McGraw-Hill, New York.

Chatanet, A., and Lockridge, O., 1989, Comparison and butyrylcholinesterase and acetylcholinesterase, *Biochem. J.* 260:625.

Davies, P., and Maloney, A.J.F., 1976, Selective loss of central cholinergic neurons in Alzheimer's type dementia, *Nature* 288:279.

Drachman, D.A., and Leavit, J., 1974, Human memory and the cholinergic system, *Arch. Neurol.* 30:113.

Ehrsson, H., Lonroth, U., Wallin, I., Ehrnebo, M., and Nilsson, S., 1981, Degradation of chlorambucil in aqueous solution: influence of human albumin binding, *J. Pharm. Pharmacol.* 33:313.

Fenstermacher, J.D., 1992, The blood-brain barrier is not a barrier for many drugs, *NIDA Res. Mongr.* 120:108.

Fenstermacher, J.D., and Cowles, A.L., 1977, Theoretic limitations of intracarotid infusions in brain tumor chemotherapy, *Cancer Treat. Rep.* 61:519.

Fenstermacher, J.D., Gross, P., Sposito, N., Acuff, V., Petersen, S., and Gruber K., 1988, Structural and functional variations in capillary systems within the brain, *Ann. NY Acad. Sci.* 529:21.

Flippen-Anderson, J.L., He, X.S., Brossi, A., Greig, N.H., and Rapoport, S., 1993, Thiaphysovenol phenylcarbamates: x-ray structures of biologically active and inactive anticholinesterase agents, *Heterocycles* 36:79.

Genka, S., Shetty, U., Stahle, P.L., John, V., Lieberburg, I.M., Ali-Osman, F., Rapoport, S., and Greig, N.H., 1993, Development of lipophilic anticancer agents for the treatment of brain tumors by the esterification of water-soluble chlorambucil, *Clin. Exp. Metastasis* 11:131.

Greig, N.H., 1984, Chemotherapy of brain metastases: current status, *Cancer Treat. Rev.* 11:157.

Greig, N.H., 1987, Optimizing drug delivery to brain tumors, *Cancer Treat. Rev.* 14:1.

Greig, N.H., 1989a, Drug delivery to the brain by blood-brain barrier circumvention and drug modification, *in*: "Implications of the Blood-Brain Barrier and Its Modification, Vol. 1, Basic Science Studies," E.A. Neuwelt, ed., Plenum Press, New York.

Greig, N.H., 1989b, Brain tumors and the blood-tumor barrier, *in*: "Implications of the Blood-Brain Barrier and Its Modification, Vol. 2, Clinial Studies," E.A. Neuwelt, ed., Plenum Press, New York.

Greig, N.H., 1992, Drug entry into the brain and its pharmacologic manipulation, *in*: "Physiology and Pharmacology of the Blood-Brain Barrier, Handbook of Experimental Pharmacology Vol. 103," M.W.B. Bradbury, ed., Springer Verlag, Berlin.

Greig, N.H., Momma, S., Sweeney, D.J., Smith, Q.R., and Rapoport, S., 1987, Facilitated transport of melphalan at the rat blood-brain barrier by the large neutral amino acid transport system, *Cancer Res.* 47:1571.

Greig, N.H., Sweeney, D.J., and Rapoport, S., 1988, Comparative brain and plasma pharmacokinetics of chlorambucil and melphalan in the rat, *Cancer Chemother. Pharmacol.* 21:1.

Greig, N.H., Soncrant, T.T., Shetty, U., Momma, S., Smith, Q.R., and Rapoport, S., 1990a, Brain uptakes and anticancer activities of vincristine and vinblastine are restricted by their low cerebrovascular permeability and binding to plasma constituents, *Cancer Chemother. Pharmacol.* 26:263.

Greig, N.H., Ries, L., Yancik, R., and Rapoport, S., 1990b, Increasing annual incidence of primary malignant brain tumors in the elderly, *J. Natl. Cancer Inst.* 82:1621.

Greig, N.H., Genka, S., and Rapoport, S., 1990c, Delivery of vital drugs to the brain for the treatment of brain tumors, *J. Controlled Release* 11:61.

Greig, N.H., Genka, S., Daly, E.M., Sweeney, D.J., and Rapoport, S., 1990d, Physicochemical and pharmacokinetic parameters of seven lipophilic chlorambucil esters designed for brain penetration, *Cancer Chemother. Pharmacol.* 25:311.

Greig, N.H., Daly, E.M., Sweeney, D.J., and Rapoport, S., 1990e, Pharmacokinetics of chlorambucil tertiary butyl ester, a lipophilic chlorambucil derivative that achieves and maintains high concentrations in brain, *Cancer Chemother. Pharmacol.* 25:320.

Greig, N.H., Pei, X.F., Soncrant, T.T., Ingram, D.K., and Brossi, A., 1995a, Phenserine and ring C hetero-analogues: drug candidates for the treatment of Alzheimer's disease, *Med. Res. Rev.* 15:3.

Greig, N.H., Pei, X.F., Soncrant, T.T., De Micheli, E., Ingram, D.K., Holloway, H.W., Deutsch, J., and Brossi, A., 1995b, Phenserine, a selective and long-acting acetylcholinesterase inhibitor for the treatment of Alzheimer's disease: in vivo pharmacokinetics and pharmacodynamics in the rat, *J. Am. Geriatr. Soc.* in press.

Hardebo, J.E., and Owman, C., 1991, Enzymatic barrier mechanisms for neurotransmitter monoamines and their precurrsors at the blood-brain barrier, *in:* "Pathophysiology of the Blood-Brain Barrier," B.B. Johansson, C. Owman, and H. Widner, eds. Elsevier, Amsterdam.

He, X.S., Greig, N.H., Rapoport, S., Brossi, A., Li, Y.Q., and Yu, Q.S., 1992, Thiaphysovenine and carbamate analogues: a new class of potent inhibitors of cholinesterases, *Med. Chem. Res.* 2:229.

Iijima, S., Greig, N.H., Garofalo, P., Spangler, E.L., Heller, B., Brossi, A., and Ingram, D.K., 1992, The long-acting cholinesterase inhibitor heptyl-physostigmine attenuates the scopolamine-induced learning impairment of rats in a 14-unit T-maze, *Neurosci. Lett.* 144:79.

Iijima, S., Greig, N.H., Garofalo, P., Spangler, E.L., Heller, B., Brossi, A., and Ingram, D.K., 1993, Phenserine: a physostigmie derivative that is a long-acting inhibitor of cholinesterase and demonstrates a wide dose range for attenuating a scopolamine-induced learning impairment of rats in a 14-unit T-maze, *Psychopharmacol.* 112:415.

Ikari, H., Spangler, E., Greig, N.H., Pei, X-F., Brossi, A., Speer, D., Patel, N., Ingram, D.K., 1995, Maze learning in aged rats is enhanced by phenserine, a novel anticholinesterase. *Neuro Report* 6:481.

Ingram, D.K., Spangler, E.L., Iijima, S., Kuo, H., Bresnahan, E.L., Greig, N.H., and London, E.D., 1994, New pharmacological strategies for cognitive enhancement using a rat model of age-related memory impairment, *Annals NY Acad. Sci.* 717:16.

Iversen, L.L., Bentley, G., Dawson, G., Freedman, S.B., Harley, E.A., Iversen, S.D., Rupniak, N.M.J., Tye, S., Tagella, P.G., and Rugarli, P.L., 1991, Heptylphysostigmine - a novel acetylcholinesterase inhibitor: biochemical and behavioral pharmacology, *in*: "Cholinergic Basis of Alzheimer's Disease," R. Becker, and E. Giacobini, eds., Birkhauser, Boston.

Jones, D.R., Hall, S.D., Jackson, E.K., Branch, R.A., and Wilkinson, G.R., 1988, Brain uptake of benzodiazepines: effects of lipophilicity and plasma protein binding, *J. Pharmacol. Exp. Ther.* 245:816.

Knapp, M.J., Knopman, D.S., Solomon, P.R., Penlebury, W.W., Davies, C.S., and Gracon, S.I., 1994, A 30-week randomized controlled trial of high-dose tacrine in patients with Alzheimer's disease, *JAMA* 271:985.

Knapp, S., Wardlow, M.L., Albert, K., Waters, D., and Thal, L.J., 1990, Correlation between plasma physostigmine concentrations and percentage of acetylcholinesterase inhibition over time after controlled release of physostigmine in volunteer subjects, *Drug Metab. Disp.* 19:400.

Leo, A., Hansch, C., and Elkins, D., 1971, Partition coefficients and their uses, *Chem. Rev.* 71:525.

Levin, V.A., 1980, Relation of octanol/water partition and molecular weight to rat brain capillary permeability, *J. Med. Chem.* 23:682.

Massoulie, J., Sussman, J., Bon, S., and Siman, I., 1993, Structure and functions of acetylcholinesterase and butyrylcholinesterase, *Prog. Brain Res.* 98:139.

Oliverio, V., 1976, Pharmacology of the nitrosoureas: an overview, *Cancer Treat. Rep.* 60:703.

Posner, J., 1977, Management of central nervous system metastases Semin, *Oncol.* 4:81.

Quinn, D.M., 1987, Acetylcholinesterase: enzyme structure, reaction dynamics, and virtual transition states, *Chem. Rev.* 87:955.

Reese, T., and Karnovsky, M., 1967, Fine structural localization of a blood-brain barrier to exogenous peroxidase, *J. Cell Biol.* 34:207.

Somani, S.M., and Kahlique, A., 1986, Distribution and pharmacokinetics of physostigmine in rat after intramuscular administration, *Fund. Appl. Toxicol.* 6:327.

Soncrant, T.T., Raffaele, K.C., and Asthana, A., 1993, Memory improvement without toxicity during chronic low dose intravenous arecoline in Alzheimer's disease, *Psychopharmacol.* 112:421.

Soreq, H., and Zakut, H., 1993, "Human Cholinesterases and Anticholinesterases," Academic Press, New York.

Spangler, E.L., Rigby, P., and Ingram, D.K., 1986, Scopolamine impairs learning performance of rats in a 14-unit T-maze, *Pharmacol. Biochem. Behav.* 25:673.

Spangler, E.L., Chachich, M.E., Curtis, N.J, and Ingram, D.K., 1989, Age-related impairment in complex maze learning in rats: relationship to neophobia and cholinergic antagonism, *Neurobiol. Aging* 10:133.

Takada, T., Vistica, D.T., Greig, N.H., Rapoport, S., and Smith, Q.R., 1992, Rapid high-affinity transport of nitrogen mustard amino acid across the blood-brain barrier, *Cancer Res.* 52:2191.

Teraski, T., and Pardridge, W.M.,1988, Restricted transport of 3'-azido-deoxythymidine and dideoxynucleosides through the blood-brain barrier, *J. Infect. Dis.* 158:630.

Torrence, P., Kinjo, J., Khamnei, S., and Greig, N.H., 1993, Synthesis and pharmacokinetics of a dihydropyridine chemical delivery system for the antiimmunodeficiency virus agent dideoxycytidine, *J. Med. Chem.* 36:529.

Tyson, G., Fenstermacher, J.D., and Davis, R., 1989, Vascular factors affecting drug delivery to brain tumors, *Basic Life Sci.* 50:115.

Whitehouse, P.J., Price, D.L., Struble, R.G., Clark, A.W., Coyle, J.T., and De Long, M.R., 1982, Alzheimer's disease and senile dementia: a quantitative study, *Science* 215:1237.

Yu, Q.S., Liu, C., Brzostowska, M., Chrisey, L., Brossi, A., Greig, N.H., Atack, J.R., Soncrant, T.T., Rapoport, S., and Radunz, H.E., 1991, Physovenines: efficient synthesis of (-)- and (+)-physovenine and synthesis of carbamate analogues of (-)-physovenine. Anticholinesterase activity and analgesic properties of optically active physovenines, *Helv. Chim. Acta* 74:761.

# CARRIER-MEDIATED DRUG TRANSPORT AT THE BLOOD-BRAIN BARRIER AND THE POTENTIAL FOR DRUG TARGETING TO THE BRAIN

Quentin R. Smith

Laboratory of Neurosciences
National Institute on Aging
National Institutes of Health
Bethesda, Maryland 20892

## INTRODUCTION

Many drugs cannot be used for the treatment of central nervous system diseases because they do not reach therapeutic concentrations in brain at acceptable doses. Though multiple factors influence the amount of drug that eventually reaches active sites in brain, one of the most important is transport across the blood-brain barrier. The blood-brain barrier generally impedes the uptake of large, hydrophilic, or charged compounds into the nervous system (Rall and Zubrod, 1962; Fenstermacher, 1989; Jezequel, 1992). In addition, it limits the brain access of drugs that bind tightly and restrictively to plasma proteins (Robinson and Rapoport, 1992).

Nowhere is this limitation more evident than in the treatment of central nervous system infections and tumors. Many antibiotic (Meulemans et al., 1988), antiviral (Terasaki and Pardridge, 1988), and chemotherapeutic drugs (Greig et al. 1990a) show restricted uptake into brain and fail to attain therapeutic concentrations in the nervous system after peripheral administration. Even when infections or tumors compromise the local integrity of the barrier, drug accumulation can be limited by surrounding brain tissue, which has an intact blood-brain barrier and serves as a diffusional sink (Levin et al., 1980). Active efflux pumps and catabolic enzymes also contribute to barrier function. For example, the anticancer drugs, vincristine and vinblastine, are transported out of brain into the circulation by the multiple drug resistance pump, P-glycoprotein (Cordon-Cardo et al., 1989; Tatsuta et al., 1992; Schinkel et al., 1994). Similarly, AZT (zidovudine) which is used in the treatment of AIDS is removed from brain and cerebrospinal fluid by a probenecid-sensitive efflux system (Dykstra et al., 1993; Wong et al., 1993).

Various strategies have been devised to improve brain delivery of poorly penetrating drugs (Table 1). Of these, the most frequently used method is to develop lipophilic analogs that more readily cross the membranes of the blood-brain barrier by passive diffusion.

Surveys of central nervous system drugs suggest that a $\log_{10}$ octanol/water partition coefficient (Log $P_{octanol/water}$) of 1-2 is optimal for brain distribution and activity (Hansch et al., 1987; Jezequel, 1992). However, with lipophilic analogs, the benefits achieved in barrier permeability can be offset, in some instances, by enhanced plasma protein binding and peripheral distribution such that the actual quantity of drug taken up into brain does not change. Such was found by Greig et al. (1990b,c) in studies of eight lipophilic esters of chlorambucil. Only one showed significantly improved brain delivery. Further, drug modification can alter receptor binding and activation. Thus, each chemical modification must be considered carefully when designing new agents for brain penetration.

In some instances drug structure cannot be modified sufficiently to obtain good brain penetration by passive diffusion without unduly compromising biological activity. In such cases, an alternative mechanism, that offers several useful possibilities, is to design drugs that are shuttled into brain by carrier-mediated transport. Over 20 carrier transporters have been identified at the cerebral capillaries of the blood-brain barrier that mediate the brain uptake of various essential nutrients, vitamins, peptides and hormones (Smith, 1992). By this route, it may be possible to obtain both enhanced brain uptake and selective brain targeting, to the extent that drugs are designed with specific affinity for the cerebrovascular transporters. This method has used to deliver L-DOPA to brain for the treatment of Parkinson's disease (Wade and Katzman, 1975; Alexander et al., 1994) and α-methyl-DOPA for the treatment of hypertension (Markovitz and Fernstrom, 1977). Comparable methods are being developed by Friden et al (1993) and Pardridge et al. (1994) for drug delivery to brain via receptor- or absorptive-mediated transcytosis.

**Table 1.** Strategies to augment brain uptake and accumulation of novel hydrophilic drugs

| | |
|---|---|
| 1. | Preparation of lipid soluble analogs or prodrugs |
| 2. | Intracarotid infusion to maximize brain arterial concentration |
| 3. | Direct injection into brain or CSF |
| 4. | Blood-brain barrier modification<br>a. Osmotic Opening<br>b. Specific Solutes |
| 5. | Design of drugs that are taken up into brain by carrier-, receptor-, or absorptive-mediated transport systems |
| 6. | Inhibition of active removal or breakdown |

In this paper, I review work that we have been conducting on the development and brain pharmacokinetics of new anticancer amino acid drugs with high affinity for the large neutral amino acid transporter of the blood-brain barrier. New agents are desperately needed for brain tumor chemotherapy that both readily penetrate the blood-brain barrier and work in combination with current agents (e.g., BCNU, CCNU, and procarbazine) to maximize tumor cell kill and minimize development of drug resistance (Greig et al., 1990a). In addition, initial results are presented on the development of selective blockers of peripheral drug uptake and distribution that may be of use to help channel active drug to brain. In the future it may be possible, with enhanced design and development, to produce new central nervous system pharmaceutics that are selectively transported by the carriers of the blood-brain barrier. This will initiate a new generation of drug targeting, much like that being attempted now with antibodies (Pietersz and Krauer, 1994), to develop "magic bullets" against brain disease. Selective delivery is important to minimize toxicity and maximize therapeutic potential.

## CEREBROVASCULAR LARGE NEUTRAL AMINO ACID TRANSPORTER

Of the carriers at the blood-brain barrier, the transporter for large neutral amino acids displays several properties that make it well suited as a brain drug delivery vector. First, it has both a large maximal transport capacity ($V_{max}$ ~40-60 nmol/min/g) and an appreciable binding affinity (affinity ≈ $1/K_m$; $K_m$ = 10-200 $\mu$M)(Smith et al., 1985,87) so that rapid rates of blood-brain barrier exchange can be obtained ($K_{in} > 10^{-3}$ ml/s/g) with half times for brain equilibration of < 15 min for high affinity substrates (Smith, 1992). Second, it possesses fairly simple structural requirements for binding and accepts a wide variety of amino acid substrates (Smith et al, 1989; Smith, 1992). And third, though the carrier mediates the brain uptake of essential amino acids that are required for brain metabolism, transient disruption in brain supply, as might occur with drug overdose, does not produce irreversible brain damage. This latter possibility may limit use of the cerebrovascular glucose transporter (GLUT-1) as a drug delivery vector because of the critical need of the brain for a continuous supply of D-glucose for brain energy metabolism.

Brain uptake via the cerebrovascular large neutral amino acid transporter is sodium independent, stereospecific and saturable (Momma et al., 1987). The carrier exhibits many of the properties of the classical L system transporter of Oxender and Christensen including inhibition by 2-aminobicyclo[2.2.1]-heptane-2-carboxylic acid (BCH) (Aoyagi et al., 1988; Smith, 1991). Substrate affinity for the cerebrovascular transporter, however, exceeds that of the L system in most tissues by 10-100 fold. Therefore, it is likely that the blood-brain barrier transporter represents one of a family of L system transporters that is selectively expressed in tissues. Such has been found for the cationic amino acid transporter (System $y_1^+$, CAT-1), the gene for which was recently cloned and shown to be highly expressed in cerebral microvessels (Stoll et al., 1993) but not in liver (Kim et al., 1991). Subsequent studies have identified a second member of the cationic amino acid transporter gene family (CAT-2) that is selectively expressed in liver (Closs et al., 1993). Final evaluation awaits identification and cloning of the L-system transporter gene family.

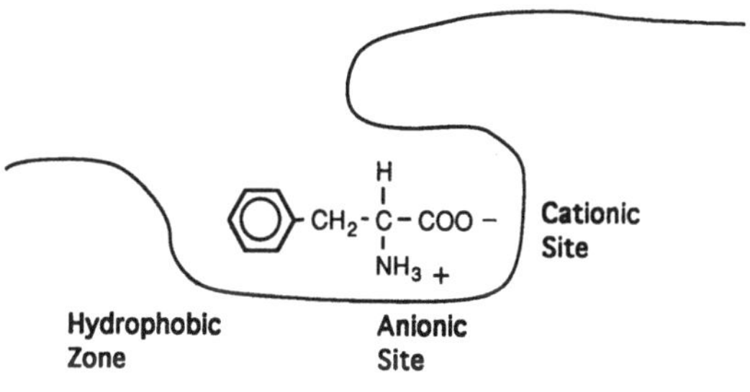


**Figure 1.** Model of the binding site of the cerebrovascular large neutral amino acid transporter in relation to the amino acid, phenylalanine.

Fig. 1 shows a model of the binding site of the cerebrovascular large neutral amino acid transporter. For affinity, substrates must contain both an unsubstituted α-amino acid functional group and a neutral, uncharged side chain (Smith et al., 1989). The transporter prefers L-amino acids with large, hydrophobic side chains, such as L-leucine and L-phenylalanine. The importance of side-chain hydrophobicity is shown by the apparent relation over >4-orders of magnitude (Fig. 2) between transport affinity ($1/K_m$) and amino acid octanol/water partition coefficient, a measure of side-chain hydrophobicity.

Marked variability is tolerated in side chain structure with such diverse compounds as O-benzyl-L-tyrosine, L-β-1-naphthylalanine, and L-2-amino-nonanoic acid all exhibiting high affinity for the carrier ($K_m$ or $K_i < 10\ \mu M$; apparent affinity > 100 $mM^{-1}$) (Smith et al, 1989; Takada et al., 1992). L-triiodothyronine (L-$T_3$) also expresses high affinity for the cerebrovascular $L_1$ system ($K_i = 1.0 \pm 0.1\ \mu M$, apparent affinity = 1000 $mM^{-1}$) though brain uptake of L-$T_3$ is mediated by the cerebrovascular thyroid hormone transporter (Hokari and Smith, 1994). Several fairly large amino acid analogs bind to the carrier, but evidence suggests that maximal transport capacity decreases with amino acid size (Takada et al., 1992).

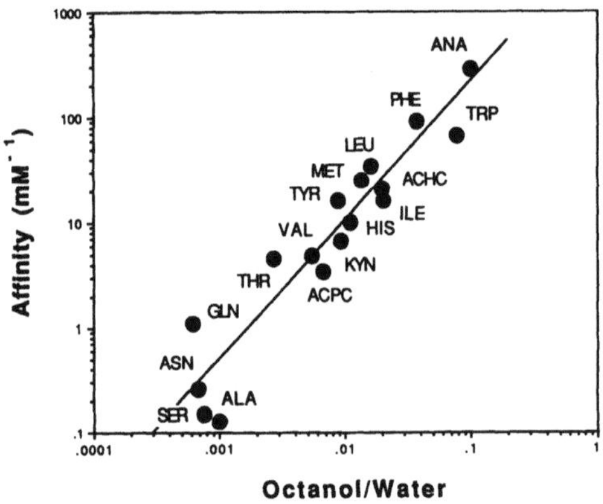


**Figure 2.** Relation of transport affinity to octanol-water partition coefficient for 17 amino acids that are taken up into brain by the cerebrovascular large neutral amino acid transporter. ANA = D,L-2-amino-1,2,3,4-tetrahydro-2-naphthoic acid, KYN = L-kynurenine, ACHC = 1-aminocyclohexanecarboxylic acid, ACPC = 1-aminocyclopentanecarboxylic acid. (Adapted from Smith, 1992).

With regard to drug targeting, amino acid transport has been suggested to be up-regulated at the blood-tumor barrier based on measurements of *in vivo* tumor uptake of radiolabeled amino acids in humans (Bergstrom et al., 1987; Langen et al., 1991) and in animals (Cornford et al., 1992). Several studies have shown that brain tumor uptake of radiolabeled amino acid can be reduced by competing amino acids and does not correlate with barrier disruption as measured using $^{68}$Ga-EDTA. The enhanced uptake may not reflect simply increased protein synthesis or metabolism, because it is also seen with nonmetabolizable amino acid tracers. The blood-tumor barrier may also express transporters that are not usually present at the blood-brain barrier and this may provide an additional means of drug targeting to critical cancer sites.

## SURVEY OF TRANSPORT AFFINITY FOR ANTICANCER AMINO ACID DRUGS

The success of the carrier approach for brain delivery depends in large part on the transport affinity and capacity of the cerebrovascular carriers for critical drugs. A number of L-amino acid anticancer agents are known that might be expected, based on structure, to express affinity for the cerebrovascular $L_1$ system transporter. In a preliminary study, we screened five such agents using the *in situ* rat brain perfusion technique (Takada et al., 1991).

The agents studied were melphalan, azaserine, L-6-diazo-5-oxo-norleucine (DON), acivicin, and buthionine sulfoximine. Melphalan is an nitrogen mustard alkylating agent, whereas azaserine, L-DON and acivicin are glutamine antagonists/antimetabolites. Buthionine sulfoximine is a glutathione synthesis inhibitor that has been used to enhance alkylating drug cytotoxicity and to limit development of drug resistance.

Each drug was found to express affinity for the cerebrovascular $L_1$ system transporter and reduced L-[$^{14}$C]leucine uptake into brain. However, transport affinity for each agent was low ($K_m$ or $K_i$ = 90-4700 $\mu$M, affinity < 0.21-11 $mM^{-1}$), only a fraction of that for L-phenylalanine ($K_m$ = 11 ± 1 $\mu$M, affinity = 91 ± 8 $mM^{-1}$), and was predicted to limit drug uptake to the nervous system (Takada et al. 1991). Consistent with this, brain uptake studies for these agents have shown only limited blood-brain barrier penetration (Greig et al., 1987; Williams et al., 1990; Fekete et al., 1990; Groothuis et al., 1992). Further, melphalan was found by Greig et al. (1988) to exhibit activity against intracerebral Walker-256 tumor after peripheral drug injection only at toxic doses. This suggested that if carrier-mediated transport were to be used to augment brain delivery, new agents would have to be developed that were targeted specifically with high affinity to the transporters of the nervous system.

## IDENTIFICATION OF A HIGH-AFFINITY ANTICANCER AMINO ACID

Knowledge that transport affinity for the cerebrovascular large neutral amino acid carrier is strongly influenced by side-chain hydrophobicity (Fig. 2) was used to initiate a search for hydrophobic analogs of melphalan and other anticancer amino acid drugs that would show improved delivery to brain. One such compound that was identified, D,L-2-amino-7-bis[(2-chloroethyl)-amino]-1,2,3,4-tetrahydro-2-naphthoic acid (D,L-2-NAM-7), is an analog of melphalan with 2 additional methylene carbons on the side chain to form a tetralene ring (Fig. 3).

Based on the presence of the two additional methylene groups, D,L-2-NAM-7 was predicted to have a 5-10 fold greater side-chain hydrophobicity than melphalan. The compound had previously been shown by Haines et al. (1987) to express high affinity for the $L_1$ system transporter of L1210 leukemia cells.

Kinetic studies using the *in situ* brain perfusion technique (Takada et al., 1992) demonstrated that D,L-2-NAM-7 also exhibits high affinity for the large neutral amino acid transporter of the blood-brain barrier ($K_m$ = 0.19 ± 0.02 $\mu$M, affinity = 5260 $mM^{-1}$). Indeed, D,L-2-NAM-7 has the highest affinity of any agent that we have tested against the transporter. Perfusion studies demonstrated that the blood-to-brain transfer coefficient ($K_{in}$) for D,L-2-NAM-7 uptake at tracer concentrations is 20-40-fold greater than that for L-melphalan. Furthermore, brain influx of D,L-2-NAM-7 was shown to be saturable, sodium independent and inhibitable by BCH, indicating transport by the cerebrovascular large neutral amino acid system(Fig. 4).

| Compound | Km ($\mu M$) | Affinity ($mM^{-1}$) |
|---|---|---|
| 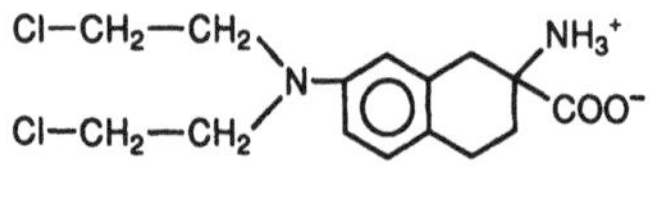 D,L-2-NAM-7 | 0.19 ± 0.02 | 5260 |
| 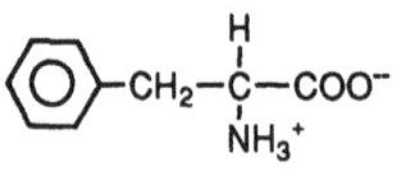 L - PHENYLALANINE | 10.5 ± 0.9 | 95 |
| 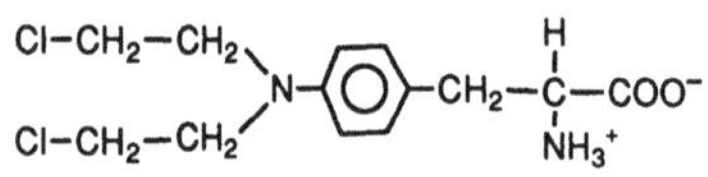 L-MELPHALAN | 150 ± 60 | 7 |

**Figure 3.** Chemical structures, half-saturation constants and affinities of D,L-2-NAM-7, L-phenylalanine and L-melphalan for the cerebrovascular large neutral amino acid transporter of the blood-brain barrier. (From Takada et al., 1992).

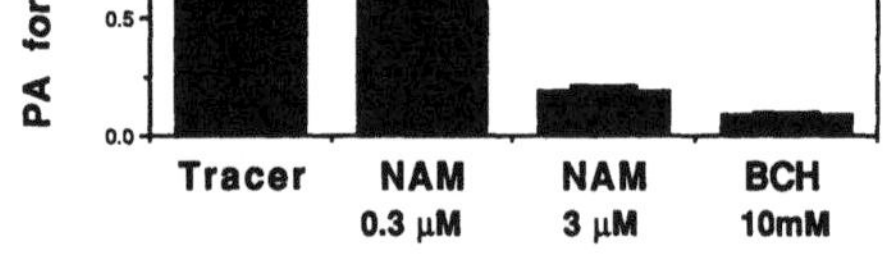


**Figure 4.** Inhibition of tritiated D,L-2-NAM-7 uptake into brain following addition of unlabeled D,L-2-NAM -7 (0.3 or 3 $\mu$M) or BCH (10 mM) to saline perfusate. From Takada et al. (1992). The uptake of tritiated D,L-2-NAM-7 was expressed as a cerebrovascular permeability-surface area product (PA) which was measured using the in situ rat brain perfusion technique.

In more recent studies, we have examined the *in vivo* pharmacokinetics of D,L-2-NAM-7 in brain and plasma in awake rats following intravenous administration (Fig. 5). The results demonstrate that D,L-2-NAM-7 is rapidly taken up into brain and reaches levels ~70% of those in plasma within 15 min after injection (Kohmo et al., 1993). The brain:plasma area-under-the-curve (AUC) ratio for D,L-2-NAM-7 exceeds that for melphalan by >20 fold. Like melphalan, primary deposition of NAM is to kidney and liver. The central nervous system appears to receive ~1% of the injected dose. Breakdown of D,L-2-NAM-7 is primarily to the monohydroxy and dihydroxy derivatives with a plasma half life for intact compound of ~60 min (Fig. 5).

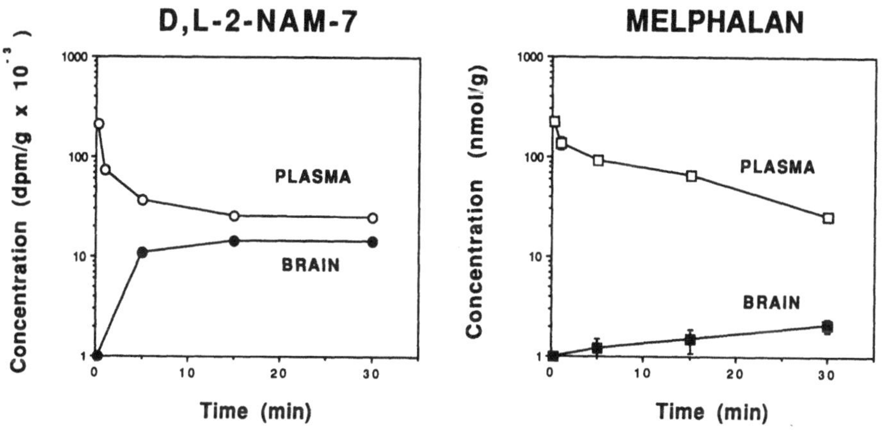


**Figure 5.** Time courses of tritiated D,L-2-NAM-7 and melphalan in brain and plasma after intravenous injection in the rat. Each point represents a mean for n=3-5 animals. Brain and plasma concentrations were measured by HPLC and reflect only intact, nondechlorinated compound. The monohydroxy and dihydroxy breakdown products represented less than 30% of total brain radioactivity for D,L-2-NAM-7 (From Greig et al., 1987 and Kohmo et al., 1993).

D,L-2-NAM-7 was also shown by Haines et al (1987) to have enhanced antitumor activity and reduced myelosuppressive activity *in vitro* as compared to melphalan. In studies using intracerebrally implanted Walker-256 carinosarcoma, tumor uptake of D,L-2-NAM-7 exceeded that into surrounding tissue by ~2 fold (Fig. 6). The mechanism for the enhanced tumor uptake, whether carrier-mediated or passive diffusion, was not determined.

## NAM ISOMERS AND DRUG TARGETING

An additional goal of our work is to target drug delivery to brain, either through the development of drugs that are selective for the transport carriers of the blood-brain barrier or the development of drugs that can be used to block peripheral uptake and distribution of active compound. In this regard, a series of structural isomers of NAM were synthesized and tested

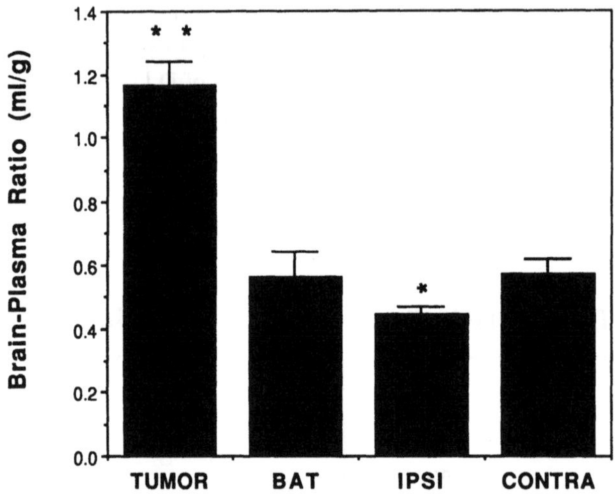


**Figure 6.** Brain tumor uptake of D,L-2-NAM-7. Each point represents a mean ± SEM for n=3-5 animals. Concentrations were determined at 15 min after injection. ** $P < 0.01$. * $P < 0.05$. BAT = Brain adjacent to tumor, Ipsi = Ipsilateral cortex or striatum, Contra = contralateral cortex or striatum (Kohmo et al., 1993).

**D,L-2-NAM-7**

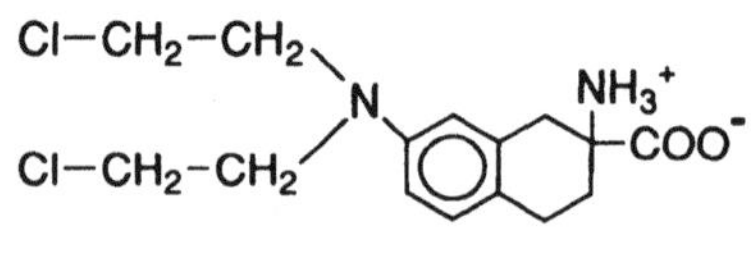

**D,L-2-NAM-6**

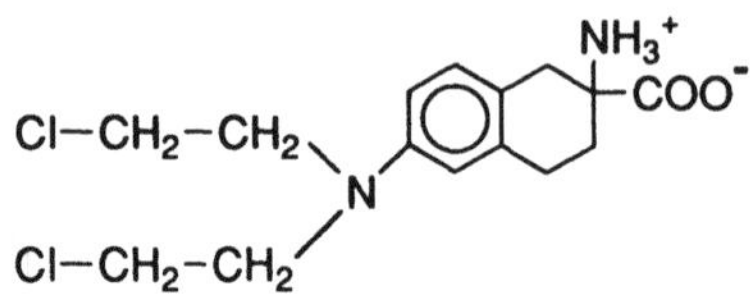

**D,L-1-NAM-7**

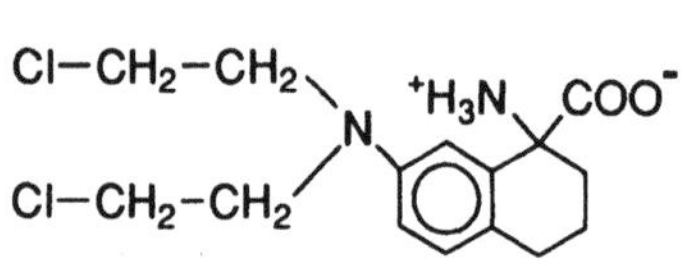

**Figure 7.** Structure of D,L-2-NAM-7, D,L-2-NAM-6 and D,L-1-NAM-7 isomers.

for activity against brain L-[$^{14}$C]leucine uptake by the cerebrovascular large neutral amino acid transporter.

The NAM analog with the amino acid functional group on carbon-1 of the tetrahydronaphthalene ring (D,L-1-NAM-7) exhibited very low affinity for the transporter ($K_i$ = 730 $\mu$M, affinity = 0.004 $mM^{-1}$; Takada et al., 1992), less than 0.1% of that for the high affinity isomer.

**Table 2.** Brain:plasma concentration ratios of D,L-2-NAM-7 and D,L-2-NAM-6 isomers 15 min after intravenous injection in rats.

| Tissue | D,L-2-NAM-7 | D,L-2-NAM-6 |
|---|---|---|
| Cortex | 0.55 ± 0.03 | 0.05 ± 0.01* |
| Cerebellum | 0.61 ± 0.02 | 0.07 ± 0.01* |
| Muscle | 0.58 ± 0.04 | 0.51 ± 0.05 |
| Heart | 2.01 ± 0.11 | 1.89 ± 0.24 |
| Liver | 6.76 ± 0.37 | 6.37 ± 0.45 |
| Kidney | 14.04 ± 0.85 | 21.09 ± 1.47* |

Values are means ± SEM for n=5. *$P < 0.05$.

Similarly, transfer of the nitrogen mustard from carbon 7 to carbon 5 or 6 of the tetrahydronaphthalene ring also reduced transport affinity of >100 fold. The results suggest that the transporter binding site has a very specific 3-dimensional conformation that is optimal for drug association and binding. Future studies with rigid amino acid structures, such as NAM, will be of use to further define the exact 3-dimensional conformation of the binding pocket.

The ability of the 6- and 7-nitrogen mustard isomers of NAM to be taken up into brain and peripheral tissues was examined following intravenous injection in rats (Table 2). As expected from the affinity studies, the 6-isomer demonstrated minimal uptake into brain. However, in peripheral tissues the uptake of the two isomers was virtually identical. Further, the plasma concentration time course did not differ significantly between the two compounds. This suggests that the two isomers are taken up in peripheral tissues by the same transport systems, but that the two compounds differ in transport at the blood-brain barrier. This may allow the 6-isomer, in the dechlorinated form, to be coadministered with D,L-2-NAM-7 to block D,L-2-NAM-7 uptake into peripheral tissues, thereby channeling active nitrogen mustard drug to brain.

## SUMMARY

Our studies demonstrate that high affinity drugs can be designed that are rapidly shuttled into brain by the large neutral amino acid carrier of the blood-brain barrier. The brain uptake of D,L-2-NAM-7 exceeds that of the low affinity analog, L-melphalan, by 20-40 fold.

In the future, it may be possible to design drugs that show not only increased brain uptake but also selective brain delivery via the saturable carriers of the blood-brain barrier. The distinction between the NAM 6 and 7 nitrogen mustard isomers is a first step in this direction. The high affinity of D,L-2-NAM-7 may also be of value for the selective labeling and isolation of the $L_1$ system transporter (Takada et al., 1992). Finally, it must be stressed that a similar approach for enhanced brain drug delivery can be taken with the other barrier transporters and preliminary work in that direction has been initiated by Allen (1994).

## ACKNOWLEDGMENTS

The author gratefully acknowledges the contributions of Dr. Peter Crooks, University of Kentucky to the identification of the NAM isomers.

## REFERENCES

Alexander, G.M., Schwartzman, R.J., Grothusen, J.R., and Gordon, S.W. , 1994, Effect of plasma levels of large neutral amino acids and degree of parkinsonism on the blood-brain transport of levodopa in naive and MPTP parkinsonian monkeys, *Neurology* 44:1491.

Allen, D.D., 1994, Assessment of blood-brain barrier choline transport using the in situ brain perfusion technique, *Soc. Neurosci. Abstr.* 20:1334.

Aoyagi, M., Agranoff, B.W., Washburn, L.C., and Smith, Q.R., 1988, Blood-brain barrier transport of 1-aminocyclohexanecarboxylic acid, a nonmetabolizable amino acid for in vivo studies of brain transport, *J. Neurochem.* 50:1220.

Bergström, M., Ericson, K., Hagenfeldt, L., Mosskin, M., von Holst, H., Norén, G., Eriksson, L., Ehrin, E., and Johnström, P., 1987, PET study of methionine accumulation in glioma and normal brain tissue: Competition with branched chain amino acids, *J. Comput. Assist. Tomogr.* 11:208.

Cordon-Cardo, C., O'Brien, J.P., Casals, D., Rittman-Grauer, L., Biedler, J.L., Melamed, M.R., and Bertino, J.R., 1989, Multidrug resistance gene (P-glycoprotein) is expressed by endothelial cells at blood-brain barrier sites, *Proc. Natl. Acad. Sci, USA.* 86:695.

Closs, E.I., Albritton, L.M., Kim, J.W., and Cunningham, J.M.., 1993, Identification of a low affinity, high capacity transporter of cationic amino acids in mouse liver, *J. Biol. Chem.* 268:7538.

Cornford, E.M., Young, D., Paxton, J.W., Finlay, G.J., Wilson, W.R., and Pardridge, W.M., 1992, Melphalan penetration of the blood-brain barrier via the neutral amino acid transporter in tumor-bearing brain, *Cancer Res.* 52:138.

Dykstra, K.H., Arya, A., Arriola, D.M., Bungay, P.M., Morrison, P.F., and Dedrick, R.L., 1993, Microdialysis study of zidovudine (AZT) transport in rat brain, *J. Pharmacol. Exp. Ther.* 267:1227.

Fekete, I., Griffith, O.W., Schlageter, K.E., Bigner, D.D., Friedman, H.S., Groothuis, D.R., 1990, Rate of buthionine sulfoximine entry into brain and xenotransplanted human gliomas, *Cancer Res.* 50:1251.

Fenstermacher, J.D., 1989, Pharmacology of the blood-brain barrier, in: "Implications of the Blood-Brain Barrier and Its Manipulation, Vol. 1," E.A. Neuwelt, ed., Plenum, New York.

Friden, P.M., Walus, L.R., Watson, P., Doctrow, S.R., Kozarich, J.W., Bäckman, C., Bergman, H., Hoffer, B., Bloom, F., and Granholm A.C., 1993, Blood-brain barrier penetration and in vivo activity of NGF conjugate, *Science* 259:373.

Greig, N.H., Momma, S., Sweeney, D.J., Smith, Q.R., and Rapoport, S.I., 1987, Facilitated transport of melphalan at the rat blood-brain barrier by the large neutral amino acid carrier system, *Cancer Res.* 47:1571.

Greig, N.H., Sweeney, D.J., and Rapoport, S.I., 1988, Comparative brain and plasma pharmacokinetics and anticancer activities of chlorambucil and melphalan in the rat, *Cancer Chemother. Pharmacol.* 21:1.

Greig, N.H., Genka, S., and Rapoport, S.I., 1990a, Delivery of vital drugs to the brain for the treatment of brain tumors, *J. Control. Release* 11:61.

Greig, N.H., Daly, E., Sweeney, D.J., and Rapoport, S.I., 1990b, Pharmacokinetics of chlorambucil-tertiary butyl ester, a lipophilic chlorambucil derivative that achieves and maintains high concentrations in brain, *Cancer Chemother. Pharmacol.* 25:320.

Greig, N.H., Genka, S., Daly, E.M., Sweeney, D.J., and Rapoport, S.I., 1990c, Physiochemical and pharmacokinetic parameters of seven lipophilic chlorambucil esters designed for brain penetration, *Cancer Chemother. Pharmacol.* 25:311.

Groothuis, D.R., Lippitz, B.E., Fekete, I., Schlageter, K.E., Molnar, P., Colvin, O.M., Roe, C.R., Bigner, D.D., and Friedman, H.S., 1992, The effect of an amino acid-lowering diet on the rate of melphalan entry into brain and xenotransplanted glioma, *Cancer Res.* 52:5590.

Haines, D.R., Fuller, R.W., Ahmad, S., Vistaca, D.T., and Marquez, V.E., 1987, Selective cytotoxicity of a system L specific amino acid nitrogen mustard, *J. Med. Chem.* 30:542.

Hansch, C., Bjorkroth, J.P., and Leo, A., 1987, Hydrophobicity and central nervous system agents: On the principle of minimal hydrophobicity in drug design, *J. Pharm. Sci.* 76:663.

Hokari, M., and Smith, Q.R., 1994, Thyroid hormones express high affinity for both the thyroid hormone and large neutral amino acid transporters of the blood-brain barrier, *Soc. Neurosci. Abst.* 20:1335.

Jezequel, S.G., 1992, Central nervous system penetration of drugs: Importance of physiochemical properties. in: "Progress in Drug Metabolism, Vol. 13," G.G. Gibson, ed., Taylor & Francis, London.

Kim, J.W., Closs, E.I., Albritton, L.M., and Cunningham, J.M., 1991, Transport of cationic amino acids by the mouse ecotropic retrovirus receptor, *Nature (London)* 352:725.

Kohmo, Y., Greig, N.H., and Smith, Q.R., 1993, Pharmacokinetics and brain tumor uptake of D,L-NAM - a high affinity anticancer amino acid, *Soc. Neurosci. Abstr.* 19:1897.

Langen, K.J., Roosen, N., Coenen, H.H., Kuikka, J.T., Kuwert, T., Herzog, H., Stöcklin, G., and Feinendegen, L.E., 1991, Brain and brain tumor uptake of L-3-[123I]iodo-α-methyl tyrosine: Competition with natural L-amino acids. *J. Nuc. Med.* 32:1225.

Levin, V.A., Patlak, C.S., and Landahl, H.L., 1980, Heuristic modeling of drug delivery to malignant brain tumors, *J. Pharmacokin. Biopharm.* 8:257.

Markovitz, D.C., and Fernstrom, J.D., 1977, Diet and uptake of Aldomet by the brain: competition with natural large neutral amino acids, *Science* 197:1014.

Meulemans, A., Vicart, P., Mohler, J., and Vulpillat, M., 1988, Determination of antibiotic lipophilicity with a micromethod: application to brain permeability in man and rats, *Chemotherapy* 34:90.

Momma, S., Aoyagi, M., Rapoport, S.I., and Smith, Q.R., 1987, Phenylalanine transport across the blood-brain barrier as studied with the in situ brain perfusion technique, *J. Neurochem.* 48:1291.

Pardridge, W.M., Kang, Y.S., and Buciak, J.L., 1994, Transport of human recombinant brain-derived neurotrophic factor (BDNF) through the rat blood-brain barrier in vivo using vector-mediated peptide drug delivery, *Pharm. Res.* 11:738.

Pietersz, G.A., and Krauer, K., 1994, Antibody-targeted drugs for the therapy of cancer, *J. Drug Targeting* 2:183.

Rall, D.P., and Zubrod, C.G., 1962, Mechanisms of drug absorption and excretion - Passage of drugs in and out of CSF, *Ann. Rev. Pharmacol.* 2:109.

Robinson, P.J., and Rapoport, S.I., 1992, Transport of drugs, in: "Handbook of Experimental Pharamcology, Vol. 103, M.W.B. Bradbury, ed., Springer-Verlag, Heidelberg.

Schinkel, A.H., Smit, J.J.M., van Tellingen, O., Beijnen, J.H., Wagenaar, E., van Deetmer, L., Moi, C.A.A.M., van der Valk, M.A., Robanus-Maandag, E.C., te Riele, H.P.J., Berns, A.J.M., and Borst, P., 1994, Disruption of the mouse mdr1a P-glycoprotein gene leads to a deficiency in the blood-brain barrier and to increased sensitivity to drugs, *Cell* 77:491.

Smith, Q.R., 1991, The blood-brain barrier and the regulation of amino acid uptake and availability to brain, *Adv. Exp. Med. Biol.* 291:55.

Smith, Q.R., 1992, Drug delivery to brain and the role of carrier-mediated transport, *Adv. Exp. Med. Biol.* 331:83.

Smith, Q.R., Takasato, Y., Sweeney, D.J., and Rapoport, S.I., 1985, Regional cerebrovascular transport of leucine as measured by the in situ brain perfusion technique, *J. Cereb. Blood Flow Metabol.* 5:300.

Smith, Q.R., Momma, S., Aoyagi, M., and Rapoport, S.I., 1987, Kinetics of neutral amino acid transport across the blood-brain barrier, *J. Neurochem.* 49:1651.

Smith, Q.R., Aoyagi, M., and Rapoport, S.I., 1989, Structural specificity of the cerebrovascular large neutral amino acid transporter, *Soc. Neurosci. Abstr.* 15:1025.

Stoll, J., Wadhwani, K.C., and Smith, Q.R., 1993, Identification of the cationic amino acid transporter (System y+) of the rat blood-brain barrier. *J. Neurochem.* 60:1956.

Takada, Y., Greig, N.H., Vistica, D.T., Rapoport, S.I., and Smith, Q.R., 1991, Affinity of antineoplastic amino acid drugs for the large neutral amino acid transporter of the blood-brain barrier, *Cancer Chemother. Pharmacol.* 29:89.

Takada, Y., Vistica, D.T., Greig, N.H., Purdon, D., Rapoport, S.I., and Smith, Q.R., 1992, Rapid high-affinity transport of a chemotherapeutic amino acid across the blood-brain barrier, *Cancer Res.* 52:2191.

Tatsuta, T., Naito, M., Oh-hara, T., Sugawara, I., and Tsuruo, T., 1992; Functional involvement of P-glycoprotein in blood-brain barrier, *J. Biol. Chem.* 267:20383.

Terasaki, T., and Pardridge, W.M., 1988, Restricted transport of 3'-azido-3'-deoxythymidine and dideoxynucleosides through the blood-brain barrier, *J. Infect. Dis.* 158:630.

Wade, L.A., and Katzman, R., 1975, Synthetic amino acids and the nature of L-DOPA transport at the blood-brain barrier. *J. Neurochem.* 25:837.

Williams, M.G., Earhart, R.H., Bailey, H., and McGovern, J.P., 1990, Prevention of central nervous system toxicity of the antitumor antibiotic acivicin by concomitant infusion of an amino acid mixture, *Cancer Res.* 50:5475.

Wong, S.L., van Belle, K., and Sawchuk, R.J., 1993, Distributional transport kinetics of zidovudine between plasma and brain extracellular fluid/cerebrospinal fluid in the rabbit: Investigation of the inhibitory effect of probenecid utilizing microdialysis, *J. Pharmacol Exp. Therap.* 264:899.

# BLOOD-BRAIN BARRIER MANIPULATION: CURRENT STATUS OF LABORATORY AND CLINICAL STUDIES

Edward A. Neuwelt

Oregon Health Sciences University
3181 SW Sam Jackson Park Road - L603
Portland, Oregon 97201-3098
USA

## INTRODUCTION

Chemotherapy has been relatively ineffective in the treatment of central nervous system (CNS) metastases of systemic cancers (i.e., breast cancer, small cell cancer of lung, lymphoma, and germ cell tumors), while clinical regression, even complete remission, of the tumor has occurred at the non-CNS systemic sites. The failure of response of such CNS metastatic tumors led us to focus on the issue of drug delivery to tumor. Besides the classic pharmacologic issues of drug delivery to tumors (i.e., blood flow, drug concentration, time of exposure, etc.), tumors within the CNS can be affected by problems of delivery because of the blood-brain barrier (BBB)[1,2]. The old concept that malignant tumors in the CNS have no effective barrier, as emphasized, for instance, by enhancement on computerized tomographic (CT) scan, has been shown to be incorrect. Numerous studies report variability in barrier permeability within malignant tumors. Even the well-vascularized, actively proliferating edge of the tumor, the brain adjacent to tumor (BAT), is known to be variable and complex in terms of barrier integrity.

The role of the BBB in the limitation of delivery of various agents to CNS tumors[1,3], infections and enzyme deficiencies has been reviewed in a recent monograph. In several experimental brain tumor models, blood-brain barrier disruption (BBBD) by the bolus administration of hyperosmotic mannitol, has resulted in increased drug delivery to tumor, BAT, and brain distant to tumor (BDT)[4]. Our animal studies have demonstrated that the BBB is relatively intact in small tumors and more leaky in large tumors[5]. Thus in larger, more permeable tumors, BBBD enhanced delivery may be less relevant to tumor as opposed to BAT, while in smaller, less permeable, tumors the importance of BBBD will be accentuated. Osmotic opening of the BBB decreases the tumor to normal brain drug ratio, which is a point of concern of several investigators. In order to increase delivery effectively to the generally less permeable proliferating edge around tumor, which may contribute to therapeutic failure, the entire vascular distribution supplying the tumor and surrounding brain needs to be disrupted. This may be

*New Concepts of a Blood–Brain Barrier*
Edited by J. Greenwood *et al.*, Plenum Press, New York, 1995

particularly advantageous when multiple tumor foci are present, as is generally the case with breast and lung tumor metastases. Technical details including anesthesia can be very important[6].

Radiation remains the primary therapy for malignant brain tumors, especially gliomas[7]. However, as pointed out in a recent lead article in the *New England Journal of Medicine*[8] and the accompanying editorial[9]. The unacceptable neurotoxicity associated with brain irradiation, especially in children, makes the search for more effective forms of chemotherapy to the CNS a priority. Additionally, our group recently published a review of the neurobehavioral sequelae of cranial radiation in adults[10], concluding that the true incidence of radiation-induced encephalopathy has probably been significantly underestimated. Therefore, alternatives for increasing drug delivery to the CNS, such as intraarterial (i.a.) administration, high-dose systemic therapy, methods of specific targeting and drug administration with BBBD, continue to be a major focus. An important consideration that must not be neglected is the characteristic steep dose-response of most sensitive tumors *in vivo*. Animal studies suggest that even a 20% increase in drug delivery is significant. Any influence that increases or decreases drug delivery must be treated as crucial. Only when adequate drug delivery is achieved can various components of chemotherapy such as tumor sensitivity, tumor cell resistance, combinations of drugs, fractionation schemes, and drug doses be defined.

The preclinical studies of BBBD as a means to increase anti-tumor agent delivery to animal models of brain tumors has led to the clinical use of this technique in brain tumor patients. The first clinical trial of osmotic opening of the BBB to increase chemotherapy delivery was undertaken in 1979. We have seen statistically significant prolongation of survival in patients with primary CNS lymphoma and glioblastoma. Response has also been dramatic in patients with germ cell tumors and primitive neuroectodermal tumors. Extensive neuropsychologic testing continues to evaluate potential neurotoxicity, which appears minimal.

## ANIMAL STUDIES

### Monoclonal Antibodies in Tumor Therapy

Tumor-specific monoclonal antibodies (mAb) offer great potential as targeting agents for chemotherapeutic agents. We have extensive experience with the L6 mAb provided by Bristol-Myers Squibb[11]. L6 is a mouse monoclonal $IgG_{2a}$ antibody that we have shown to be reactive with a propressophysin-like cell surface protein abundant on human lung carcinoma, breast carcinoma and colon carcinoma. While the L6 mAb can be used to distinguish carcinoma cells by immunohistochemistry, in order for mAbs to effectively transport drugs or toxins into the cytoplasm after cell surface binding, the carrier must be internalized and hydrolysed for maximum antitumor activity. Unfortunately, very few internalizing, or modulating, antibodies have been described which bind solid tumors (L6 binds tumor cells specifically but does not modulate). Our collaborators at Bristol-Meyers Squibb have developed two modulating mouse mAbs, BR64 and BR96 (both $IgG_1$), that have been shown to selectively react with carcinomas of the lung, breast, ovary and colon while having low reactivity with normal human tissues. Following conjugation to a doxorubicin (ADR) derivative, both mAbs retain immunoreactivity and cytotoxic activity. The BR96 mAb has a direct cytotoxic effect on antigen-positive tumor cells *in vitro* and *in vivo*. The BR96-ADR conjugate has recently been shown to be highly effective against a number of tumor types, grown subcutaneously in the nude rat, and phase 1 clinical trials in patients with systemic tumors have begun as of January 1, 1994. We have begun studies to delivery the doxorubicin immunoconjugate by BBBD to provide a novel approach to treatment of a drug sensitive brain metastasis when using a drug with potential neurotoxicity.

In addition to chemotherapeutic drugs such as ADR, protein toxins can also be linked to antibodies that specifically recognize tumor cell surface receptors. This results in cell-specific cytotoxic reagents that are referred to as oncotoxins, mitotoxins, or immunotoxins. These bifunctional compounds are either chemical conjugates or fusion proteins between an antibody and a toxin. Antibody fusion proteins have been developed between BR96 and the *Pseudomonas* exotoxin, sFv-PE40 (PE). Protein toxins such as PE are extremely potent and very low quantities of the hybrid molecules are required for effective therapy. It is important to recognize that they are immunogenic and can elicit an anti-toxin immune response.

Many cancer types express tumor associated differentiation antigens (complexes of glycoproteins and carbohydrate antigens) to which antibodies have been developed. Antibodies of this type have been shown to bind to numerous types of carcinomas. Identification of mAbs, such as BR96, that bind strongly to carcinoma cells, rapidly internalize, and have weak reactivity with most normal human tissues, allows development of molecules that are relatively effective, non-toxic therapeutic agents.

## MR Imaging and Diagnosis of Brain Metastases

A major problem in brain tumor therapy is the lack of a noninvasive method to determine tumor type or to accurately determine the complete extent of tumor infiltration. Tumors are largely indistinguishable by current CT or magnetic resonance (MR) imaging methods. These techniques demonstrate mass effect, edema, and pathologic changes associated with tumors without showing the tumor cells themselves. The use of paramagnetic substances conjugated to tumor-specific mAbs has potential to specifically diagnose tumor type and extent.

Development of agents for the specific imaging of tumors as well as monitoring BBBD has been an ongoing goal in our laboratory. We have investigated the toxicity and pharmacology of ionic and nonionic iodinated contrast agents for use with CT. More recently, we began evaluation of MR imaging as a new technology offering greater contrast resolution and better anatomic and pathologic detail compared to CT. Enhanced imaging with the paramagnetic materials, such as gadolinium (Gd), has been used clinically to evaluate areas of BBB compromise. Due to the problem of chelating sufficient Gd to mAbs, and the toxicity of Gd conjugates, we are now exploring the use of superparamagnetic iron oxide (FeO) conjugates.

Superparamagnetic labels are preferable over other paramagnetic compounds for MR imaging because of their two- to threefold higher magnetic susceptibility, allowing *in vivo* detection at concentrations as low as 1 $\mu$g of iron/gm tissue[12]. In the past, the use of superparamagnetic magnetite particles attached to antibody has been limited because of their large size (50 to 2000 nm), which resulted in the rapid clearance of these particles from the blood by liver and spleen before they could bind to antigen. In contrast, the monocrystalline FeO nanoparticles (MIONs) developed by Weissleder and colleagues can pass through capillary fenestrae to reach the interstitial space of the systemic circulation. The iron core of these particles is a single FeO crystal, 4.6 ± 1.2 nm in diameter as determined by electron microscopy (EM). The microcrystalline iron preparations, have a stable dextran coat consisting of 25 ± 6 dextran molecules (10 kd each) that isolates the central FeO core from surrounding solutions and minimizes protein adsorption to the particle surface. The microcrystalline iron particles have a slight overall negative charge, probably due to the dextran-bound carboxylic groups. Since brain endothelium also has a slight overall negative charge, nonspecific binding of these particles to brain endothelium should be less of a problem. Weissleder has reported polyclonal human immunoglobulin labeled with polymeric FeO as a promising *in vivo* imaging agent for the localization of IgG. Weissleder et al. were also able to demonstrate marked changes in the signal intensity of infarcted myocardium using antimyosin Fab fragments bound to microcrystalline iron particles. We have demonstrated that FeO particles can be successfully delivered with BBBD.

## Differential Permeability

In preclinical studies, we found only a few chemotherapeutic agents (including cytoxan, methotrexate, etoposide, carboplatin and, most recently etoposide-phosphate) were amenable to administration with BBB opening. There was neurotoxicity associated with other agents evaluated. We observed differential increase of BBB permeability in CNS lesions and demonstrated this quantitatively in a rat brain abscess model evaluating gentamicin delivery and binding characteristics[13]. Additionally, we have demonstrated differential permeability qualitatively in a nude rat brain tumor model in that only the low-molecular weight dye fluorescein penetrated tumor, with little or no penetration of high-molecular weight agents such as Evans blue-albumin or radiolabeled antibody. These findings suggest a new therapeutic approach in that, after parenteral drug administration, subsequent delayed antidrug antibody administered after reaching peak drug delivery to brain can bind potentially toxic systemic drug without affecting intracerebral drug. We are currently evaluating methotrexate and adriamycin immune complex binding with monoclonal or polyclonal antibody in both differential permeability and drug inactivation and drug rescue studies[14].

## Sequencing of Combination BBBD Chemotherapy and Radiation

A major concern in clinical neuro-oncology therapeutics is the potential neurotoxic effect of therapeutic modalities. The neurotoxic sequelae of cranial radiation, particularly in children, has been extensively reviewed. We undertook preclinical studies to evaluate the long term effect of various sequences of chemotherapy and cranial radiation administration. An initial trial evaluated the efficacy of chemotherapy regimens in an animal intracerebral lung tumor metastasis model (LX-1). Two combinations of drugs (methotrexate and cytoxan) and (carboplatin and etoposide) were determined to be efficacious against the LX-1 subcutaneous tumor model, as shown by a 50% decrease in tumor volume over time. A current study was designed to evaluate the long term (one year) neuropathological effects of chemotherapy (either methotrexate [MTX], or carboplatin and etoposide) in combination with whole head irradiation given either 30 days prior, 24 hours prior, or 30 days after chemotherapy with or without BBB opening. Initial observations included the following: mortality (within the first 30 days after chemotherapy) was significantly higher in animals that received radiation 30 days prior to chemotherapy (although no difference between drugs administered); seizures were observed within 10 minutes following discontinuation of anesthesia in 35% of animals receiving radiation prior to MTX, though no seizures were seen with MTX alone or with carboplatin/etoposide with or without radiation; and there was a significant decrease in drug delivery (radiolabeled MTX and radiolabeled Dextran 70) when radiation was given before chemotherapy as compared to control animals (no radiation). Necropsies are now being performed at the one year time point to evaluate neuropathology.

## Virus Delivery to Intracerebral Tumors

Genetically engineered retroviruses or herpes simplex virus type I (HSV-I) with decreased neurovirulence have been proposed as novel therapy for malignant brain tumors. The primary issue with such therapy is obtaining adequate delivery of virus to tumor without toxicity to normal brain. To date, preclinical studies and a limited Phase I clinical study have relied on either intratumor inoculation of infectious toxic virus, or intratumor implantation of cells producing retroviruses that transfer toxic genes. We hypothesize that the limited volume of distribution obtained by intracerebral inoculations will limit the efficacy of such therapy[15]. Our current data support the hypothesis that delivery of virus across the BBB can be accomplished by the hypertonic mannitol BBB disruption technique. Vascular permeability to small molecules (chemotherapeutic drugs), large molecules (immunoglobulins), and even very large particles (iron oxide or radiolabeled herpes virus) is increased maximally at 15 min post infusion of hypertonic mannitol, after which it decreases rapidly, and returns to pre infusion levels within 2 hours. This method has been well characterized in animals and humans.

We are examining new methods for the delivery, to intracerebral tumor xenografts in the nude rat, of two strains of high titer ($10^{10}$ pfu/ml), replication compromised recombinant HSV-1 and adenovirus bearing the lacZ marker gene. We are evaluating virus infection, gene expression, and toxicity and efficacy after delivery to intracerebral tumor by direct inoculation or convection enhanced diffusion (CED), compared with osmotic blood-brain barrier disruption (BBBD). We hypothesize that CED will enhance local delivery, but only osmotic BBB disruption will result in global delivery.

## CLINICAL OSMOTIC BLOOD-BRAIN BARRIER PROGRAM

The previous, ongoing and proposed tumor biology and molecular genetics investigation described above have had a direct clinical application in treating patients with malignant brain tumors. A number of the specific preclinical studies were a direct result of observations made and concerns raised in our clinical protocols of osmotic opening of the blood-brain barrier. The issues of tumor resistance, specific tumor targeting, potential neurotoxicity and treatment modality sequencing spur our basic science investigation. Specifically, further investigation is underway in animal models to reevaluate previously studied chemotherapeutic agents, initially thought too neurotoxic to use in conjunction with osmotic BBBD. The neurotoxicity may be species specific. Additionally, as tumor resistance to chemotherapeutic agents remains a significant impediment to durable response, a broader armamentarium of anti-tumor agents is needed. Further, treatments with tumor-specific targeting is necessary to reduce the risks of systemic and neurotoxicity. Tumor-specific targeting may also lead to improved imaging techniques can enhance our ability to identify tumor involvement, response to treatment, neurotoxicity (if any), and possibly aid in tumor diagnosis.

We undertook the first clinical osmotic opening of the blood-brain barrier in 1979 while at the University of Texas Southwestern Medical School and since then have performed osmotic opening of the blood brain barrier over 3,000 times in approximately 300 patients without a procedurally related death[16]. Patients have ranged in age from 18 months to 72 years and have had a myriad of tissue diagnoses including: glioblastoma multiforme, anaplastic astrocytoma, primitive neuroectodermal tumor, disseminated germ cell tumor, primary central nervous system lymphoma and brain metastases. Dramatic responses to our combination chemotherapy have been seen particularly in non-radiated patients with primary CNS lymphoma, germ-cell tumor and medulloblastoma. Our goal is to complement our laboratory investigation with continued integration of laboratory and clinical investigation in a national multicenter clinical blood-brain barrier program, thus increasing patient accrual to further evaluate the efficacy of treatment regimens which we have employed at the Oregon Health Sciences University.

## CHEMOTHERAPY REGIMENS

In our initial clinical series of 38 patients[17] with glioblastoma (all who had received surgery and cranial radiation) who received methotrexate-cytoxan plus osmotic BBBD, we demonstrated a significant prolongation of survival (17.5 months) in 27 evaluable patients (using the Cox proportional hazards model). This was compared to conventional therapy used in two historical controls from the same institution (14 patients treated with surgery and radiation, and 8 patients treated with surgery, radiation and chemotherapy) where survival was 12.8 and 11.4 months respectively. Neither neuroradiologic nor neuropsychologic abnormalities consistent with progressive leukoencephalopathy were seen. We have also evaluated 30 consecutive non-acquired immune deficiency syndrome patients with primary CNS lymphoma treated with a methotrexate-cytoxan regimen in conjunction with osmotic opening of the BBB[18]. Thirteen patients received cranial radiation 1 to 9 months before referral (Group 1). Seventeen patients received initial BBB disruption chemotherapy with subsequent radiation only for tumor progression or recurrence (Group 2). The difference in median survivals from diagnosis - 17.8 months for Group 1 and 44.5 months for Group 2- was statistically significant. Group 1 survival was comparable with the 20-month median survival of a historical series of patients (n=208) treated with radiotherapy $\pm$ chemotherapy. Group 2 patient survival represented an advance in the survival of this disease and was associated with preservation of cognitive function in 6/7 nonirradiated complete responders observed for 1 to 7 years. Patient toxicity was manageable and the survival data suggests that 30-40% of such patients may be cured without the attendant neuropsychological sequelae of cranial radiation. Additionally, the risk of neurotoxicity was evaluated in eight non-AIDS primary CNS lymphoma patients surviving disease free more than one year following completion of osmotic BBB opening and the methotrexate-cytoxan (and procarbazine, decadron) regimen. Trends in neuropsychologic assessment results between baseline and follow-up (1 to 7 years) were analyzed for all eight nonradiated survivors. This serial assessment design addressed the specific issue of neurotoxic risk potential of treatment, when confounding factors of tumor persistence and/or recurrence and cranial radiation were ruled out. This long term follow-up suggests that chemotherapy can be given in conjunction with osmotic BBB opening in nonradiated patients without cognitive manifestations of neurotoxicity.

We have recently completed a review of a broad Phase II trial of carboplatin and etoposide administered in conjunction with osmotic BBBD. This regimen was carried out in 34 patients with glioblastoma multiforme (n=3), malignant astrocytoma (n=8), malignant astrocytoma-oligodendroglioma (n=1), primitive neuroectodermal tumor (n=4), disseminated CNS germ cell tumors (n=6), CNS lymphomas (n=7) and metastatic carcinomas (n=5). The major toxicity observed in patients treated with multiple courses of this regimen was reversible myelosuppression and irreversible high-frequency hearing loss. Twenty two patients had measurable disease and nine objective responses were observed in these patients. It appears that carboplatin and etoposide with BBBD is an active regimen in treatment of malignant astrocytomas, primitive neuroectodermal tumors, CNS lymphoma and disseminated CNS germ cell tumors. However, such therapy is associated with significant high-frequency hearing loss. Based on the above results and relatively minimal toxicity of these combination chemotherapy agents, we feel that additional combinations of the four chemotherapeutic agents (methotrexate, cytoxan, carboplatin and etoposide) may prove more efficacious. We began a combination chemotherapy regimen of methotrexate and cytoxan and escalating doses of etoposide (to a maximum of 200mg/m$^2$/infusion) in six CNS lymphoma patients to date with minimal hematologic toxicity and excellent radiographic responses in the nonradiated patients included.

We have recently received approval to proceed with a Phase I study to evaluate the toxicity, maximum tolerated dose and possible efficacy of etoposide-phosphate in conjunction with osmotic BBB opening in patients with primary or metastatic brain tumors. Etoposidephosphide is an investigational new drug (IND). The chemical properties of etoposide, including its formulation in bile salts due to its poor solubility, requires it to be administered slowly i.v. to avoid precipitation in the blood and resultant hypertension. However, the increased solubility of etoposide-phosphate, the phosphorylated form of etoposide, permits administration of increased concentrations of the drug systemically and intraarterial administration. We have shown in preclinical studies that intraarterial administration of chemotherapeutic agents increases drug delivery over an intravenous route when either route is used in conjunction with BBBD. Our preclinical rodent studies of etoposide-phosphate showed efficacy against subcutaneous LX-1 tumors line (human small cell lung cancer) and no adverse side effects observed with intraarterial doses of 200 - 400mg/m$^2$ when administered with BBBD.

## MONOCLONAL ANTIBODIES

Based on our preclinical studies of differential permeability and monoclonal antibody studies in the normal, nude and LX-1 tumor-bearing rat models, we evaluated eight patients with malignant brain tumors comparing uptake in tumor of an iodinated contrast with CT scanning and the uptake of low molecular weight ($^{99m}$Tc-glucoheptonate) and high molecular weight ($^{99m}$Tc-albumin) imaging agents measured by radionuclide brain scans[19]. The agent $^{99m}$Tc-labeled albumin was chosen for evaluation because its molecular weight (68,000) was similar to that of the most clinically promising monoclonal antibody fragment (the IgG Fab monomeric fragment). The radionuclide brain scans in these patients showed uptake of the high molecular weight marker in tumor being much less than the low molecular weight radionuclide. A clinical implication of these studies was that the success of monoclonal antibody therapy in the treatment of malignant brain tumors may require techniques to increase blood-brain and blood-tumor barrier permeability to protein.

Additionally, we evaluated the delivery of melanoma-specific radiolabeled monoclonal antibody (mAb) Fab fragments in a pilot study of three patients with melanoma metastatic to the CNS[20]. Tumor samples demonstrated excellent reactivity by immunohistochemistry with Fab 96.5, specific for a 97,000-molecular- weight melanoma antigen (p97) and/or Fab 48.7, specific for a melanoma- associated proteoglycan antigen. All three patients received $^{131}$I-tumor-specific Fab (5 to 7 mg, 1 mCi/mg) intravenously, and on a separate occasion, two received $^{131}$I nonspecific Fab (5 to 7 mg, 1 mCi/mg). There was no uptake of either antibody into the region of the tumor (as documented by gamma camera brain images). However, there was increased uptake in the blood-brain barrier (BBB) modified areas in all three patients when radiolabeled tumor-specific mAb was administered intravenously in conjunction with osmotic BBB opening.

## SUMMARY

In summary, the Blood-Brain Barrier program began in 1979 with preclinical animal studies to evaluate drug delivery and imaging. Clinical trials ensued following the conventional sequence of Phase I (toxicity) and Phase II (efficacy) studies. These studies are currently being expanded to a multiinstitutional setting in preparation for Phase III randomized trials. Concomitantly, a number of issues have arisen from the clinical trials which have stimulated further laboratory studies, such as the sequencing of chemotherapy and radiation. The issue of maximizing drug delivery using monoclonal antibodies as targeting agents, the delivery of chemotherapy and the global delivery of genetic material are other therapeutic options being pursued. Thus, we plan to expand manipulation of the blood-brain barrier for continued application in treatment of patients with malignant brain tumors, and as a potential therapeutic technique for treatment of infections and genetic and degenerative disorders of the central nervous system.

## ACKNOWLEDGEMENTS

This work was supported by the Veterans Administration Merit Review Grant, and the National Institutes of Health, grants CA31770 and NS27757.

The author wishes to express sincere appreciation to Gail Engles for assistance in preparing this manuscript.

## REFERENCES

1. E.A. Neuwelt and P. Barnett, Blood-brain barrier disruption in the treatment of brain tumors: Animal studies, *in:* Implications of the blood-brain barrier and its manipulation: Vol II Clinical aspects, E.A. Neuwelt, ed., pp 107-194, Plenum Press, New York (1989).
2. S.I. Rapoport, Blood-brain Barrier in Physiology and Medicine, Raven Press, New York (1976).
3. N.H. Greig, Brain tumors and the blood-tumor barrier, *in*: Implications of the blood-brain barrier and its manipulation: Vol I Basic science aspects, E.A. Neuwelt ed., Plenum Press, New York (1989).
4. R.G. Blasberg, D. Groothuis, and P. Molnar, A review of hyperosmotic blood-brain barrier disruption in seven experimental brain tumor models, *in*: Pathophysiology of the Blood-brain Barrier, B.B. Johansson, C. Owman, and H. Widner, eds., Volume 14:197-220, Elsevier, New York, (1990).
5. P.A. Barnett, S. Roman-Goldstein, F. Ramsey, C. McCormick, G. Sexton, J. Szumowski, and E.A. Neuwelt, Differential permeability and quantitative MR imaging of a human lung carcinoma brain xenograft in the nude rat, *Am. J. Path.* 146:436 (1995)
6. M.K. Gumerlock and E.A. Neuwelt, The effects of anesthesia on osmotic blood-brain barrier disruption, *Neurosurgery* 26:268 (1990).
7. M.D. Walker, S.B. Green, D.P. Byar, E. Alexander, Jr., U. Batzdorf, W. H. Brooks, W.E. Hunt, C.S. MacCarty, M.D. Mahaley, Jr., J. Mealey, Jr., G. Owens, J. Ransohoff, J.T. Robertson, W.R. Shapiro, K.R. Smith, Jr., C.B. Wilson, and T.A. Strike, Randomized comparisons of radiotherapy and nitrosoureas for the treatment of malignant glioma after surgery, *N. Engl. J. Med.* 303:1323 (1980).
8. P.K. Duffner, M.E. Horowitz, J.P. Krischer, H.S. Friedman, P.C. Burger, M.E. Cohen, R.A. Sanford, R.K. Mulhern, H.E. James, C.R. Freeman, F.G. Seidel, and L.E. Kun, Postoperative chemotherapy and delayed radiation in children less than three years of age with malignant brain tumors, *N. Engl. J. Med.* 328:1725 (1993).
9. J.C. Allen, What we learn from infants with brain tumors, (Editorial), *N. Engl. J. Med.* 328:1780 (1993).
10. J.R. Crossen, A. Garwood, E. Glatstein, and E.A. Neuwelt, Neurobehavioral sequelae of cranial irradiation in adults: A review of radiation induced encephalopathy, *J. Clin. Oncol.* 12:627 (1994).
11. E.A. Neuwelt, P.A. Barnett, K.E. Hellström, I Hellström, C.I. McCormick, and F.L. Ramsey, The effect of blood-brain barrier on intact and fragmented monoclonal antibody localization in intracerebral human carcinoma xenografts, *J. Nuclear Medicine* 35:1831 (1994).
12. E.A. Neuwelt, R. Weissleder, G. Nilaver, R.A. Kroll, S. Roman-Goldstein, J. Szumowski, M.A. Pagel, R.S. Jones, L.G. Remsen, C.I. McCormick, E.M. Shannon, and L.L. Muldoon, Delivery of virus-sized iron oxide particles to rodent CNS neurons, *Neurosurgery* 34:777 (1994).
13. J.M. Nazzaro, M. Pagel, L. Rosenbaum, and E.A. Neuwelt. A new model of systemic drug rescue based on permeability characteristics of the blood-brain barrier in intracerebral abscess-bearing rats, *J. Neurosurg.* 74:467 (1991).
14. R.A. Kroll, M.A. Pagel, J.J. Langone, G.J. Sexton, and E.A. Neuwelt, Differential permeability of the blood-tumor barrier in intracerebral tumor-bearing rats: Anti-drug antibody to achieve systemic drug rescue, *Therapeutic Immunol.* 1:333 (1995).
15. E.A. Neuwelt, M.A. Pagel, A.I. Geller, and L.L. Muldoon, Gene replacement therapy in the central nervous system; Viral vector mediated therapy of global neurodegenerative disease, *Behav. Brain Sci.* 18: 1 (1995).
16. E.A. Neuwelt, Blood-brain barrier disruption in the treatment of brain tumors: clinical implications, *in:* Implications of the blood-brain barrier and its manipulation: Vol. 2. Clinical implications, E.A. Neuwelt, ed., Plenum Press, New York, (1989).

17. E.A. Neuwelt, J. Howieson, E.P. Frenkel, D. Specht, R. Weigel, C.G. Buchan, and S.A. Hill, Therapeutic efficacy of multiagent chemotherapy with drug delivery enhancement by blood-brain barrier modification in glioblastoma, *Neurosurgery* 19:573 (1986).
18. E.A. Neuwelt, D. Goldman, S.A. Dahlborg, J. Crossen, F. Ramsey, S.M. Goldstein, R. Braziel, and B. Dana, Primary CNS lymphoma treated with osmotic blood-brain barrier disruption: Prolonged survival and preservation of cognitive function, *J. Clin. Oncol.* 9:1580 (1991).
19. E.A. Neuwelt, H.D. Specht, and S.A. Hill, Permeability of human brain tumor to $^{99m}$Tc-glucoheptonate and $^{99m}$Tc-Albumin: Implications for monoclonal antibody therapy, *J. Neurosurg.* 65:194 (1986).
20. E.A. Neuwelt, D. Specht, S. Larson, K. Krohn, K. Hellström, I. Hellström, S.A. Dahlborg, and P. Barnett. Increased delivery of tumor-specific monoclonal antibodies to brain after osmotic blood-brain barrier modification in patients with melanoma metastatic to the CNS, *Neurosurgery* 20:885 (1987).

# PERIPHERAL TISSUE GRAFTS AS PORTALS INTO BRAIN FOR SOLUTES AND LEUKOCYTES

S. Ishihara, S-I. Tsubaki, L. Chang and M. W. Brightman

Laboratory of Neurobiology, NINDS
Bld. 36, Rm. 2A - 21
National Institutes of Health
Bethesda MD 20892

## INTRODUCTION

The blood-brain barrier (BBB) to ions and hydrophilic molecules can be focally and permanently bypassed at sites where peripheral tissues have been transplanted to brain (Rosenstein and Brightman, 1984; Tsubaki et al., 1986; Wakai et al., 198). A fragment of superior cervical ganglion or of skeletal muscle inserted onto the choroid plexus of the IV cerebral ventricle or upon the dorsal, pial surface of the medulla become portals of entry for cirulating solutcs. Thc present work demonstrates that the grafts also create focal sites through which blood-borne, exogenous cells, can be brought into the interstitium of an otherwise normal brain. A major conclusion is that, in this relationship between graft and brain, the route taken by solutes differs from that taken by cells.

## SOLUTE ENTRY

An intrusive but focal and permanent means of bringing protein into brain is circumvention of the BBB with grafts of peripheral tissues placed upon or within the brain. When a fragment of superior cervical ganglia (SCG) is inserted into the IV ventricle of mice, a portion of it survives for at least 12 months. Although there is attrition with time, of the graft's vessels, which are normally permeable to hydrophilic molecules, some of the capillaries persist. The survival of at least a portion of the graft may be partly attributed to the rapid formation of anastomoses between the intrinsic capillaries of the ganglion and those of the nearby choroid plexus and the medulla. If the graft is on the pial surface, subarachnoid vessels or those from the adjacent brain parenchyma may anastomose with capillaries of the graft (Fig. 1). Vascularization is rapid. Within 9 to 16 hours after transplantation, the microvessels of the ganglion can be filled with biologically inert India ink that had been infused into the systemic circulation (Fig. 1). Sympathetic ganglia and

*New Concepts of a Blood–Brain Barrier*
Edited by J. Greenwood *et al.*, Plenum Press, New York, 1995

skeletal muscle grafts to brain not only permit the passage of solutes, but also create sites of entry for circulating leukocytes. The following discussion will emphasize the difference in the route of solute entry from that of cell penetration (Fig. 6).

Unless otherwise stated, the bar on each figure represents 500 $\mu$m.

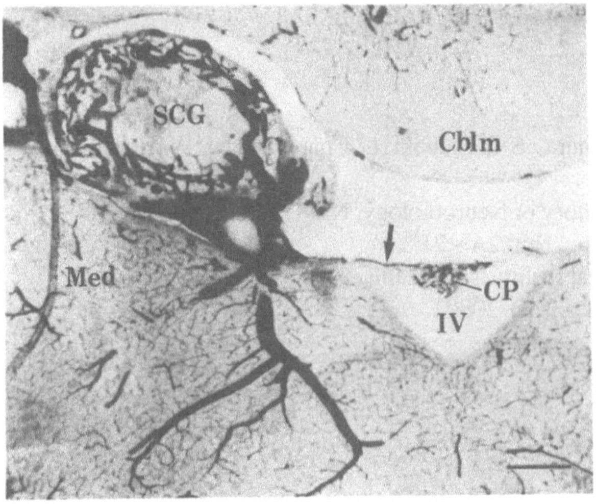


**Figure 1.** Vessels of a superior cervical ganglion (SCG) allograft, inserted between cerebellum (Cblm) and medulla (Med), have anastomosed with vessels of medulla. Graft is within subarachnoid space and separated from IV ventricle by the tela choriodea (arrow) of choroid plexus (CP). Donor was 3 weeks old; host rat was adult. Graft is 48 hours old.

Skeletal muscle was selected to circumvent the BBB, because it is readily accessible, becomes occasionally well innervated, is rapidly vascularized and persists indefinitely (Wakai, et al., 1986). Muscle capillaries are normally permeable to the tracer, horseradish peroxidase (HRP), which rapidly enters the brain adjacent to the graft. Does HRP enter brain from the graft or directly from brain vessels rendered permeable by the foreign graft tissue near them?

The question was resolved by reducing the circulation time of the protein; the results, obtained from SCG grafts, presumably pertain to muscle grafts as well. At circulation times of about 1 minute, HRP was confined to the vessels and interstitial clefts of the SCG graft. At longer periods, HRP entered the medulla. This time dependency signified that the HRP spread from the graft's vessels into its interstitial clefts and then into the clefts of brain. Had the medullary vessels been rendered permeable by the graft, HRP entry into graft and brain would have been simultaneous (Rosenstein and Brightman, 1984).

The vascular permeability of muscle and of SCG grafts to the ventricular surfaces simulates only crudely the permeability of circumventricular organs, such as the area postrema. The degree of solute entry at the site of the SCG grafts has been measured by quantitative autoradiography of a $^{14}$C-labeled small, neutral amino acid, a-amino isobutyric acid, which rapidly entered graft and surrounding brain (Tsubaki et al., 1987). The permeability of vessels in grafts that had been placed on the medulla's surface was greater than that of grafts placed within the brain parenchyma. The permeability also diminished with time, presumably because, by 4 weeks after transplantation, there had been an appreciable attrition of vessels. The permeability of even parenchymal graft vessels was approximately 20 to 40 times greater than barrier vessels (Tsubaki et al., 1987). Nevertheless, the permeability of vessels in the grafts was only a fraction of that in circumventricular organs, such as the pineal gland and neural lobe of the pituitary gland (Table 1) (Gross, et al., 1987). Blood flow, as measured by radioactive antipyrine, was comparable to that of barrier vessels and those of the pineal gland; however, the flow through the vessels of another circumventricular organ, the neural lobe, was 5 to 6 fold greater than that in the SCG grafts and normal pineal gland. There was some decrement in blood flow by 4 weeks, probably too, because vessels were lost with time (Table 1).

**Table 1.** Comparison of CVO with SCG Graft

| Tissue | Blood Flow<br>μl / g / min | Blood - Brain Transfer Δ<br>μl / g / min |
|---|---|---|
| ¥ Pineal Gland | 943 | 710 ± 100 |
| ¥ Neural Lobe | 5,707 ± 528 | 781 ± 94 |
| * SCG Graft | | |
| 1 week | 1,250 ± 110 | 44.8 ± 5.1 |
| 4 weeks | 900 ± 90 | 24.5 ± 2.6 |
| © BBB Region (Brainstem) | | |
| 1 week | 1,290 ± 60 | 0.86 ± 0.09 |
| 4 weeks | 1,460 ± 70 | 0.75 ± 0.06 |

¥ CVO - Circumventricular organ. Values adapted from Gross et al., 1987.

* SCG - Superior cervical ganglion allografted to IV ventricle ependyma; 1 and 4 weeks refers to age of graft. Values from Tsubaki et al., 1987.

© BBB - Region of brain with blood - brain barrier, e.g., brain stem.

^ - Blood to brain brain transfer of the circulating amino acid, a-aminoisobutyric acid.

Values ± standard deviation

## CELL ENTRY

Monocytes only enter mature CNS in appreciable numbers when it is perturbed, as in the chronic demyelination brought on by an inherited deficiency in galactosylceramidase. As a means of replacing the enzyme, bone marrow, containing enzymatically normal monocytes, has been injected peritoneally. The monocytes proliferate, cross the BBB and enter the CNS of the affected mice, thereby introducing the deficient enzyme and reversing the demyelinating process (Hoogebrugge et al., 1985).

Would an atopic and, therefore, foreign piece of autogeneic or isogeneic muscle, placed on the brain surface, be sufficient to attract exogenous circulating macrophages (Mø) to the graft site and thence into an otherwise normal, undisturbed brain? This question was approached by collecting peritoneal macrophages, stimulating them with phorbol myristate acetate and tagging them with a lipid-soluble, fluorescent dye with a long half-life of more than 100 days. After free dye was removed, the stimulated, tagged Mø were injected as a bolus of about 2-3X$10^6$ cells into the axillary artery, which eventually becomes confluent with the vertebral arteries (Ishihara et al., 1993). The infused Mø were permitted to circulate for 2 hours before aldehyde fixation. Five weeks prior to the infusion of Mø, an isogeneic graft of skeletal muscle had been inserted into the IV ventricle of adult host rats.

### Adhesion of Macrophages to Brain Endothelium

Before circulating monocytes can enter brain, they must first adhere to the endothelium of cerebral vessels (Fig. 5). The adhesion is mediated by molecules on the plasmalemma of endothelial cells and Mø. Leukocyte function antigen (LFA-1) on lymphocytes is one such molecule that binds to its intercellular adhesion molecule (I-CAM-1) ligand on endothelial cells

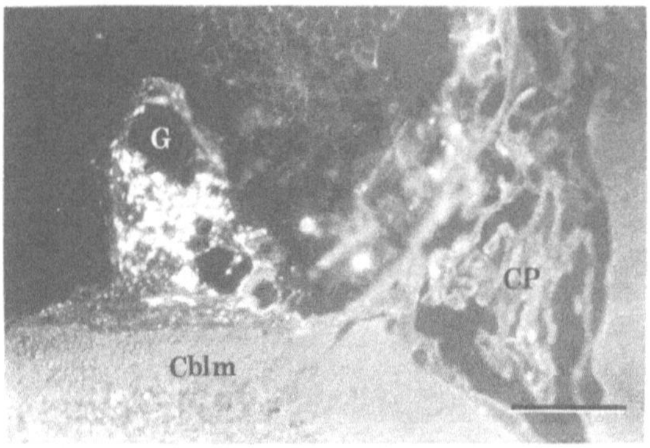


**Figure 2.** Muscle graft (G) rests on the cerebellum (cblm) and near, but not in contact with choroid plexus (CP) of IV ventricle at this level. The large, brightly fluorescent bodies in graft are degenerating muscle cells. Some of the smaller ones are labeled Mø.

(Marlin and Springer, 1987; Smith et al., 1989). It is likely that other cell surface molecules of the selectin family, e.g., endothelial leukocyte adhesion molecule or E-selectin (Carlos et al., 1991; Luscinska et al., 1991) are more relevant than I CAM-1 as a mediator of Mø-endothelial adhesion. L-selectin on leukocytes (Spertini et al., 1992) may also be involved. However, available antibodies that recognized these antigens in rats were not available at the time. Adhesion may also be influenced by a local stimulation of brain endothelium with cytokines and a peripheral activation of monocytes (Male et al., 1990). Our in vivo approach includes both of these components: the supply of factors, possibly including cytokines, to adjacent brain capillaries by an indwelling muscle graft, and the prior activation of Mø which are then infused into peripheral blood.

Activated, dye-tagged Mø were attracted to medulla capillaries in the vicinity of the graft and adhered to the luminal face of their endothelium (Fig. 4). Activation by phorbol ester might include an enhanced adhesion of Mø to endothelium as it does for T lymphocytes so stimulated (Haskard et al., 1986). Despite the flushing of salt solution and fixative through the blood vessels of brain and graft at a pressure of approximately 100 to

110 mm Hg for 20 minutes at a rate of about 0.5 to 1.0 ml/minute, the adhesive bond between Mø and endothelium was sufficiently strong to enable the Mø to remain attached. In contrast, the red blood cells had been swept out of the lumen of the same vessels.

Circulating leukocytes enter brain by moving across the endothelial cells of brain vessels near the grafts. A primary interaction required for the transcapillary passage of such cells is adhesion to the lumenal face of the endothelium. One component of adhesion is the LFA-1 molecule, a member of the integrin family of adhesion receptors, present on the surface of leukocytes, including Mø (Springer, 1990). LFA-1 was expressed on both non-activated and activated Mø in our preparations. The second component of adhesion is the counter - receptor for LFA-1, the intercellular cell adhesion molecule, I CAM-1, situated on endothelial (Tamatani et al., 1990) and other cells. The I CAM-1 molecule on the lumenal surface of capillary endothelium selectively binds to the LFA-1 molecule on the surface of Mø. I CAM-1 expression is induced during inflammation (Dustin et al., 1988) through the intermediation of interferon-g and interleukin-1. The inflammation mediated by these cytokines results in the infiltration of mononuclear cells into those regions where I CAM-1 appears on endothelial cells (Munro, '89). The muscle graft appears to induce I CAM-1, visualized immunohistochemically, on brain vessels near the graft (Ishihara et al. 1993).

LFA-1 and I CAM-1 are only two of several, cell surface molecules that promote adhesion between leukocytes and endothelial cells (Hakkert et al., 1991). In the rat brain near a muscle graft, LFA-1 may not be the major adhesion molecule. As mentioned, it is expressed constitutively in Mø that had not been activated with phorbol ester as well as in those cells that had been stimulated, yet many more activated Mø penetrated brain than did non-activated cells. Accordingly, molecules other than LFA-1, must have been responsible for the adhesion between Mø and endothelium. The probable mediators are a group of lectin-like proteins, the selectins, situated on both endothelium (E-selectins) (Springer 1990) and leukocytes (L-selectins) (Spertini et al., 1992). E-selectin, also designated as endothelial leukocyte adhesion molecule, may contribute to adhesion of the isogeneic Mø in vivo as it does to neutrophils and monocytes in vitro (Hakkert et al., 1991; Luscinskas et al., 1991). The L-selectin on the surface of monocytes, stimulated with cytokines, contributes to adhesion of the monocytes to human derived endothelium in vitro. It is likely that the Mø, activated with phorbol ester, likewise expressed L-selectin on their surface, but this likelihood has yet to be substantiated. These reports also suggest that, during the five weeks in which the grafts were in place, the intrinsic immune cells that had been attracted to the graft released cytokines that promoted the adhesion and transcapillary passage of the exogenous Mø.

**Roles of Cytokines**

Because cytokines probably mediate the host's inflammatory reponses to the muscle grafts, intrinsic inflammatory cells were immunostained for their content of tumor necrosis factor (TNF)-a and gamma interferon (INF-g). The two cytokines enhance the expression of I CAM-1 on endothelial cells in vitro (Pober et al., 1986). Another reason for attempting to detect TNF-a in the graft itself, is that this cytokine is released by Mø and mediates angiogenesis (Leibovich et al., 1989). Although TNF-b antibody immunostained the muscle grafts, TNF-a did not. Antibody to transforming factor (TGF)-b 1-3 did not stain endogenous lymphocytes and monocytes that had been carried to the graft site. The failure to stain may not have been due to the absence of this antigen because the radial glia of cerebellum, known to contain the antigen (Unsicker et al., 1991), was not stained in our sections by the antibody used.

Cytokines affect both endothelial and immune cells. What is alluded to as "perturbation"of the vasculature in brain adjacent to the graft, may also be mediated by cytokines. The intracerebral injection of IFN-g leads to the recruitment of monocytes and Mø to the site of injection (Sethna and Lampson, 1991). In vitro, endothelial cells become more adhesive for neutrophils after having been stimulated with IL-1 and phorbol ester (Schleimer and Rutledge, 1986). TNF, IL-1 and IFN-g elevate and sustain for 72 hours, the expression of I CAM-1 on endothelial cells (Pober et al., 1986). An additional consequence of TNF and IFN-g release is that they appear to induce the migration of monocytes to sites where they have stimulated I CAM-1 in vessels (Munro et al., 1989).

The means by which large solutes, like the glycoprotein HRP, enter brain from a muscle graft is by moving across permeable vessels of the graft into the graft's interstitial compartment and thence into the interstitial clefts of the brain which are confluent with

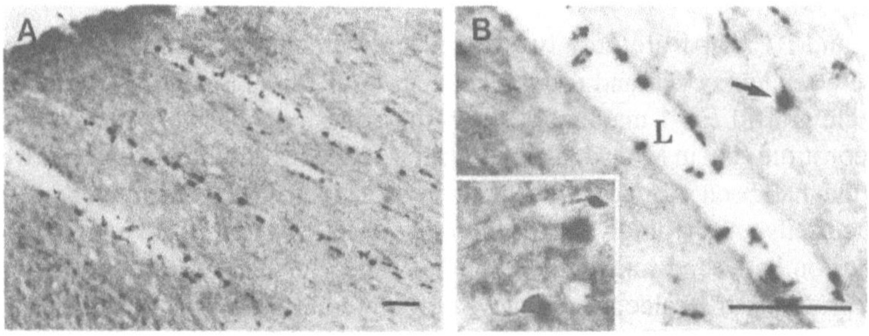


**Figure 3A.** Exogenous, fluorescent macrophages, activated by phorbol ester, adhere to luminal face of 3 blood vessels in medulla to graft.

**B.** Similar vessel in medulla with adherent macrophages, some of which have passed across vessel to enter brain parenchyma. **Inset:** Some macrophages have entered the parenchyma near graft.

those of the graft (Fig. 6). The route taken by activated Mø into brain is not from the graft first, but directly across perturbed barrier vessels in the vicinity of the graft.

The requirements for the entry of circulating Mø into normal brain are twofold: the insertion of an autologous or isogeneic muscle graft onto the brain's surface and the activation of the Mø prior to their introduction into the bloodstream. Soluble tropic factors emanating from the muscle graft probably reached the nearby brain through the grafts' extracellular compartment. Having entered the brain, the soluble factors could affect the blood vessels' interactions with circulating cells.

The penetration of Mø into a gelfoam graft and the subjacent medulla provided a baseline with which to compare the entry around muscle tissue. The entry into muscle may be due, only in small part, to its presence as a foreign body. The acellular, gelatin xenograft is a foreign body, yet the entry of Mø into the gelfoam and nearby medulla was much less than that associated with the muscle graft. Many non-activated Mø were attracted to the muscle but only a few to the gel and still fewer to the brain near the gel.

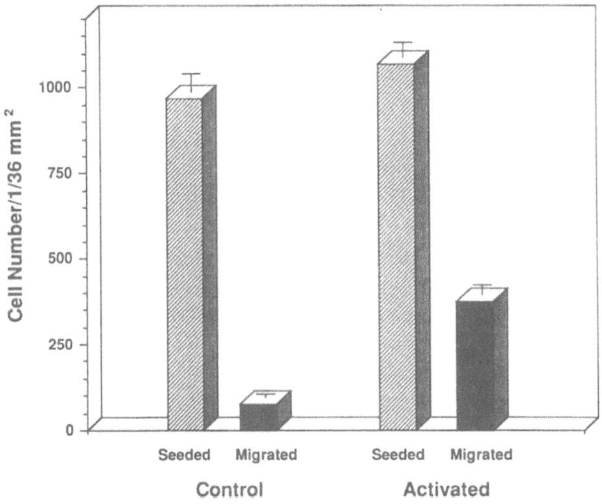


**Figure 4.** In vitro migration of Mø through a 3 $\mu$m pore size filter. About 35 % of Mø activated by phorbol ester migrated in 24 hours; only about 7 % of non-activated cells migrated in the same length of time. The lower chamber receiving the cells did not contain a chemotactic agent.

## Effects of Cell Passage on Cell Layer Permeability

Having adhered to brain endothelium, the Mø were in a position to cross the vessels and enter the interstitial clefts of the brain. The influence of the phorbol ester activation was demonstrated in vitro by placing the cells in the top chamber of a bipartite, "trans-well" separated from its lower chamber by a synthetic filter perforated with 3 $\mu$m pores. The passage of Mø, maintained at 37° C, across the filter over a 3 day period was enhanced by activating the Mø with phorbol ester. Five-fold more activated cells than non-activated cells were able to migrate into the lower compartment which did not have a chemotactic agent in its medium (Fig. 4).

In vivo, two requirements must be met for the passage of cells across endothelium: (i) the cells must be motile, (ii) the endothelium must react by forming pores or enclosures through which the cells can migrate, or by permitting their tight junctions to be sufficiently deformed to accommodate the motile cells. Polymorphonuclear leukocytes are able to enter brain when they are engulfed by endothelial evaginations where a-bungarotoxin is topically applied, as a chemotactic agent, to the pial surface (Faustmann and Dermietzel, 1985). The rapid transcapillary passage of Mø in vivo, suggests an active response on the part of the endothelium: intercellular junctions that might have opened, evaginations of the endothelium that engulfed Mø, then released them at the abluminal surface and endothelial pores that formed, through which the cells could move (Faustmann and Dermietzel, 1985). The possibility, that altered tight junctions between endothelial cells can be a pathway for Mø exudation, has yet to be examined.

A "tight" endothelium or epithelium that separates two fluid compartments must maintain its imperviousness to ions and macromolecules during the migration of cells across it, if it is to maintain a concentration gradient (Milks et al., 1986). The degree of impermeability of an endothelial or epithelial monolayer is readily monitored by measuring the electrical resistance (R) across it. The greater the degree of tightness, the higher the R. Does the passage of cells across such monolayers compromise the barrier property of the monolayer? The evidence is contradictory. The passage of leukocytes across a monolayer of an intestinal epithelial cell line is by way of disrupted tight junctions and there results a leakage of macromolecules during the passage (Nash et al., 1987).

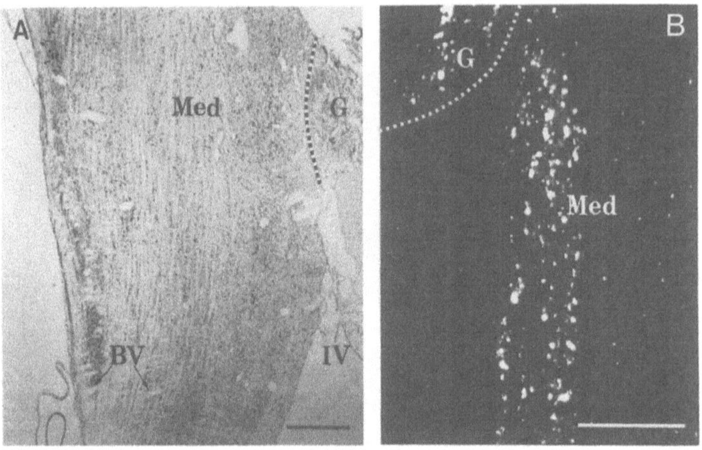


**Figure 5** **A.** Medulla (Med) with muscle graft (G) on its dorsal surface facing the IV ventricle (IV). Whereas many of the blood vessels (BV) course perpendicularly to the long axis of the medulla, the axon bundles around them run parallel to it.

**B.** Exogenous, isogeneic macrophages that are fluorescent, have left the vessels of the medulla near graft (G) and have migrated parallel to the axon fascicles. Bar = 100 $\mu$m

A limitation of the in vitro method is that, in a relatively impervious monolayer of epithelial cells with a high R, a small hole will lead to a large drop in R (Fuller et al., 1986). For example, in a high R (1,000 $\Omega$) strain of a canine kidney epithelial cell line, MDCK, a hole that is only 0.5% of the monolayer's area leads to a fall in R of ~ 75%, whereas the decrement across a monolayer of a low R (100 $\Omega$) strain is only ~ 3%. A few openings in what was a highly impermeable tight junction can likewise lead to a large drop in R across a monolayer (Martinez-Palomo et al., 1980).

It would be expected from the above considerations, that the passage of leukocytes through the tight junctions of a monolayer would be accompanied by a drop in R. However, the R may not fall during such migration of cells. Leukocytes, in vitro, may traverse a confluent layer of endothelial cells derived from human unmbilical vein and elsewhere, by moving across the cell layer either parajunctionally or through their tight junctions. The R across a monolayer of MDCK cells remained stable during the migration of neutrophils across it. The monolayer also retained its impermeability to macromolecules and, when viewed by light microscopy, appeared intact (Milks et al., 1986); but such an impression is convincing only when verified by an R that is sufficiently high.

There is further evidence that an endothelial monolayer can be crossed by leukocytes without lowering the layer's permeability. The contact between endothelial cell and leukocyte is apparently so closely maintained during the passage of the leukocyte that there is no leakage either of serum albumin or of ions, detectable by a lowered R across the endothelium (Huang et al., 1988). Monolayers of brain endothelial cells have been established with a trans-layer R of ~ 600 $\Omega\cdot cm^2$ (Dehouck et al., 1990; Rubin et al., 1991). These tight layers should provide a sufficiently high, initial R, to demonstrate whether cell passage across the particularly impermeable endothelium of brain alters the R and other properties of the barrier endothelium.

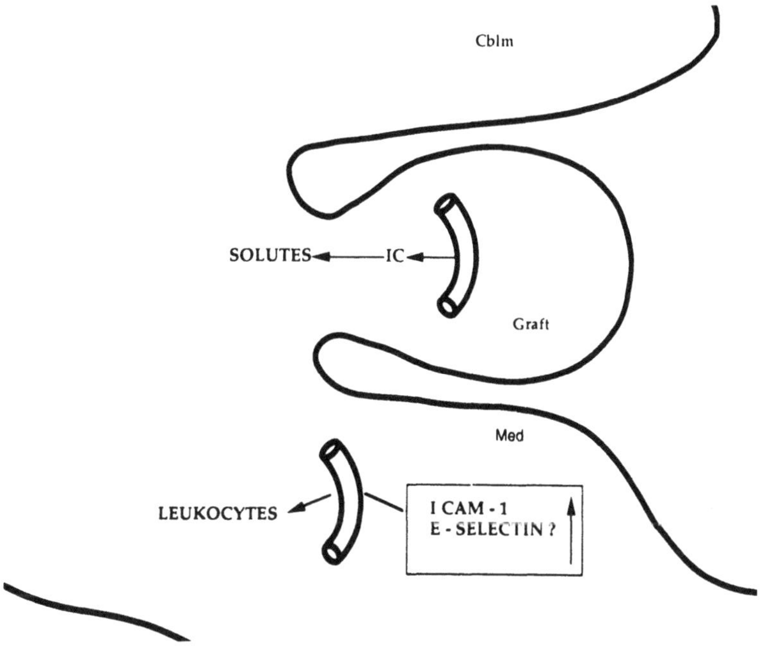


**Figure 6.** Portals of entry into brain for solutes and for circulating cells are different. The pathway for solutes is across permeable vessels in graft, into the graft's interstitial clefts which are confluent with the clefts of the adjacent brain. The route taken by exogenous macrophages is not via the graft directly, but across vessels constituting the blood-brain barrier. The graft enhances the expression, on brain vessels near graft, of I CAM-1 and, probably, lectin-like adhesion molecules.

## Migration of Exogenous Macrophages in Brain.

In the in vivo arrangement of muscle graft to brain, the extravasated Mø, once having traversed the endothelium, migrated between the fiber tracts of the medulla. The longitudinal spread of activated Mø was most likely due to their migration rather than to their local, transendothelial exudation along the length of reactive blood vessels. The basis for this inference is the course of medullary vessels, which is at right angles to the longitudinal axis of the fiber tracts (Fig. 5A). Had the Mø traversed the vessels along their length and had not then migrated further, the cells would have been arranged more or less perpendicularly to the long axis of the medulla. Instead, many of the cells were aligned parallel to the long axis of the axon fascicles of the medulla (Fig. 5B). Migration of Mø within the parenchyma of the medulla is also suggested by the elongation of extruded Mø (Fig. 3B, Inset), a shape likely to have been imposed by the longitudinally disposed axon bundles, among which the cells moved. The in vitro experiments demonstrated that activated Mø can migrate for a limited distance in the absence of a chemotactic signal (Fig. 4). It is emphasized that, although activated Mø were able to enter the medulla in appreciable numbers near the graft, their more distant dissemination would require a chemotactic influence beyond the site of the graft.

The conditions necessary for the successful penetration of cells into brain were indicated by several control experiments. Thus, non-activated Mø did not invade the graft-bearing brain in any appreciable number. Viable muscle grafts were required for cell entry; substitution of muscle with an acellular piece of gel foam did not result in discernible cell entry of cells into brain. A very few Mø did enter the gel foam, which appeared to be avascular, so that the Mø that did infiltrate the gel may have come from the leptomeninx or medulla. Abrading the brain surface, by inserting a piece of muscle then immediately removing it, was insufficient to attract Mø. It was inferred from these control experiments that both a viable graft and activated Mø were requirements for the invasion of brain by the cells.

The muscle graft itself does not act as a reservoir for Mø which then move into the brain. Rather, the muscle cells, some of which continue to degenerate during their 5 week residence, may release factors that attract circulating Mø and promote their passage across vessels in adjacent brain.

## CONCLUSIONS

Several conclusions may be drawn from these and other experiments. (1) At graft sites, solutes enter brain by crossing the permeable vessels of the graft into its interstitial compartment which is confluent with that of the brain. (2) Circulating cells do not enter brain by way of the graft but via nearby brain vessels that have been altered by the graft to become more adherent to the cells and assist in bringing them from blood to brain. (3) In vitro observations suggest that exogenous cells may be delivered to brain near a graft without opening the blood-brain barrier. (4) Monocytes or Mø which can be altered to provide needed factors, could be brought into a region of the CNS that is selected by inserting a sterile, small tissue graft upon or within it.

## ACKNOWLEDGEMENT

The authors are greatly indebted to the helpful criticisms and suggestions of Dr. Lois A. Lampson.

## REFERENCES

Carlos, T.N., N. Kovach, B. Schwartz, M. Rosa, B. Newman, E. Wayner, C. Benjamin, L. Osborn, R. Lobb, and J. Harlan. 1991. Human monocytes bind to two cytokine-induced adhesive ligands on cultured human endothelial cells: endothelial-leukocyte adhesion molecule-1 and vascular cell adhesion molecule-1. *Blood* 77 : 2266.

Dehouck MP, Meresse S, Delorme P, Fruchart J-C, Cecchelli R (1990) An easier, reproducible, and mass - production method to study the blood-brain barrier in vitro. *J Neurochem* 54 : 1798.

Dustin, M.L., and T.A. Springer. 1988. Lymphocyte function-associated antigen (LFA-1) interaction with intercelular adhesion molecule-1 (ICAM-1) is one of at least three mechanisms for lymphocyte adhesion to cultured endothelial cells. *J Cell Biol* 107 : 321.

Faustmann, P.M., and Dermietzel R. 1985. Extravasation of polymorophonuclear leukocytes from the cerebral microvasculature. Inflammatory response induced by alpha-bungarotoxin. *Cell Tissue Res* 242: 399.

Fuller, S.D., and Simons, K. 1986. Transferrin receptor polarity and recycling accuracy in "tight" and "leaky" strains of Madin-Darby canine kidney cells. *J. Cell. Biol.* 103: 1767.

Gross, P.M., Blasberg, R.G., Fenstermacher, J.D. and Patlack, C.S. 1987. The microcirculation of rat cirumventricular organs and pituitary gland. *Brain Res. Bull.* 18 : 73.

Hakkert, B.C., T.W. Kuijpers, J.F.M. Leeuwenberg, J.A. van Mourik, and D. Roos. 1991. Neutrophil and monocyte adherence and migration across monolayers of cytokine-activated endothelial cells; contribution of CD18, ELAM-1, and VLA-4. *Blood* 78 : 2721.

Haskard, D., D. Cavender, and M. Ziff. 1986. Phorbol ester-stimulated T lymphocytes show enhanced adhesion to human endothelial cell monolayers. *J Immunol* 137 : 1429.

Hoogebrugge,P.M., Suzuki,K., Poorthuis, B.J.H.M., Kobayashi,T. Wagemaker, G. and van Bekkum, D.W. 1988. Donor-derived cells in the central nervous system of twitcher mice after bone marrow transplantation. *Science* 294: 1035.

Huang, A.J., Furie, M.B., Nicholson, S.C., Fischbarg, J., Kiebovitch, L.S., and Silverstein, S.C. 1988. Effects of human neurophil chemotaxis across human endothelial cell monolayers on the permeability of these monolayers to ions and macromolecules. *J Cellular Physiology* 135: 355.

Ishihara, S., Sawada, M., Chang, L., Kim, J-M., and Brightman, M.W. 1993. Brain vessels near muscle autografts are sites for entry of isogeneic macrophages into brain. *Exper Neurol* 124: 219.

Leibovich, S.J., P.J. Polverini, H.M. Shepard, D.M. Wiseman, V. Shively, and N. Nuseir. 1987. Macrophage-induced angiogenesis is mediated by tumour necrosis factor. *Nature* 329 : 630.

Luscinskas F.W., M.I. Cybulsky, J-M. Kiely, C.S. Peckins, V.M. Davis, and M.A. Gimbrone. 1991. Cytokine-activated human endothelial monolayers supportenhanced neutrophil transmigration via a mechanism involving both endothelial-leukocyte adhesion molecule-1 and intercellular adhesion molecule. *J Immunol* 146 : 1617.

Male, D., Pryce, G. Hughes, C. and Lantos, P. 1990. Lymphocyte migration into brain modelled in vitro: Control by lymphocyte activation, cytokines, and antigen. *Cell Immunol* 127: 1.

Marlin,S.D., and Springer,T.A. 1987. Purified intercellular adhesion molecule-1 (ICAM-1) is a ligand for lymphocyte function-associated antigen-1 (LFA-1). *Cell* 51 : 813.

Martinez-Palomo, A., Meza, I., Beatyu, G, Cereijido, M. 1980. Experimental modulation of occluding junctions in a cultured transporting epithelium. *J Cell Biol* 87: 736.

Milks, L.C., Conyers, G.P., Cramer, E.B. 1986. The effect of neutrophil migration on epithelial permeability. *J Cell Biology* 103: 2729.

Munro, J.M., J.S. Pober, and R.S. Cotran. 1989. Tumor necrosis factor and interferon-g induce distinct patterns of endothelial activation and associated leukocyte accumulation in skin of Papio anubis. *Am J Pathol* 135 : 121.

Nash, S., Stafford, J., Madara, J.L. 1987. Effects of polymorphonuclear leukocyte transmigration on the barrier function of cultured intestinal epithelial monolayers. *J Clin Invest* 80: 1104.

Pober, J.S., M.A. Gimbrone, L.A. Lapierre, D.L. Mendrick, W. Fiers, R. Rothlein, and T. Springer. 1986. Overlapping patterns of activation of human endothelial cells by interleukin 1, tumor necrosis factor and immune interferon. *J. Immunol.* 137 : 1893.

Rosenstein, J., and Brightman, M.W. 1983. Circumventing the blood brain barrier with autonomic ganglion transplants. *Science* 221 : 879.

Rubin, L.L., Hall, D.E., Porter, S., Barbu, K. Cannon, C., Horner, H.C., Janatpour, M, Liaw, C.W., Manning, K., Morales, J., Tanner, L.I., Tomaselli, K.J. and Bard, F. 1991. A cell culture method of the blood-brain barrier *J Cell Biol* 115: 1725

Schleimer, R.P., and B.K. Rutledge. 1986. Cultured human vascular endothelial cells acquire adhesiveness for neutrophiles after stimulation with interleukin 1, endotoxin, and tumor-promoting phorbol diesters. *J. Immunol.* 136 : 649.

Sethna, M.P. and Lampson, L.A. 1991. Immune modulation within the brain: recruitment of inflammatory cells and increased major histocompatability antigen expression following intracerebral injection of interferon-g. *J. Immunol.* 34 : 121.

Smith, C.W., Marlin, S.D., Rothlein, R. C., Toman,C., and Anderson,D.C. 1989. Cooperative interactions of LFA-1 and Mac-1 with intercellular adhesion molecule-1 in facilitating adherence and transendothelial migration of human neutrophils in vitro. *J. Clin. Invest.* 83 : 2008.

Spertini, O., Luscinskas,F.W., Gimbrone,M.A. Jr., and Tedder, T.F. 1992. Monocyte attachment to activated human vascular endothelium in vitro is mediated by leukocyte adhesionmolecule-1 (L-selectin) under nonstatic conditions. *J. Exp. Med.* 175 : 1789.

Springer, T.A. 1990. Adhesion receptors of the immune system. *Nature* 346: 425.

Tamatani, T., and M. Miyasaka. 1990. Identification of monoclonal antibodies reactive with the rat homolog of ICAM-1, and evidence for a differential involvement of ICAM-1 in adherence of resting vs activated lymphocytes to high endothelial cells. *Internat. Immunol.* 2 : 165.

Tsubaki, S.I., Brightman, M.W., Nagakawa, H., Owens, E. and Blasberg, R.G. 1987. Local blood flow and vascular permeability of autonomic ganglion-transplants in the brain. *Brain Res.* 424 : 71.

Unsicker, K., Flanders,K.C., Cissel,D.S., Lafayatis R., and Sporn,M.B. 1991. Transforming growth factor beta isoforms in the adult rat central and peripheral nervous system. *Neuroscience* 44 : 613.

Wakai, S., Meiselman, S. E., and Brightman, M.W. 1986. Muscle grafts as entries for blood-borne proteins into the extracellular space of the brain. *Neurosurgery* 18 : 548.

# CNS TRANSPLANTS AND THE HOST IMMUNE RESPONSE: THE BLOOD-BRAIN BARRIER AND IMMUNOLOGICAL PRIVILEGE WITHIN THE MAMMALIAN BRAIN

Richard D. Broadwell[1], Belinda J. Baker[2], and William F. Hickey[3]

[1] Office of Research Integrity
United States Public Health Service
Rockville, Maryland 20852, U.S.A.
[2] Laboratory of Molecular Medicine and Neuroscience
National Institutes of Health
Bethesda, Maryland 20894, U.S.A.
[3] Department of Pathology
Dartmouth School of Medicine
Lebanon, New Hampshire 03756, U.S.A.

## INTRODUCTION

Survival of fetal mammalian brain tissue or of cell suspensions grafted into the adult mammalian brain is dependent upon the graft developing a vascular network supplied with host blood and failure of the host immune system to reject the graft as foreign tissue. The mammalian CNS has long been suspected to be a site of "immunological privilege" where constituents of the peripheral immune system do not venture. This suspected privilege is attributed to the absence of easily identifiable antigen presenting cells (e.g., cells expressing MHC class II antigen constitutively), the absence of lymphatic drainage, and the presence of the blood-brain barrier (for reviews see Head and Billingham, 1985; Widner and Brundin, 1988; Sloan et al., 1991).

The concept of the mammalian CNS as a privileged site in which grafted tissue can fluorish without rejection by the host immune system was envisioned, although not specifically stated as such, by Medawar (1948), whose study of skin allografts transplanted to the brain and elsewhere led to the discovery of "immunological tolerance". The rejection of an allograft, regardless of its location, is a direct consequence of the afferent and efferent arms of the host immune system acting in a coordinated fashion to promote rejection. The efferent arm is represented by effector cells such as T-helper and T-cytotoxic lymphocytes, activated monocytes (e.g., macrophages), and antibodies, all of which are conveyed in the blood and seek the foreign tissue. Lymphatic drainage and a blood supply are prerequisites for a host immune response; if one or both is absent, some degree of immunological privilege is likely to prevail, and the foreign tissue or antigen may survive for an extended period of time if not indefinitely. The afferent arm of the immune system is associated with foreign antigen presentation and subsequent activation of the effector cells. Antigen presenting cells sample the foreign antigen and process it intracellularly for presentation on their cell surfaces to T-lymphocytes in the lymph nodes and spleen.

The discussion that follows focuses on the hypothesis that neither the blood-brain barrier (BBB) nor the "immunological privilege" of the CNS is absolute. Immunological privilege afforded by the CNS is greater for allografts than for xenografts.[1] The privilege is relative at best and may relate moreso to antigen presentation, or the absence thereof, than to the absence of lymphatic drainage and the presence of a BBB. This topic is not without considerable significance given the current clinical and research attention human fetal CNS grafting has received to ameliorate Parkinson's disease and perhaps other neuro-degenerative disorders.

## POTENTIAL ANTIGEN PRESENTING CELLS WITHIN THE CNS

The BBB is not absolute. Blood-borne macromolecules readily enter the CNS by suspected intracellular pathways through the BBB endothelium, a process referred to as transendothelial transfer or transcytosis, and by patent extracellular pathways circumventing the BBB (Broadwell and Banks, 1993). The latter includes the subarachnoid space, pial surface and the half dozen, well-documented circumventricular organs such as the median eminence and area postrema; each of these sites possesses blood vessels permeable to blood-borne macromolecules and permits endogenous antibodies as large as IgM (500,000 Daltons) and complement unrestricted access to the brain parenchyma and cerebrospinal fluid (Broadwell and Sofroniew, 1993). Antibodies and complement do engage in cellular destruction, and the CNS sites fostering extracellular entry of these serum proteins could jeopardize the immunological privilege afforded by the CNS.

In the normal adult rat brain, the circumventricular organs, subarachnoid space, and pial surface contain populations of microglia, macrophages, and cells expressing major histocompatibility complex (MHC) class II antigens (Figure 1A-C). The macrophages and MHC class II$^+$ cells are similar morphologically and may represent one and the same cell type. Microglia and macrophages, although distinctly different morphologically, claim direct lineage from bone marrow monocytes; microglia and macrophages, but not cells demonstrably expressing MHC class II antigens, also are associated with BBB vessels (Figure 1D,E).

Microglia are abundant throughout the brain parenchyma, whereas macrophages and MHC class II$^+$ cells, revealed immunohistochemically in the normal adult rat CNS, are not. Hickey and Kimura (1988) have published compelling data to suggest perivascular microglia (hereafter referred to as perivascular cells) are antigen presenting cells in the CNS. These monocyte/macrophage elements represent a distinct subpopulation of CNS cells. Compatible with the role of proposed antigen presentation, perivascular cells and macrophages are interpreted as highly mobile cells capable of moving bidirectionally between the blood and intracerebral perivascular space, perhaps passing through BBB endothelia.

---

[1] An allograft is associated with differences in major and/or minor histocompatibility complex antigens between donor and host of the same species. A xenograft represents tissue transplanted between donor and host each of a different species. A syngeneic graft is tissue transplanted between genetically identical donor and host of the same inbred strain and species.

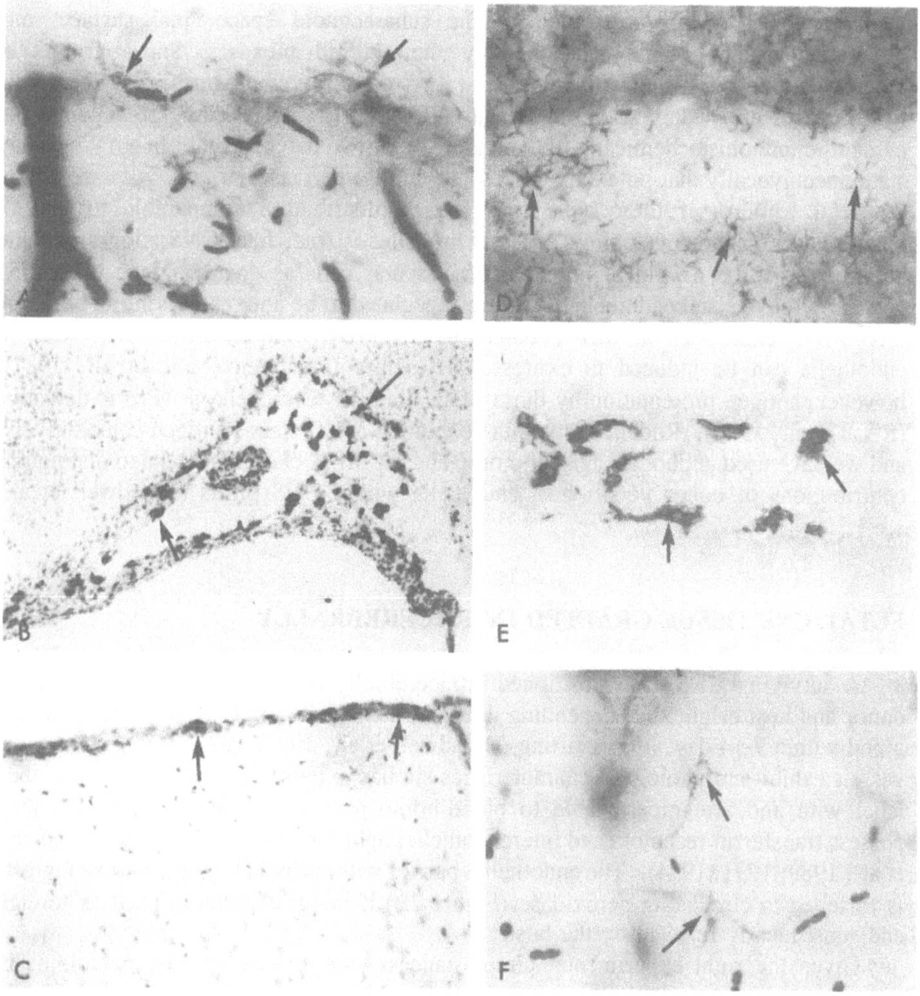

**Figure 1.** Within the normal CNS, potential antigen presenting cells revealed immunohistochemically or enzyme cytochemically include microglia (A,D; arrows), macrophages (B,E; arrows), and MHC class $II^+$ cells (C; arrows) on the pial surface (A-C) or associated with parenchymal BBB vessels (D,E); the perivascular cells in D and E also label with blood-borne peroxidase (F; arrows), which enters the CNS extracellularly by circumventing the blood-brain barrier (see Broadwell and Sofroniew, 1993).

The cells occupying the subarachnoid space, pial surface, and perivascular clefts have a phagocytic function and label with intravenously administered horseradish peroxidase that enters the brain extracellularly through sites deficient in a BBB (Figure 1F). Once the blood-borne peroxidase has gained access to the extracellular spaces within the brain, it is conveyed through the perivascular clefts by the pulsatile activity of arterioles for widespread exposure to perivascular phagocytes (Broadwell and Sofroniew, 1993). Cells inhabiting the subarachnoid space, pial surface, and perivascular clefts and which serve as phagocytes or antigen presenters can be looked upon as providing a crucial line of defense within the brain once the BBB is breached pathologically or circumvented extracellularly.

Recognition and distribution of potential antigen presenting cells within the developing rat brain is noticeably different from that we see in the adult CNS. In the fetal rat a few days before birth, microglia are present throughout the CNS with macrophages appearing restricted to the subarachnoid space, pial surface, and circumventricular organs, most notably the choroid plexus. Small groups of macrophages occupy the white matter at two days following birth, and individual macrophages are widely scattered around BBB vessels. MHC class $II^{+}$ cells are not evident immunohistochemically in the fetal CNS or two day neonatal brain. We cannot state unequivocally that potential antigen presenting, perivascular cell types are within the fetal CNS or if these cell types have a distribution comparable to that of macrophages in the developing CNS. If the latter is true, fetal CNS allografts could be expected to be devoid of potential perivascular cells as donor antigen presenting cells. Astrocytes and endothelial cells are speculated to be antigen presenting based on their MHC expression *in vitro* (Fontana et al., 1984; Male et al., 1987). Cerebral endothelia can be induced to express MHC class II antigen (Male et al., 1987); however, antigen presentation by these MHC class $II^{+}$ endothelia *in vitro* is defective (Risau et al., 1990). Rodent endothelia do express MHC class I antigen constitutively, and we have used antibodies against mouse and rat MHC class I antigens to distinguish contributions of donor versus host endothelia within CNS grafts (Broadwell et al., 1991, 1994).

## FETAL CNS TISSUE GRAFTED INTRACEREBRALLY

A surviving CNS graft positioned intracerebrally is vascularized with vessels of donor and host origin and, depending upon the size of the graft, is perfused with host blood within 7-14 days after grafting (Broadwell et al., 1987, 1991, 1994). The graft vessels exhibit morphological characteristics similar to those of host BBB vessels; they label with and are impermeable to blood-borne peroxidase (Figure 2A), and they possess transferrin receptors and interendothelial tight junctional complexes (Broadwell et al., 1989, 1991, 1994). The endothelial barrier within CNS allografts and xenografts is forfeited to circulating peroxidase (Figure 2B) if the graft is recognized as foreign and subsequently rejected by the host.

Given the right experimental circumstances with various grafting paradigms in rodent models, fetal CNS allografts and xenografts placed within the adult brain do survive and exhibit a BBB for a period of several weeks to months or indefinitely. The right circumstances include: immunosuppressing immunocompetent hosts with such drugs as Cyclosporin A; athymic or nude (T-lymphocyte deficient) rodents serving as the host; using highly inbred rodents of different strains as donor and host; and, employing donor and host of the same species that are MHC and minor HC compatible. Available evidence suggests the inbred PVG, PVG(AO), and AO rat strains are ideal rodent models for intracerebral allograft investigations (Table 1). These rat strains as donor and host represent excellent controls for assessing allograft survival when the host brain and fetal CNS allograft are analyzed immunohistochemically with the panel of monoclonal antibodies listed in Table 2 (Mason et al., 1986; Sloan et al., 1991; Broadwell et al., 1994). For example, a PVG donor CNS allograft placed within the brain parenchyma or cerebral ventricular system of a PVG(AO) host for 30 days exhibits viable neurons throughout the graft when stained immunohistochemically with OX-7 antibody against the neuronal cell surface marker Thy 1.1. The PVG allografts also express MHC class I immunoreactivity which, with OX-27 antibody against the PVG rat MHC class I antigen, is restricted to donor PVG blood vessels within the graft (Figure 2C); no significant immunoreactivity is evident within the allograft when immunohistochemical staining is performed with OX-6 antibody against MHC class II

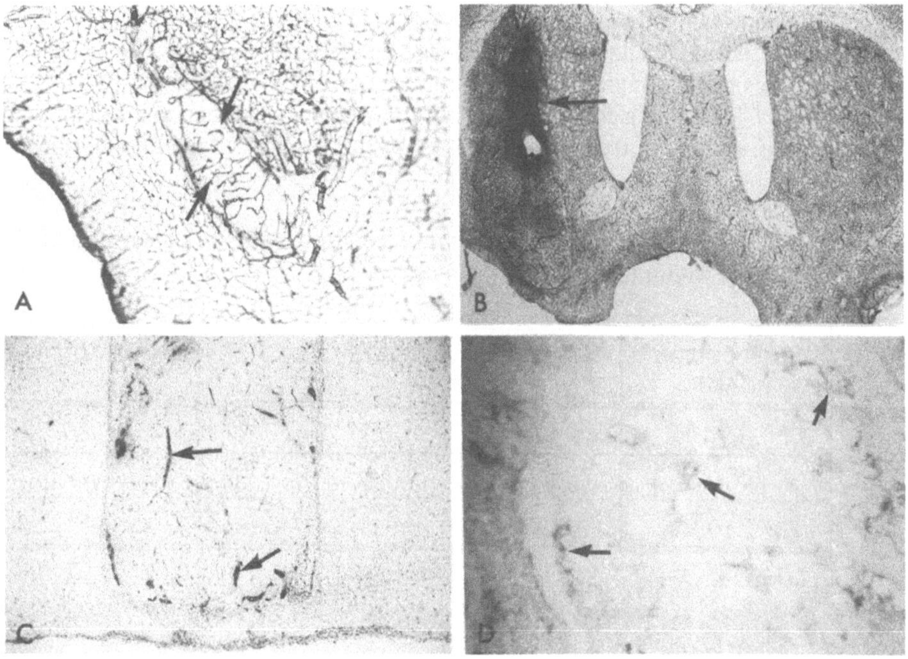


**Figure 2.** Healthy, intracerebrally positioned, CNS grafts possess BBB vessels (A; arrows) that label with blood-borne peroxidase delivered intravenously to the host. A CNS allograft rejected by the host immune system looses its BBB to circulating peroxidase (B; arrow). Allograft BBB vessels of donor rat origin (C; arrows) are identified immunohistochemically with an antibody that recognizes only the donor rat MHC class I antigen expresssed constitutively by these endothelia. In a fetal rat CNS graft placed within the brain of an adult athymic mouse (a xenograft), perivascular microglia (D; arrows) of host mouse origin express MHC class I antigen that is recognized immunohistochemically with antibody against that antigen.

antigen, with OX-1+30 antibodies against leucocyte common antigen, with W3/25 antibody against CD4 antigen, and with OX-8 and OX-10 antibodies against CD8 antigen.

CNS xenografts (fetal rat) placed intracerebrally in adult, homozygous, athymic mice survive quite nicely. Antibodies against MHC class I antigen of the donor and against that of the host demonstrate graft blood vessels are derived from both donor and host brains with donor vessels predominating (Broadwell et al., 1991). Perivascular cells within the rat xenograft also stain immunohistochemcially with antibody against host mouse MHC class I antigen (Figure 2D) (Broadwell et al., 1994). This histological observation suggests that although T-lymphocytes may be absent from the immune system of athymic mouse hosts, potential antigen presenting cells within the host CNS are very much in evidence, having entered the rat xenograft directly from either the systemic circulation or surrounding host brain. Potential perivascular, antigen presenting cells can be induced to express MHC class II antigen when exposed to gamma interferon and tumor necrosis factor *in vivo* (Hickey et al., 1992).

CNS allografts contributed by fetal rat donors of the same outbred strain as the host (e.g., Wistar or Sprague Dawley rats) demonstrate a disrupted BBB to blood-borne peroxidase (Figure 2B) and histological evidence signalling a host rejection response (Broadwell et al., 1994; Bergman et al., 1994), which can be highly variable ranging from no apparent rejection to mild or demonstrable rejection. The rejection may be initiated by differences between donor and host in the minor histocompatibility complex antigens as opposed to the MHC antigens. Immunohistochemical application of our panel of monoclonal antibodies (Table 2) to the allografts and host brains at 30 days post-grafting reveals scattered pockets of viable Thy 1.1$^{+}$ neurons within the grafts (Fig. 3A); this picture contrasts sharply with that of a healthy, surviving CNS allograft in which Thy 1.1 staining for neuronal viability is homogeneous and indistinguishable from Thy 1.1 staining of the surrounding host brain. Additionally, allografts between

**Table 1.** Inbred rat strains that serve as controls for assessing survival of CNS allografts within the host brain parenchyma.

| COLLOQUIAL NAME | RAT STRAIN | MHC | non-MHC |
|---|---|---|---|
| PVG | PVG-RT1$^{c}$ | C | PVG |
| PVG(AO) | PVG-RT1$^{u}$ | U | PVG |
| AO | AO-RT1$^{u}$ | U | AO |

1) PVG differs from PVG(AO) at the MHC loci only.
2) PVG(AO) has the same MHC antigens as AO but has the non-MHC antigens of PVG.
3) PVG differs from AO at both MHC and non-MHC loci.

**Table 2.** Rat monoclonal antibodies for immunohistochemical assessement of allograft survival versus rejection.

| MONOCLONAL ANTIBODY | DETECTS | MONOCLONAL ANTIBODY | DETECTS |
|---|---|---|---|
| OX-26 | CNS Blood Vessels (rat transferrin receptor) | OX-27 | Rat RT1A$^{c}$ (PVG MHC Class I) |
| OX-18 | Rat MHC Class I Monomorphic RT1A | OX-7 | Adult Neuronal Components (rat Thy 1.1 molecule) |
| OX-6 | Rat MHC Class I Monomorphic RT1B | OX-1+30 | Leucocyte Common Antigen CD45 |
| W3/25 | Thymocytes, CD4$^{+}$ T-cells, Macrophages | OX-8/OX-10 | CD8$^{+}$ T-cells, NK cells |

donor and host rats of the same outbred strain may exhibit mild to dense immunoreaction product for MHC classes I and II antigens, leucocyte common antigen, and CD4 and CD8 antigens (Figure 3B-F). The host rejection of outbred allografts may not be obvious in sections prepared for routine histology and counterstained with cresyl violet or thionine (Broadwell et al., 1994). Support for this allograft rejection is provided ultrastructurally by tissue sampled for conventional electron microscopy (Figures 4 and 5).

## IMMUNOLOGICAL PRIVILEGE OF THE MAMMALIAN BRAIN TO CNS ALLOGRAFTS IS RELATIVE

That immunological privilege is not absolute within the adult mammalian brain is evidenced by the host rejection of CNS xenografts and of CNS allografts from fetal donor to adult host of the same outbred strain. Three additional examples of CNS allograft rejection deserve brief consideration to emphasize that immunological privilege in the adult CNS is relative at best. Two of the examples involve fetal CNS allografts placed within the third cerebral ventricle of adult hosts. First, intraventricular CNS allografts between fetal donor AKR inbred mice and adult host outbred hypogonadal mutant mice survive well for several months but then exhibit demonstrable signs of rejection at nine to twelve months (Broadwell et al., 1994); in this example, both donor and host mice possess the same MHC loci genetically but differ in their *minor* HC loci, which also may contribute to the host rejection response. Second, while CNS allografts from inbred fetal PVG rats do survive when located within the brain parenchyma of inbred adult host AO

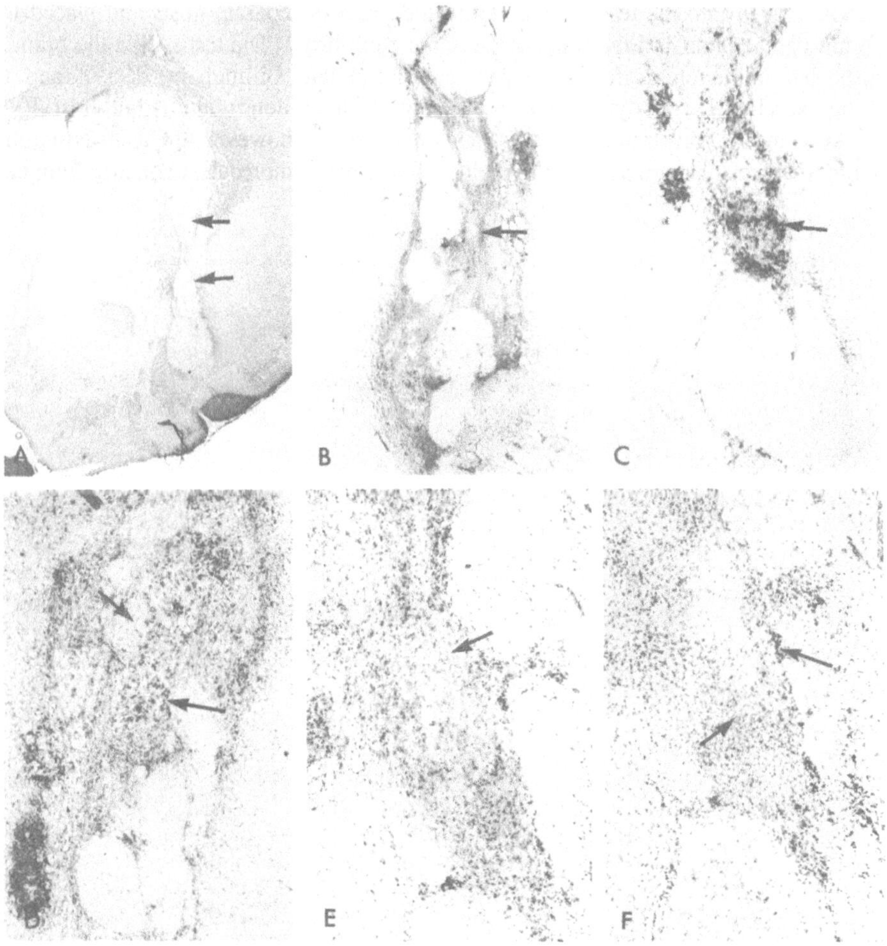


**Figure 3.** CNS allografts between outbred fetal donor and adult host Wistar rats exhibit a host rejection (arrows) observed within 30 days using immunohistochemistry and the monoclonal antibodies listed in Table 2. (A) spotty Thy 1.1 neuronal staining; (B) MHC class $I^+$ cells; (C) MHC class $II^+$ cells; (D) leucocyte common antigen$^+$ cells; (E) $CD4^+$ and (F) $CD8^+$ T-lymphocytes and other cell types.

rats, the allografts fail to survive generally when placed into the third ventricle of the AO host (Sloan et al., 1990; Broadwell et al., 1994); inbred PVG and AO rats are disparate genetically at both MHC and minor HC loci (Table 1). These data suggest that as a sanctuary for CNS allografts, the cerebral ventricular system is less tolerant than the brain parenchyma to foreign antigen. If this is so, contributing factors could include the direct access of blood-borne macromolecules and antigen presenting cells to the cerebrospinal fluid-filled subarachnoid space and ventricular system (Broadwell and Sofroniew, 1993) and drainage of the cerebrospinal fluid through deep cervical lymph nodes (Cserr and Knopf, 1992). The drainage of cerebrospinal fluid to the cervical nodes could convey host antigen presenting cells that occupy the subarachnoid space, cerebral ventricles and other locations within the brain as well as convey donor allograft cells. Following intracerebral injection, cell suspension grafts can move in the retrograde direction to take up residence within the subarachnoid space and on the pial surface (Broadwell et al., 1994). We also have reported that donor allograft cells expressing MHC class I leave the host brain, perhaps by way of the cerebral circulation, for sequestration within the host spleen and peripheral lymph nodes (Broadwell et al., 1994).

Lastly, Baker and Broadwell (1993) have drawn attention to the fate of fetal CNS allografts placed into the testis compared to the brain in separate hosts and placed into the brain parenchyma and the testis of the same adult host. The testis, like the brain, is a suspected immunologically privileged site (Head and Billingham, 1985), and the grafting paradigms employed involve the inbred, fetal donor and adult host PVG, PVG(AO), and AO rat strains (Table 1). Unlike the brain, however, the testis is supplied with blood vessels that are permeable to blood-borne macromolecules, contains lymphatic

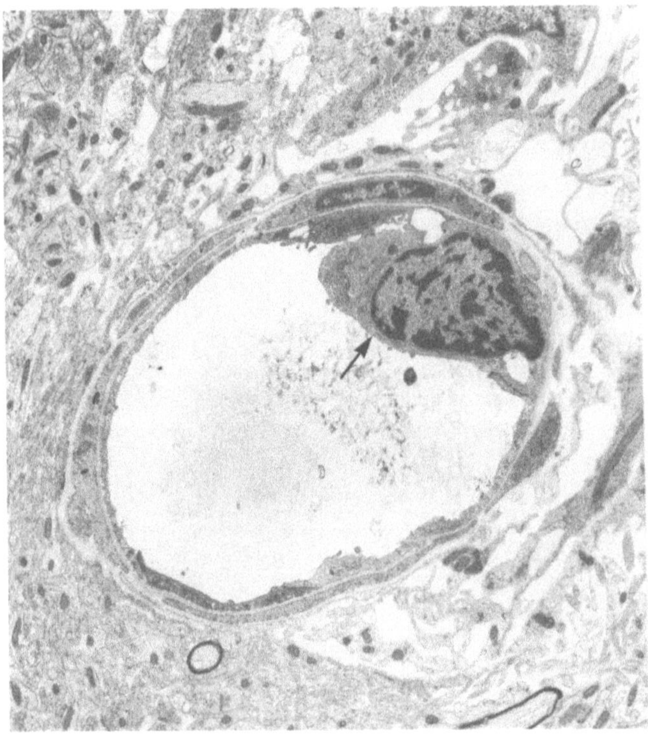

**Figure 4.** A Wistar rat CNS allograft in the early stage of being rejected shows a host Wistar rat lymphocyte (arrow) attached to the luminal surface of a BBB endothelial cell within the allograft.

channels, and exhibits antigen presenting cells that constitutively express, and demonstrably so by immunohistochemistry, MHC class II antigens. Some CNS allografts within the testis do survive, but the survival rate is significantly less (in particular instances no greater than 50%) than the success rate encountered with a similar CNS allograft placed within the brain parenchyma of separate but identical hosts. CNS allografts that do survive in the testis are well-vascularized with BBB vessels indigenous

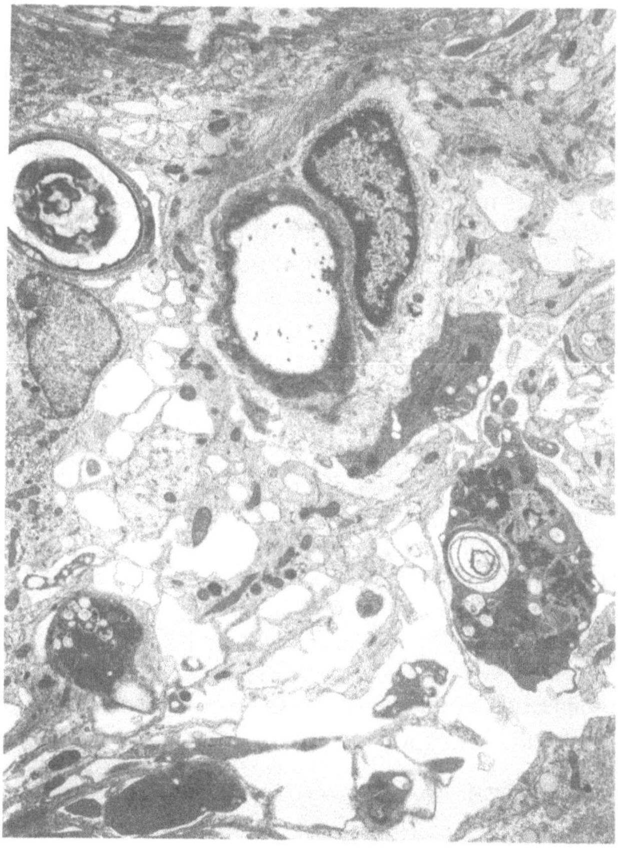

**Figure 5.** A Wistar rat allograft experiencing a prominent Wistar host rejection exhibits a disrupted parenchyma and phagocytic cells (compare to Figure 4).

to the donor brain and display little immunohistochemical evidence for rejection using the monoclonal antibodies listed in Table 2. This survival conceivably is related to locally secreted, but as yet unidentified, immunosuppressant factors inherent to the testis (Maddocks and Setchell, 1990). Rejecting CNS allografts in the testis exhibit populations of MHC class $I^+$ and class $II^+$ cells and $CD4^+$ and $CD8^+$ T-lymphocytes and other cell types. CNS allografts to the brain followed three to four weeks later by a concomitant CNS allograft (from the same inbred donor strain) to the host testis are rejected only if the CNS allograft to the testis is first undergoing rejection (Baker and Broadwell, 1993). The results suggest: (1) "immunological privilege" in the testis is not comparable to that in the brain; (2) the rejection response in immunocompetent hosts possessing an intact effector cell arm of the immune system is a manifestation of significant and successful antigen presentation, which appears to be greater for the testis than for the brain; and (3) when the host is sensitized to a level that an effector cell immune response is triggered, the BBB is not a deterrent to the host rejection of an allograft within the brain parenchyma.

## SUMMARY AND CONCLUSIONS

CNS xenografts and allografts within the host mammalian brain possess a BBB that is compromised readily by the host immune response for graft rejection; hence, the BBB is not a significant factor in determining immunological privilege of the CNS or in protecting CNS grafts from a host immune response. Reported immunological privilege of the CNS, testis, and perhaps other suspected sites is relative and by no means absolute. The relative immunological privilege of the CNS may be a consequence of physical factors like the presence of a blood-brain barrier, the absence of lymphatic channels, and the absence of a major population of MHC class $II^+$ cells, all of which may combine with the presence of chemical factors (i.e., a local production and secretion of immunosuppressant factors) for inhibiting foreign antigen presentation. Antigen presentation of foreign CNS grafts in the mammalian brain to the host immune system may be accomplished by host perivascular cells and MHC class $II^+$ cells occupying the subarachnoid space, pial surface, and circumventricular organs and/or by graft cells that gain entry to the blood or cerebrospinal fluid. Indeed, the escape of allograft cells from the host brain may provide a means by which the host immune system can become aware of, but potentially tolerant to, CNS allografts occupying this "immunologically privileged" site. The rejection of intracerebrally located CNS allografts between donor and host of the same outbred strain suggests that when grafting to an "immunologically privileged" site the minor HC antigens can be just as important if not potentially more important than the MHC antigens. Survival or rejection of a CNS allograft within the brain of an immunocompetent mammalian host very likely is determined by the success of antigen presentation to the host immune system; this success is influenced by the physical and chemical factors alluded to above. To insure allograft survival in the clinical arena, immunosuppression of the human host is recommended prior to the grafting of human fetal CNS tissue.

## ACKNOWLEDGEMENTS

These studies were conducted when RDB and BJB were affiliated with the University of Maryland at Baltimore and were supported by NINDS/NIH grant NS18030 (to RDB) and NS27321 (to WFH).

## REFERENCES

Baker, B.J., and Broadwell, R.D., 1993, Balls and brains: I. Immunological privilege for CNS grafts, *Soc. Neurosci. Abstr.* 19:1740.

Bergman, H., Nagy, J.I., and Granholm, A-C., 1995, Intracranial transplantation and survival of tuberomammillary histaminergic neurons, *Neurosci.* 64:61.

Broadwell, R.D., Baker, B.J., Ebert, P.S., and Hickey, W.F., 1994, Allografts of CNS tissue possess a blood-brain barrier: III. Neuropathological, methodological, and immunological considerations, *Micros. Res. and Tech.* 27:471.

Broadwell, R.D., and Banks, W.A, 1993, A cell biological perspective for the transcytosis of peptides and proteins through the mammalian blood-brain fluid barriers, *in*: "The Blood-Brain Barrier Molecular and Cellular Biology," W.M. Pardridge, ed., Raven Press, New York.

Broadwell, R.D., Charlton, H.M., Balin, B.J., and Salcman, M., 1987, Angioarchitecture of the CNS, pituitary gland, and intracerebral grafts revealed with peroxidase cytochemistry, *J. Comp. Neurol.* 260:47.

Broadwell, R.D., Charlton, H.M., Ebert, P.S., Hickey, W.F., Shirazi, Y., Villegas, J., and Wolf, A.L., 1991, Allografts of CNS tissue possess a blood-brain barrier: II. Angiogenesis in solid tissue and cell suspension grafts, *Exp. Neurol.* 112:1.

Broadwell, R.D., Charlton, H.M., Ganong, W.F., Salcman, M., and Sofroniew, M.V., 1989, Allografts of CNS tissue possess a blood-brain barrier: I. Grafts of medial preoptic area in hypogonadal mice, *Exp. Neurol.* 105:135.

Broadwell, R.D., and Sofroniew, M.V., 1993, Serum proteins by-pass the blood-brain barrier for extracellular entry to the central nervous system, *Exp. Neurol.* 120:245.

Cserr, H.F., and Knopf, P.M., 1992, Cervical lymphatics, the blood-brain barrier and the immunoreactivity of the brain: A new review, *Immunol. Today* 13:507.

Fontana, A., Fierz, W., and Wekerle, H., 1984, Astrocytes present myelin basic protein to encephalitogenic T-cell lines, *Nature* 307:273.

Head, J.R., and Billingham, R.E., 1985, Immunologically privileged sites in transplantation immunology and oncology, *Perspect. Biol. Med.* 29:115.

Hickey, W.F., and Kimura, H., 1988, Perivascular microglial cells of the CNS are bone marrow derived and present antigen in vivo, *Science* 239:290.

Hickey, W.F., Vass, K., and Lassmann, H., 1992, Bone marrow-derived elements in the central nervous system: An immunohistochemical and ultrastructural survey of rat chimeras, *J. Neuropath. Exp. Neurol.* 51:246.

Maddocks, S., and Setchell, B.P., 1990, Recent evidence for immune privilege in the testis, *J. Reprod. Immunol.* 18:9

Male, D.K., Pryce, G., and Hughes, C.C.W., 1987, Antigen presentation in brain: MHC induction on brain endothelium and astrocytes compared, *Immunol.* 60:453.

Mason, D.W., Charlton, H.M., Jones, A.J., Lavy, C.B.D., Puklavec, M., and Simmonds, S.J., 1986, The fate of allogeneic and xenogeneic tissue transplanted into the third ventricle of rodents, *Neurosci.* 19:685.

Medawar, P.B., 1948, Immunity to homologous grafted skin. III. The fate of skin homografts transplanted to the brain, to subcutaneous tissue and to the anterior chamber of the eye, *Brit. J. Exp. Path.* 29:58.

Risau, W., Engelhardt, B., and Wekerle, H, 1990, Immune function of the blood-brain barrier: Incomplete presentation of protein (auto-)antigens by rat brain microvascular endothelium in vitro, *J. Cell Biol.* 110:1757.

Sloan, D.J., Baker, B.J., Puklavec, M., and Charlton, H.M., 1990, The effect of site of transplantation and histocompatibility differences on the survival of neural tissue transplanted to the CNS of inbred rat strains, *Prog. Brain Res.* 82:141.

Sloan, D.J., Wood, M.J., and Charlton, H.M., 1991, The immune response to intracerebral neural grafts, *Trends Neurosci.* 14:341.

Widner, H., and Brundin, P., 1988, Immunological aspects of grafting in the mammalian central nervous system. A review and speculative synthesis. *Brain Res. Rev.* 13:287.

# THE BLOOD-BRAIN BARRIER—NO BARRIER TO A DETERMINED LYMPHOCYTE

David Male

Department of Neuropathology
Institute of Psychiatry
DeCrespigny Park
London SE5 8AF
UK

## INTRODUCTION

Lymphocytes continuously patrol the tissues of the body in their search for pathogenic micro-organisms, viruses, and infected cells. The central nervous system is not excluded from this process. Although lymphocyte traffic into the CNS is normally limited, when immune reactions do develop, the rate of inward migration increases rapidly[1]. Leukocyte migration occurs across the endothelium of small venules. In these areas haemodynamic shear forces are relatively low, endothelial surface charge is low and specific adhesion molecules are expressed on the luminal aspect of the endothelium which mediate leukocyte migration. The control of these processes is equally dependent on which lymphocyte populations are available in the blood and on the state of activation of the endothelium.

By analogy with other tissues, the migration is a multistep process, which involves a variety of different adhesion molecules and their ligands as well as cytokines and chemokines originating from the local tissues[2]. The principle steps are:

1) Slowing and rolling of leukocytes along the venular endothelium, mediated principally by the interactions of selectins with carbohydrates on surface glycoproteins.
2) Triggering of the leukocytes, which may be mediated by cytokines, chemotactic agents, chemokines or direct cell/cell interaction.
3) Adhesion of leukocytes to the endothelium mediated by the interactions of integrins with cellular adhesion molecules (CAMs).
4) The initation of a programme of leukocyte diapedesis across the endothelium in which both cells types are active.

The research described in this article was carried out to determine which adhesion molecules are involved in the interactions between the leukocytes and the brain endothelium, how these adhesion molecules are induced on the endothelium and what are the characteristics of migratory populations of lymphocytes.

*New Concepts of a Blood–Brain Barrier*
Edited by J. Greenwood *et al.*, Plenum Press, New York, 1995

## CHARACTERISTICS OF LYMPHOCYTES WHICH MIGRATE ACROSS BRAIN ENDOTHELIUM IN VITRO

It has been noted both in vivo and in vitro, that lymphocytes which have been activated by their specific antigen or by mitogens, adhere to brain endothelium more effectively than do resting lymphocytes[3]. In vitro, mitogen-activated cells do not migrate rapidly across the endothelium, despite binding efficiently. In contrast T cell lines which have been stimulated with antigen and then pulsed with IL-2 exhibit high levels of migration at 2-4 days after the IL-2 pulse which subsequently wanes as the cells fall back into resting phase[4].

In vivo, it has been shown that lymphocyte activation favours migration to the CNS and many of the cells which are seen in lesions of acute experimental encephalomyelitis have recently been in cell cycle[5].

What characteristics determine whether a lymphocyte can cross CNS endothelium? Antibody blocking studies have shown that lymphocyte adhesion to the endothelium can be inhibited most effectively with anti-LFA-1 and less effectively with anti-VLA-4[6]. Anti-LFA-1 inhibits the binding of lymphocytes to cytokine-activated (TNF-$\alpha$, or IFN-$\gamma$) endothelium, and to a much lesser extent to unstimulated endothelium. In contrast anti-VLA-4 has a marginal effect on adhesion to activated endothelium and no effect on adhesion to resting endothelium. Antibodies to LFA-1 and VLA-4 can block synergistically[6]. These data imply that a ligand for LFA-1 on the endothelium plays a major role in controlling lymphocyte adhesion while lymphocyte VLA-4 makes only a minor contribution to adhesion.

Evidence discussed below, implies that the ligand on brain endothelium for LFA-1 is ICAM-1 and the endothelial ligand for VLA-4 is VCAM-1. This is also the case with non-CNS endothelium, although non-CNS endothelia apparently express higher levels of ICAM-2, an alternative ligand for LFA-1. We have scanned over 150 surface molecules on normal and cytokine induced CNS endothelium by two-dimensional gel electrophoresis and to date we have found no evidence for a brain-specific adhesion molecule.

LFA-1 and VLA-4 are inducible integrins expressed on various leukocyte populations. Since these molecules appear to be most important for adhesion to brain endothelium, their presence on lymphocytes is a necessary prerequisite for migration. Whether a lymphocyte expresses these integrins is determined by events occuring ouside the CNS.

Using CD4+ rat T cell lines specific for PPD, it has been shown that VLA-4 is increased at 2-3 days following cellular stimulation with IL-2, and LFA-1 is induced at 2-6 days after stimulation[4]. This corresponds to the time when cells have the highest migratory potential in vitro. Similarly, mitogen-activation of mixed populations of peripheral lymph node cells for 24 hours causes the induction of a new population of LFA-$1^+$,VLA-$4^-$ cells and a population of LFA-$1^+$,VLA-$4^+$ cells. Removal of LFA-$1^+$ cells from such a mixed population substantially reduces the number of adherent lymphocytes.

Taken together this implies that T cell activation induces enhanced expression of LFA-1 and VLA-4. This occurs at a time when lymphocytes would be released into the circulation following activation by antigen in secondary lymphoid tissues. These cells now have the opportunity to interact with their ligands on endothelium in any tissue of the body.

Where they will migrate depends on the expression of appropriate ligands on the endothelium in different vascular beds, and on the local release of cytokines which can stimulate lymphocytes to put their integrins into a high affinity state (see below).

## INDUCTION OF ADHESION MOLECULES ON CNS ENDOTHELIUM

The receptor on the endothelium for VLA-4 is thought to be VCAM-1 and the receptors for LFA-1 are ICAM-1 and ICAM-2. Both VCAM-1 and ICAM-1 can be induced on CNS endothelium by the cytokines TNF-$\alpha$ and IFN-$\gamma$. TNF-$\alpha$ is produced by mononuclear phagocytes, including microglia, and IFN-$\gamma$ is exclusively generated by activated T cells.

In vitro, VCAM-1 is absent from unstimulated endothelium whereas ICAM-1 is normally present on a proportion of the cells detectable by immunofluorescence and ELISA. VCAM-1 is induced 5-48 hours after endothelial activation; while LFA-1 appears slightly later at 8-96 hours (Figure 1). The levels of VCAM-1 are invariably lower than those of ICAM-1, and we have been able to immunoprecipitate ICAM-1 from endothelial lysates but not VCAM-1. Removal of the cytokines causes a decline in expression.

ICAM-1 has also been identified on endothelia in the plaques of multiple sclerosis[7] and in the lesion of acute experimental allergic encephalomyelitis[8].

The presence of ICAM-1 on CNS endothelia would allow lymphocytes to bind to the endothelium prior to migration. Of course, the expression of ICAM-1 will only be of any significance if populations of circulating cells are available which express LFA-1 or CR3, the ligands for ICAM-1.

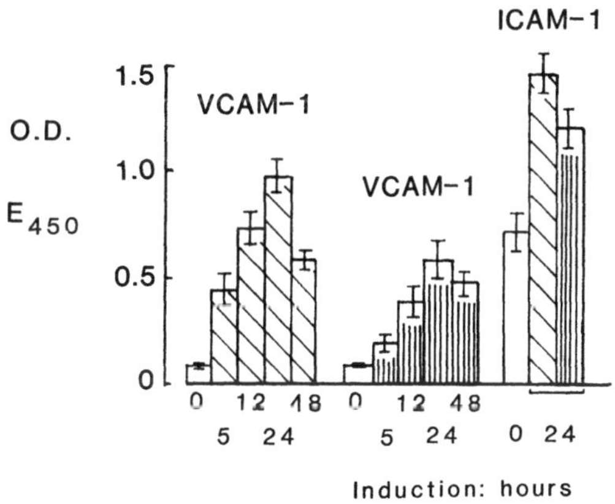


**Figure 1.** Induction of ICAM-1 and VCAM-1 on brain endothelium in vitro following stimulation by 25ng/ml TNF-$\alpha$ (cross hatch) or 100 units/ml rat IFN-$\gamma$ (vertical hatch) over a period of 0 - 48 hours, measured by an enzyme immunoassay[9]

## CONTROL OF ADHESION AND MIGRATION

Another important molecule is the one recognised by the monoclonal antibody 4A2[10]. It is inducible by IFN-$\gamma$, but not TNF-$\alpha$, on CNS endothelium, and is maximally expressed 24 - 48 hours after endothelial activation, at a time when the endothelium is most efficient at supporting lymphocyte adhesion. This molecule is also constitutively present on lymphocytes but is not further induced by cytokines. In vivo, it appears on microglia and brain endothelium during acute experimental allergic encephalomyelitis. Cross-linking of the molecule on lymphocytes in vitro with the 4A2 antibody enhances adhesion by up to 400%. Antibody blocking studies show that the enhanced adhesion is mediated by the integrins, and indirect evidence suggests that this occurs by upregulation of lymphocyte protein kinase-C. Integrins on lymphocytes are often present in a non-functional state, which requires activation of protein-kinase C to induce the functional integrin, and this may be acheived by stimulation with chemokines or by cross-linking of surface CD31[11]. In these experiments the 4A2 antibody has a similar effect to lymphocyte activation by these routes.

Using this information, we can now develop a theory on how lymphocyte traffic into the CNS is controlled during the initiation of an immune response:

If damage or an infection occurs in the brain parenchyma, this causes microglia to release cytokines, which induce VCAM-1 and ICAM-1 on the endothelium. This allows a small number of the activated CD4+ T cells in the circulation to migrate across the endothelium. If these cells now encounter their specific antigen, presented by the microglia or perivascular macrophages, they are stimulated to release IFN-$\gamma$, which further induces ICAM-1, VCAM-1 and 4A2-Ag. These adhesion molecules now facilitate a major inward migration of lymphocytes and mononuclear phagocytes (ICAM-1 is also a receptor for CR3, an adhesion molecule on blood monocytes).

This theory and the evidence which underlies it, all see the CNS endothelium and the circulating lymphocytes as equal partners in the control of immune reactions in the brain.

## REFERENCES


1. H.M. Wisniewski and A.S. Lossinsky, Structural and functional aspects of the interaction of inflammatory cells with the blood brain barrier in experimental brain inflammation. *Brain Pathol.* 1:89 (1991).
2. T.A. Springer, Traffic signals for lymphocyte recirculation and leukocyte emigration: the multistep paradigm, *Cell* 76:301 (1994).
3. D.K. Male, G. Pryce, C.C.W. Hughes C and P.L. Lantos, Lymphocyte migration into brain modelled in vitro: control by lymphocyte activation, cytokines and antigen. *Cell. Immunol.* 127:1 (1990).
4. G. Pryce, J. Greenwood, D. Male and I. Campbell, Time-dependent T cell migration across normal rat cerebral endothelium in vitro is dependent on duration of IL-2 stimulation and is LFA-1/ICAM-1 dependent. Submitted to *Int. Immunol.* (1994).
5. A.H. Cross, B. Cannella, C.F. Brosnan and C.S. Raine, Homing to central nervous system vasculature by antigen-specific lymphocytes I.Localization of $^{14}C$-labelled cells during acute and chronic relapsing experimental allergic encephalomyelitis. *Lab. Invest.* 63:162 (1990).
6. D. Male, J. Rahman, G. Pryce, T. Tamatani and M. Miyasaka, Lymphocyte migration into the CNS modelled in vitro: roles of LFA-1, ICAM-1 and VLA-4. *Immunology* 81:366 (1994).
7. R.A. Sobel, M.E. Mitchell and G. Fondren, Intercellular adhesion molecule-1 (ICAM-1) in cellular immune reactions in the human central nervous system. *Am. J. Path.* 136:1309 (1990).
8. B. Cannella, A.H. Cross and C.S. Raine, Adhesion-related molecules in the the central nervous system. Up-regulation correlates with inflammatory cell influx during experimental allergic encephalomyelitis. *Lab. Invest.* 65:23 (1991).
9. D.K. Male, G. Pryce and C.C.W. Hughes, Antigen presentation in brain: MHC induction on brain endothelium and astrocytes compared. *Immunology* 60:453 (1987).
10. N.A. Flaris, T.L. Densmore, M.C. Molleston and W.F. Hickey, Characterization of microglia and macrophages in the CNS of rats: Definition of the differential expression of molecules using standard and novel monoclonal antibodies in normal CNS and in four models of parenchymal reaction. *Glia* 7:34 (1993).
11. Y. Tanaka, S.M. Albelda, K.J. Horgan, G.A. van Seventer, Y. Shimizu, W. Newman, J. Hallam, C.A. Buck and S. Shaw, CD31 expressed on distinctive T cell subsets is a preferential amplifier of ß1 integrin-mediated adhesion. *J. Exp. Med.* 176:2457 (1992).

# THE BLOOD-RETINAL BARRIER IN IMMUNE-MEDIATED DISEASES OF THE RETINA

John Greenwood, Simon Bamforth,
Yufei Wang and Lesley Devine

Division of Clinical Science
Department of Clinical Ophthalmology
Bath Street
London EC1V 9EL

## INTRODUCTION

The retina, as an extension of the central nervous system, is potentially subject to the same systemic influences that can affect the delicate homeostasis of the brain. As a result, the neuroretina is protected from the circulation, and the content of the extracellular fluid carefully controlled, by the presence of a blood-tissue barrier. In the brain this selective cellular barrier, known as the blood-brain barrier (BBB), is formed mainly by the endothelial cells of the cerebral vasculature which posses tight junctions and negligible non-specific vesicular transport. In addition to the BBB there is also the blood-CSF barrier which is situated at the choroid plexus and comprises of a sheet of choroidal epithelium joined by tight junctions and which overlies the fenestrated vasculature beneath. This latter barrier, however, only represents a small fraction of the total surface area available for exchange between the blood and cerebral parenchyma.

In the retina, the blood-tissue barrier is also comprised of endothelial and epithelial cells but unlike that of the brain these barriers are jointly referred to as the blood-retinal barrier (BRB). Furthermore, the total surface area available for exchange between the blood and retinal parenchyma is shared more equally by the two sites (Gratton et al., 1992). The retina forms a semicircular cup lying at the posterior aspect of the eye (Figure 1) and is situated over a vascular bed with one of the highest blood-flows found in the animal. This choroidal vasculature is composed of permeable, fenestrated endothelia analogous to those in the choroid plexus of the brain. Between this leaky vascular bed and the neuroretina lies a monolayer of epithelial cells, the retinal pigment epithelium (RPE), which are joined together by tight apical junctions. These important cells are responsible for phagocytosing shed rod outer segments as well as controlling the passage of substances into and out of the avascular posterior retina (Figure 1). The anterior retina, on the other hand, is supplied with nutrients by the retinal vasculature, the endothelial cells being functionally and structurally identical to those of the majority of the CNS.

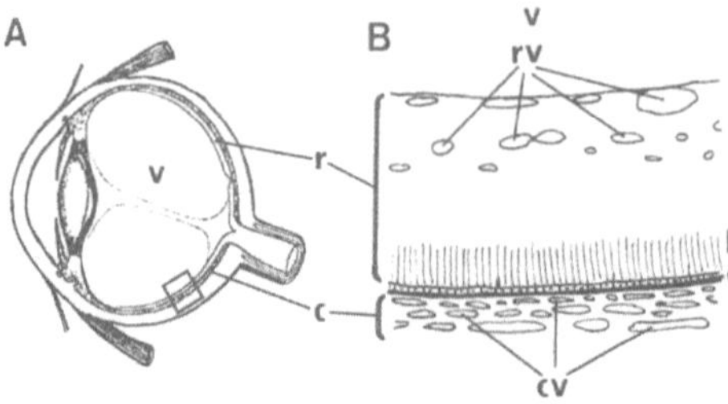


**Figure 1.** A. Diagram of horizontal section through the human eye showing the posterior position of the retina. B. enlarged boxed section from A indicating the arrangement of the anterior (retinal vasculature) and posterior (RPE) aspects of the blood-retinal barrier. v, vitreous; r, retina; rv, retinal vessels; c, choroid; cv, choroidal vessels; e, RPE; p, photoreceptor layer.

## IMMUNOLOGICAL ROLE OF THE BRB

As with the brain the retina also possesses a degree of immune privilege and traffic of leucocytes into the parenchyma is restricted by the presence of the BRB. In inflammatory diseases of the eye, such as posterior uveitis, there is active recruitment which leads to a dramatic upregulation of leucocyte influx into the retina (Caspi et al., 1993). This process can lead to a breakdown in the structural integrity of the barrier resulting in oedema, tissue damage and loss of vision. The BRB therefore plays a major part in the pathogenesis of inflammatory eye disease as it is responsible for recruiting leucocytes from the circulation and its subsequent dysfunction leads to all the clinical problems associated with oedema. In addition, there also remains the question of whether cells of the BRB are capable of inducing or perpetuating the disease by presenting antigen to circulating T cells and eliciting a proliferative response.

The role of the BRB in ocular inflammatory disease has been investigated in a number of *in vivo* studies where the animal model of posterior uveitis, experimental autoimmune uveoretinitis (EAU), has been employed. As with the animal model of multiple sclerosis, experimental autoimmune encephalomyelitis (EAE), EAU is believed to be a $CD4^{+}$ T cell mediated diseases as it can be adoptively transferred by inoculating animals with uveitogenic T cell lines (Caspi et al., 1986; Calder and Lightman, 1992). In these *in vivo* studies the route that leucocytes take in entering the retina and the ensuing damage have been investigated as well as the molecules involved in recruiting leucocytes from the circulation. This latter phenomena, however, has been investigated more successfully by employing *in vitro* techniques in which both retinal vascular endothelium and RPE are cultured and their immune properties examined.

## IN VIVO STUDIES

### Ultrastructural studies

In posterior uveitis barrier disruption and the ensuing oedema poses a serious threat to sight and as such the factors leading to increased BRB permeability have been widely investigated in the animal model EAU (Greenwood, 1992a). A major question to be addressed is whether barrier disruption occurs prior to, is concomitant with or develops after the influx of inflammatory cells. The indications are that although oedema is an early event in the disease process it only occurs at the time when leucocytes are first seen to enter the parenchyma (Lightman & Greenwood, 1992; Greenwood et al., 1994). This is generally in agreement with the studies on EAE where leucocyte infiltration and disruption of the BBB are also related events (Stoul et al., 1979; de Rosbo et al., 1985; Hawkins et al., 1990; Claudio et al., 1990).

The route by which leucocytes enter the retina and their path across the BRB remains one of the contentious issues in this field of study. Leucocytes are thought to enter the parenchyma predominantly via the retinal vasculature. The endothelia of the retinal vasculature, being in close contact with circulating immune cells, are more readily able to orchestrate the influx of these cells into the parenchyma. The RPE, on the other hand, is separated from the blood and any leucocyte crossing via this route must first be captured from the circulation by the choroidal endothelia, migrate out into the perivascular space, penetrate Bruch's membrane and finally cross the RPE. Ultrastructural evidence suggests, however, that inflammatory cells are capable of entering the retina across the posterior barrier formed by the RPE but mainly during the later stages of the disease (Lin et al., 1991; Dua et al., 1991; McMenamin et al., 1993; Greenwood et al., 1994). That this can occur is also illustrated by the finding that in some species (eg. guinea pig) which have an avascular retina, EAU can still be induced (Forrester et al., 1985).

If the lymphocytes that initiate the disease cross the vascular BRB then what path do they take? Unlike lymphocyte extravasation at non-CNS vascular beds where the passage between endothelial cells is largely unrestricted, the tight junctions of the retinal vasculature present a significant physical barrier to the infiltrating cell. If the lymphocyte does not take this route, however, the only other option is for it to pass through the body of the endothelial cell. Whichever route lymphocytes take it is likely that they utilize processes that are unique to the CNS. Careful ultrastructural studies in both EAE (Lossinsky et al., 1989; Raine et al., 1990) and EAU (McMennamin et al., 1992; Greenwood et al., 1994) suggest that lymphocytes penetrate the barrier by passing through the endothelial cell in close proximity to the junction. The perceived mechanism of passage is for the lymphocyte to probe the endothelial cell causing invagination of the luminal membrane. As the lymphocyte continues its probing the endothelial cell becomes highly attenuated with the luminal and abluminal membranes abutting. The two membranes are then able to fuse allowing for a pore to form through which the lymphocyte can pass (Figure 2). Naturally both this process and penetration through the tight junctions may occur. Indeed, at the BBB it has been suggested that extravasation occurs via both these routes and depends upon the phenotype of the leucocyte (Raine et al., 1990).

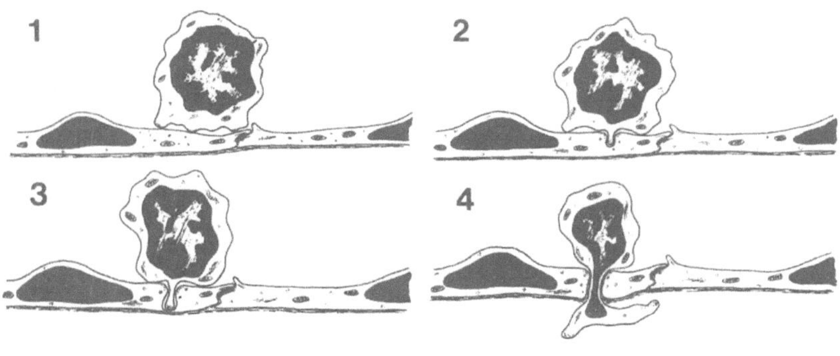


**Figure 2.** Possible route of lymphocyte extravasation across the tight vasculature of the CNS. The lymphocyte penetrates the endothelium at a parajunctional site (1-3) and passes through a pore formed by the latter (4). Proposed mechanism based on ultrastructural studies in both EAE and EAU.

## Permeability studies

There are various routes through which leakage of the BRB can arise. Firstly there are the two barrier sites; the posterior barrier formed by the RPE and the anterior barrier formed by the vascular endothelium. Secondly there is the question of whether permeability changes are the consequence of disrupted junctions, increased non-specific vesicular transport, damaged cells of the BRB or due to leakage occurring during the process of leucocyte migration. Moreover, there remains much confusion over the role that inflammatory compounds play in inducing these permeability changes.

In a number of studies of EAU in the rat we have investigated the degree of BRB disruption and attempted to identify the ultrastructural correlate of any leakage. By using tracers of different sizes introduced into the vasculature we have found that the small tracer $^{14}$C-sucrose, which is normally excluded from the parenchyma by the barrier, enters the tissue at the same time as infiltrating leucocytes (Lightman and Greenwood, 1992). $^{125}$I-albumin and horseradish peroxidase (HRP), however, does not leak into the parenchyma until later in the disease (Table 1). This size selectivity suggests that the early leakage is

**Table 1.** Blood-retinal barrier permeability and leucocyte infiltration in EAU

| Days post-immunization | Leucocyte infiltration | Sucrose permeability | HRP permeability | Retinal damage |
|---|---|---|---|---|
| 0-10 | ND | very low | ND | ND |
| 11-12 | many | increased | ND | ND |
| 13 on | large-scale | not done | increased | yes |

ND; not detected

due to the formation of channels, possibly at the tight junctions, as vesicular transport is less likely to discriminate between tracers of different sizes (Robinson and Rapoport, 1987). Whether leakage was occurring via the RPE or the vascular endothelia was determined by

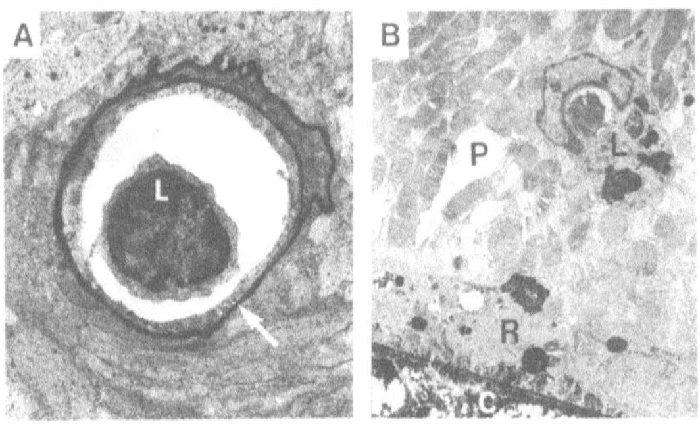


**Figure 3.** A. Retinal microvessel with adherent leucocyte (L), probably a lymphocyte, and extravasated HRP in basement membrane (arrow) and perivascular region. B. Posterior region of retina with leucocyte (L) probably a macrophage, present in the photoreceptor layer (P). RPE layer (R) is successfully excluding lanthanum from the neuroretina which has flooded the choroid (C), basal infoldings of the RPE and up to the apical junction.

using the visual tracers HRP and lanthanum. At the ultrastructural level it was found that there was a significant leakage of tracer into the perivascular region of the retina (Lightman and Greenwood, 1992; Greenwood et al., 1994; Figure 3) with the route of passage appearing to be via tight junctions (Lin et al., 1991; Greenwood et al., 1994). Only very rarely could tracer be seen extending from the choroidal circulation into the neuroretina via the RPE. Indeed, even with severe disruption of the retina the RPE was able to maintain considerable structural integrity and retain its ability to exclude tracer (Figure 3).

It has been suggested in EAE that tracer may also diffuse across the barrier during leucocyte migration (Lossinsky et al., 1989; Claudio et al., 1990). Whilst we have not been able to confirm this finding in EAU this possibility still exists. A further potential route of leakage is by irreversible damage to the RPE and endothelial cells. As has been stated above, the RPE appear to be able to maintain their structural integrity but the endothelia, where most leucocyte traffic occurs, can be severely damaged especially in the latter stages of the disease. It must be made clear, however, that the BRB, like that of the BBB, appears to possess a considerable capacity for repair. Furthermore, the leakage that occurs can vary significantly in its severity, in its distribution and in its capacity to resolve.

Unlike in the brain where the role of inflammatory agents on the BBB has been extensively researched (see chapters by Wahl and Sharma), there have been few studies investigating their effect on the permeability of the BRB (for review see Greenwood, 1992b). Of recent interest has been the role of proinflammatory cytokines in the pathogenesis of brain and retinal inflammation (Martiney et al., 1990; 1992). In studies conducted in our laboratory, these compounds have been introduced into the vitreous and we have been able to detect with $^{14}C$-mannitol any temporal changes in BRB permeability and correlate this with the induction of leucocyte influx into the retina. Administration of interleukin-1ß (IL-1ß) into the vitreous has been found to bring about a biphasic opening of the barrier at 4 and 24h after injection (Bamforth et al., 1994). The initial phase of disruption coincides with the first cohort of infiltrating leucocytes whilst the second phase corresponds with maximal leucocyte infiltration. These studies suggest that IL-1ß does not have an immediate and direct effect on the BRB but increases permeability by its induction of leucocyte infiltration and the release of other vasoactive agents. Of the other cytokines so far studied IL-2 (Hall et al., 1992) and IL-6 were found to have no effect on either

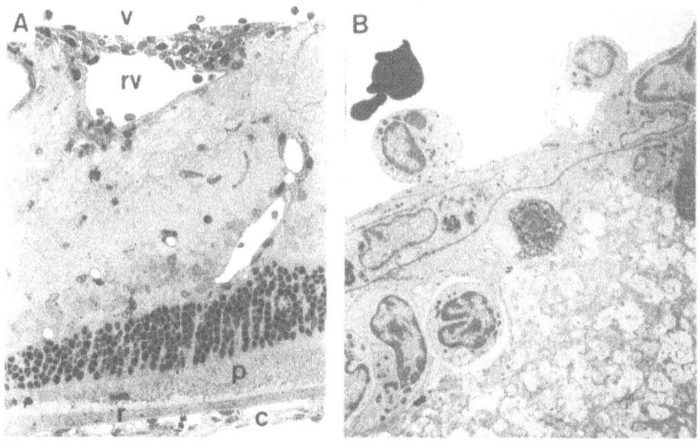


**Figure 4.** A. Toluidine blue section through the retina 48h after IL-1ß injection into the vitreous. Note the numerous leucocytes adhering to the retinal vessel (rv) walls and present in the perivascular region and vitreous (v) with few leucocytes in the photoreceptor layer (p), RPE (r) and choroid (c). B. Electron micrograph of retinal vessel and tissue with leucocytes adhering to the vessel endothelium and present in the perivascular region 72h after IL-ß injection.

barrier permeability or leucocyte infiltration whereas TNF-$\alpha$ opened the barrier but did not induce leucocyte infiltration. This latter finding, however, contradicts a report in which TNF introduced into the rabbit vitreous caused cellular infiltration (Brosnan et al., 1990).

## IN VITRO STUDIES

### Culturing cells of the BRB

The distinct advantages in experimental design afforded by tissue culture has led to the development of numerous techniques for isolating and culturing brain endothelia (see chapters by Joó and Abbott). Less common, however, is the culture of retinal endothelia and RPE. Based on an original technique by Hughes & Lantos (1986) we have developed a method for isolating and growing primary cultures of rat retinal endothelia (Greenwood, 1992c) (Figure 4A). These cells have been thoroughly characterised and are positive for the tight junction-associated protein ZO-1, P-glycoprotein (the product of the multidrug resistancy gene), von Willebrands factor, transferrin receptor, the rat endothelial epitope RECA-1 in addition to expressing a high uptake of acetylated LDL.

We have also developed a technique, based on that of Chang et al., (1991), to isolate and grown primary cultures of rat RPE (Figure 4B). These cells grow as pigmented monolayers positive for cytokeratin and the rat RPE-specific epitope RET-PE2 (Neill and Barnstable, 1990).

Using *in vitro* techniques both RPE and retinal endothelial cells have been employed to investigate the interactions between these cells and lymphocytes with a view to identifying the role these cells play in the pathogenesis of inflammatory eye disease.

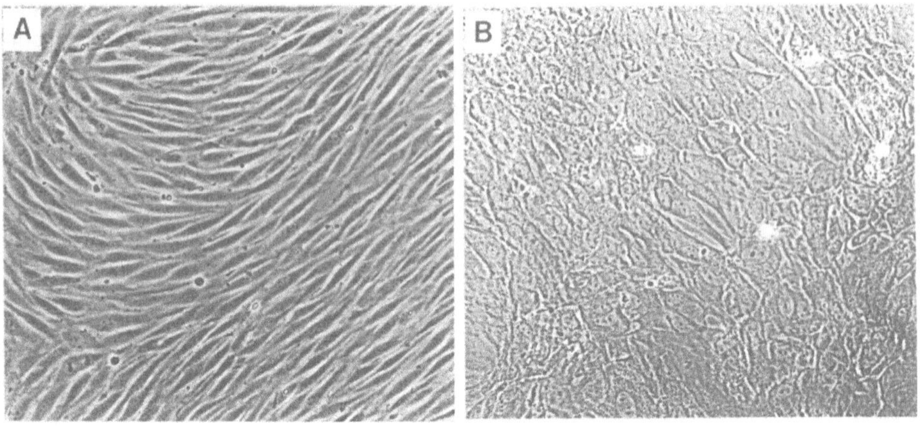


**Figure 5.** Primary cultures of (A) rat retinal endothelia and (B) retinal pigment epithelium.

### Antigen presentation

Whether the cells of the BBB and BRB are capable of presenting antigen and supporting T cell proliferation has remained a vexed question. Although cerebral (Male et al., 1987, 1988a, 1988b; McCarron et al., 1991) and retinal endothelia (Figure 6; Liversidge et al., 1990) are capable of expressing MHC class II following cytokine activation, and are able to induce class II mediated cell lysis, they are poor at inducing T cell proliferation (Pryce et al., 1989; Risau et al., 1990; Sedgewick et al., 1990; Fabry et

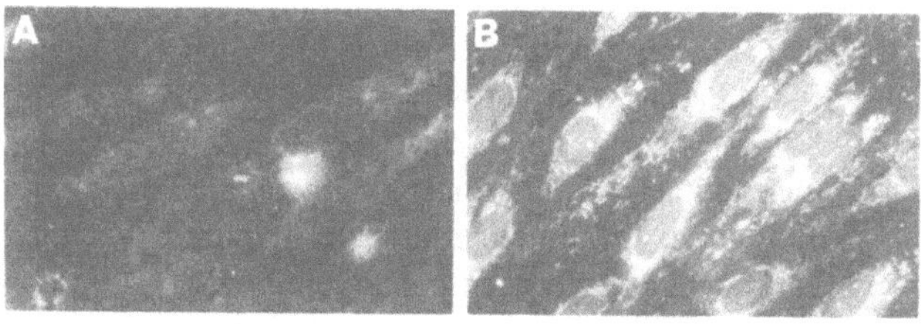

**Figure 6.** Primary cultures of retinal endothelial cells labelled with antibodies to the MHC class II antigens I-A and I-E (OX-6 and OX-17 respectively). A. Non-activated endothelia. B. IFN-γ activated endothelia.

al., 1990; McCarron et al 1991). It has therefore been suggested that these cells lack the necessary co-stimulatory factors that are expressed by "professional" antigen presenting cells.

We have found that in a conventional proliferation assay in which antigen-specific T cells in the presence of their antigen are added to confluent brain and retinal EC monolayers, no significant T cell proliferation could be induced. If, however, the ratio of endothelial cells to T cells is reduced by using subconfluent endothelial monolayers a significantly greater degree of proliferation can be achieved. This suggests that confluent endothelial cells are either producing inhibitory factors or are able to switch off the T cells thus rendering them unresponsive (Wang et al., 1995).

As with retinal endothelia, the RPE can also be induced to express MHC class II antigens (Liversidge et al., 1988) although their ability to support T cell proliferation has not yet been demonstrated. Because of their normal physiological role, it has been suggested that RPE are more likely to function as antigen presenting cells as in many respects they closely resemble macrophages. The RPE are actively engaged in phagocytosing photoreceptor outer segments and degrading them enzymatically. Furthermore, it is interesting to note that many of the uveitogenic antigens that have been identified are found to be present in the photoreceptor layer. These cells therefore posses the potential to re-express these molecules in association with MHC class II. In addition, the normal function of the RPE will mean that many of these molecules may also be deposited at the basal side of the RPE where they may come into contact with dendritic cells in the choroid that act as antigen presenting cells. Further work is underway to establish whether RPE are able to initiate T cell proliferation although in the retina, as in the rest of the CNS, the major antigen presenting cell is probably the microglia.

## Lymphocyte extravasation

For circulating lymphocytes to be recruited from the blood their passage along the vessel must first be reduced and halted. This process of capture has elicited an enormous body of research over the last decade which has led to a generalised model being proposed. Lymphocyte extravasation can thus be separated into three broad stages mediated by surface expressed molecules. The first involves transient interactions between the lymphocyte and the endothelia via lectin-like molecules and their counter receptors. These initial connections slow the lymphocyte down and allow for firmer attachments to be made and generated. The major molecules believed to be involved during this second stage are LFA-1 and VLA-4 on the lymphocyte and their receptors ICAM-1/2 and VCAM-1 respectively on the endothelia (see chapter by Male). Once adhesion has occurred the leucocyte can then proceed to migrate across the vascular wall and into the perivascular region: a process referred to as diapedesis.

One of the most striking differences between non-CNS and CNS endothelia is their ability to bind lymphocytes. The important work of Male and colleagues (Hughes et al., 1988; Male et al., 1990; Pryce et al., 1991) and others (Waldschmidt et al., 1991; McCarron et al., 1993) has clearly shown that the degree of lymphocyte adhesion to monolayers of rat brain endothelia is significantly lower than for non-brain endothelia. Indeed, in lymph nodes where the endothelium are required to capture non-activated lymphocytes the vessels (known as high endothelial veins; HEV) have a very high binding capacity (Ager and Mistry, 1988). As with brain endothelia we have shown that retinal endothelia are also much less adhesive than non-CNS endothelia (Wang et al., 1993). This finding illustrates an important functional difference between CNS and non-CNS vasculature and is likely to explain why there is thought to be such a low level of lymphocyte traffic through the brain and retina. Adhesion, which is the first mandatory stage of extravasation, is therefore minimal in normal CNS tissue. The binding of lymphocytes to both brain and retinal endothelia, however, can be significantly enhanced by activating the lymphocyte with mitogen or antigen and the endothelia with cytokines such as IL-1, IFN-$\gamma$ and IL-4. This is believed to induce, upregulate and switch to high affinity states the molecules involved in adhesion (Table 2).

**Table 2.** Percentage of ICAM-1 expression on primary cultures of retinal and brain endothelia from rat following IFN-$\gamma$ treatment (200U/ml) determined by flow cytometry. Means $\pm$ S.D.

| Duration of treatment | Retinal endothelia | Brain endothelia |
|---|---|---|
| 0 | $60.8 \pm 6.6$ | $66.8 \pm 9.7$ |
| 18 h | $86.1 \pm 3.9$ | $81.2 \pm 2.5$ |
| 3 days | $86.9 \pm 4.5$ | $83.7 \pm 2.5$ |
| 5 days | $93.9 \pm 2.9$ | $86.1 \pm 6.1$ |

What has become apparent, however, is that induction of lymphocyte adhesion does not necessarily lead to an increase in diapedesis. Thus, although mitogen activation of peripheral lymph node (PLN) lymphocytes significantly upregulates adhesion and locomotion it has no effect upon migration (Greenwood & Calder, 1993), which remains well below the levels achieved by antigen-specific T cell lines (Figure 7). This strongly suggests that the signals propagated by mitogen lack the necessary triggers for switching on the migratory phenotype. Moreover, these studies have shown that the ability of an antigen-specific T cell line to migrate through a retinal endothelial cell monolayer is independent of the antigen with an irrelevant antigen (purified protein derivative) T cell line migrating to the same degree as a retinal (S-Ag) antigen-specific T cell line.

The molecules involved in binding lymphocytes to CNS endothelia have been investigated in adhesion assays by employing antibodies with specific affinities to known surface molecules. Of crucial importance to this process is the lymphocyte integrin LFA-1. Although one of the major ligands for this molecule, ICAM-1, is expressed on CNS endothelia and is upregulated following cytokine activation it does not appear to be involved in the initial binding process. This suggests that an alternative ligand such as ICAM-2 or another, as yet unidentified molecule, may be involved. In addition to LFA-1 the integrin VLA-4 on the lymphocyte may also play a minor role in adhesion, but only to activated endothelia as its ligand, VCAM-1, is not expressed on resting cells (Male et al., 1994; Greenwood et al., 1995).

Although ICAM-1 does not appear to play a significant part in the binding of lymphocytes to brain and retinal endothelia it does play a vital role in the subsequent process of diapedesis. Using videomicroscopy to assess the degree of migration of lymphocytes across a monolayer of retinal endothelial cells, we have shown that ICAM-1 is instrumental in mediating the migration of T cell lines. The VLA-4/VCAM-1 pairing also appears to play a role in lymphocyte migration across activated endothelia which is likely to be significant in the diseased retina where cytokines and other inflammatory agents are released (Greenwood et al., 1995).

The other site through which lymphocytes may enter the retina is via the RPE. Unlike the retinal endothelium the RPE is not in direct contact with circulating lymphocytes and thus is only a barrier to immune cells that have already left the blood and entered the perivascular space of the choroid. It is therefore the choroidal endothelium that will orchestrate the extravasation of leucocytes into the back of the eye. The RPE, however, can influence this process by releasing cytokines and chemotactic agents (Elner et al., 1990; 1991; Planck et al., 1992) which will directly affect the ability of the choroidal endothelium to capture circulating cells. The role of adhesion molecules on the RPE are not therefore involved in leucocyte capture but their expression may influence the ability of the leucocyte to cross this cellular barrier. We have investigated the ability of lymphocytes to transmigrate across a monolayer of RPE cells. As with the vascular endothelial studies migration was largely controlled by the state of activation of the lymphocyte with T cell lines exhibiting a much greater degree of transmigration than unactivated or mitogen activated PLN cells. The LFA-1/ICAM-1 pairing was also found to play a significant part in mediating this process.

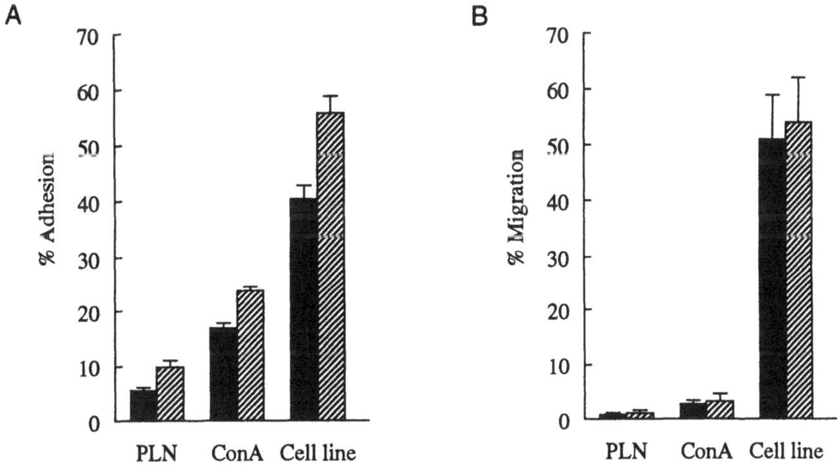


**Figure 7.** The effect of cell activation on lymphocyte adhesion to and migration across resting (closed bars) and IFN-γ activated (hatched bars) retinal endothelial monolayers. A. Activation of resting peripheral lymph node (PLN) cells with concanavalin A (ConA) or antigen (antigen-specific T cell line) significantly increases their adhesion to retinal endothelia. Activation of the endothelia with IFN-γ also increases adhesion. B. Activation of lymphocytes with ConA or the endothelia with IFN-γ does not increase the limited ($<5\%$) degree of transmonolayer migration but antigen-specific T cells exhibit significant levels of migration by 4h. Data taken from Wang et al., 1993 and Greenwood and Calder, 1993.

In recent studies in our laboratory we have been investigating the role of second messenger and signal transduction pathways in the endothelial cell during lymphocyte migration. Preliminary data would suggest that in addition to the endothelium expressing the necessary adhesion molecules, diapedesis is further actively facilitated by the endothelial cell. Whether this is by assisting in the penetration of the lymphocyte through the endothelial cell by pore formation or by unlocking the tight junctions remains to be resolved.

## CONCLUSIONS

The vascular BRB in most respects is identical to that of the BBB and much of our understanding of its pathology until recently has been extrapolated from studies in the brain. The posterior BRB, formed by the RPE, also possess some similarities with the blood-CSF barrier although significant differences do exist. The relative importance of the RPE as a site of exchange of molecules and cells is significantly greater than that of the blood-CSF barrier. Thus in this respect, the RPE is likely to influence the neuroretina to a greater extent than does the epithelium of the choroid plexus affect the brain. Because of its unique anatomical situation the vasculature of the retina offers a unique opportunity for the investigator to study the properties of the CNS vascular barrier as it is considerably more accessible than brain microvessels. Indeed, in a recent study we have been able to gain access to retinal microvessels *in vivo* and have measured their transvascular resistance (Towler et al., 1994). Finally, the role of the BRB in inflammatory diseases of the retina is central to the pathogenesis of the disease and by increasing our understanding of these processes, therapeutic intervention may ultimately be developed.

## ACKNOWLEDGEMENTS

The authors would like to thank the Leverhulme Trust, Wellcome Trust and Moorfields Eye Hospital Locally Organised Research Scheme who have supported this work.

## REFERENCES

Ager, A., and Mistry, S., 1988, Interactions between lymphocytes and cultured high endothelial cells: an in vitro model of lymphocyte migration across high endothelial venule endothelium, *Eur. J. Immunol.* 18:1265.

Bamforth, S.D., Greenwood, J., and Lightman, S.L., 1994, The effect of interleukin-1ß on blood-retinal barrier permeability of the anaesthetized rat, *J. Physiol.* 480:19P.

Brosnan, C.F,, Claudio, L., Tansey, F.A., and Martiney, J., 1990, Mechanisms of autoimmune neuropathies, *Ann. Neurol.* 27 (suppl):S75.

Calder, V.L., and Lightman, S.L., 1992, Experimental autoimmune uveoretinitis (EAU) verses experimental allergic encephalomyelitis (EAE): a comparison of T cell-mediated mechanisms, *Clin. exp. Immunol.* 89:165.

Caspi, R.R., Chan, C-C., Fujino, Y., Najafian, F., Grover, S., Hansen, C.T., and Wilder, R.L., 1993, Recruitment of antigen-nonspecific cells plays a vital role in the pathogenesis of a T cell-mediated organ-specific autoimmune disease, experimental autoimmune uveoretinitis, *J. Neuroimmunol.* 47:177.

Caspi, R.R., Roberge, F.G., MacAllister, C,G,, et al., 1986, T cell lines mediating experimental autoimmune uveoretinitis (EAU) in the rat, *J. Immunol.* 136:928.

Chang, C.W., Roque, R.S., Defoe, D.M., and Caldwell, R.B., 1991, An improved method for isolation and culture of pigment epithelial cells from rat retina, *Curr. Eye Res.* 10:1081.

Claudio, L., Kress, Y., Factor, J., and Brosnan, C.F., 1990, Mechanisms of edema formation in experimental autoimmune encephalomyelitis, *Am. J. Pathol.* 137:1033.

Dua, H.S., McKinnon, A., McMenamin, P.G., and Forrester, J.V., 1991, Ultrastructural pathology of the "barrier sites" in experimental autoimmune uveitis and experimental autoimmune pinealitis, *Br. J. Ophthalmol.* 75:391.

Elner, V.M., Strieter, R.M., Elner, S.G., Baggiolini, M., Lindley, I., and Kunkel, S.L., 1990, Neutrophil chemotactic factor (IL-8) gene expression by cytokine-treated retina pigment epithelial cells, *Am. J. Pathol.* 136:745.

Elner, S.G., Strieter, R.M., Elner, V.M., Rollins, B.J., del Monte, M.A. and Kunkel, S., 1991, Monocyte chemotactic factor gene expression by cytokine treated human retinal pigment epithelial cells, *Lab. Invest.* 64:8.

Fabry, Z., Waldschmidt, M.M., Moore, S.A., and Hart, M.N., 1990, Antigen presentation by brain microvessel smooth muscle and endothelium, *J. Neuroimmunol.* 28:63.

Forrester, J.V., Borthwick, G.M., and McMenamin, P.G., 1985, Ultrastructural pathology of S-antigen uveoretinitis, *Invest. Ophthalmol. Vis. Sci.* 26:1281.

Gratton, J., Greenwood, J., Luthert, P., and Lightman, S., 1992, A quantitative comparison of the blood-retinal and blood-brain barrier capillary distribution and density in the rat using image analysis, *J. Physiol.* 446:508P.

Greenwood, J., 1992a, The blood-retinal barrier in experimental autoimmune uveoretinitis (EAU): a review, *Curr. Eye Res.* 11 (Suppl):25.

Greenwood, J., 1992b, Experimental manipulation of the blood-brain and blood-retinal barriers, *in:* Physiology and pharmacology of the blood-brain barrier, M.W.B., Bradbury, ed., *Handbook Exp. Pharmacol.* 103:459, Springer-Verlag, New York.

Greenwood, J., 1992c, Characterization of a rat retinal endothelial cell culture and the expression of P-glycoprotein in brain and retinal endothelium in vitro, *J. Neuroimmunol.* 39:123.

Greenwood, J., and Calder, V.L., 1993, Lymphocyte migration through cultured endothelial cell monolayers derived from the blood-retinal barrier, *Immunology* 80:401.

Greenwood, J., Howes, R., and Lightman, S., 1994, The blood-retinal barrier in experimental autoimmune uveoretinitis, *Lab. Invest.* 70:39.

Greenwood, J., Wang, Y., and Calder, V.L., 1995, Lymphocyte adhesion and transendothelial migration in the CNS: the role of LFA-1, ICAM-1, VLA-4 and VCAM-1, *Immunology* in press.

Hall, A.J.H., Greenwood, J., Patel, B., Brada, M., and Lightman, S.L., 1992, The effect of histamine and interleukin-2 on the blood-brain and blood-retinal barriers of the anaesthetized rat, *J. Physiol.* 446:497P.

Hawkins, C.P., Munro, P.M.G., MacKenzie, F., Kesselring, J., Tofts, P.S., du Boulay, E.P.G.H., Landon, D.N., and McDonald, W.I., 1990, Duration and selectivity of blood-brain barrier breakdown in chronic relapsing experimental allergic encephalomyelitis studied by gadolinium-DTPA and protein markers, *Brain* 113:365.

Hughes, C.C.W., and Lantos, P.L., 1986, Brain capillary endothelial cells in vitro lack surface IgG Fc receptors, *Neurosci. Lett.* 68:100.

Hughes, C.C.W., Male, D.K., and Lantos, P.L., 1988, Adhesion of lymphocytes to cerebral microvascular cells: effects of interferon-$\gamma$ tumour necrosis factor and interleukin-1, *Immunology* 64:677.

Lightman, S., and Greenwood, J., 1992, Effect of lymphocytic infiltration on the blood-retinal barrier in experimental autoimmune uveoretinitis, *Clin. exp. Immunol.* 88:473.

Lin, W-L., Essner, E., and Shichi, H., 1991, Breakdown of the blood-retinal barrier in S-antigen-induced uveoretinitis in rats, *Graefe's Arch. Clin. Exp. Ophthalmol.* 229:457.

Liversidge, J.M., Sewell, H.F., and Forrester, J.V., 1988, Human retinal pigment epithelial cells differentially express MHC class II (HLA DP, DR and DQ) antigens in response to *in vitro* stimulated with lymphokine or purified IFN-$\gamma$, *Clin. exp. Immunol.* 73:489.

Liversidge, J., Sewell, H.F., and Forrester, J.V., 1990, Interactions between lymphocytes and cells of the blood-retina barrier: mechanisms of T lymphocyte adhesion to human retinal capillary endothelial cells and retinal pigment epithelial cells in vitro, *Immunology*, 71:390.

Lossinsky, A.S., Badmajew, V., Robson, J.A., Moretz, R.C., and Wisniewski, H.M., 1989, Sites of egress of inflammatory cells and horseradish peroxidase transport across the blood-brain barrier in a murine model of chronic relapsing experimental allergic encephalomyelitis, *Acta Neuropathol. (Berl)* 78:359.

Male, D.K, Pryce, G., and Hughes, C.C.W., 1987, Antigen presentation in brain: MHC induction on brain endothelium and astrocytes compared, *Immunology* 60:453.

Male, D.K., and Pryce, G., 1988a, Kinetics of MHC gene expression and mRNA synthesis in brain endothelium, *Immunology* 63:37.

Male, D.K., and Pryce, G., 1988b, Induction of Ia molecules on brain endothelium is related to susceptibility to experimental allergic encephalomyelitis, *J. Neuroimmunol.* 21:87.

Male, D., Pryce, G., Hughes, C., and Lantos, P., 1990, Lymphocyte migration into brain modelled in vitro: control by lymphocyte activation, cytokines, and antigen, *Cell. Immunol.* 127:1.

Male, D., Rahman, J., Pryce, G., Tamatani, T., and Miyasaka, M., 1994, Lymphocyte migration into the CNS modelled in vitro: roles of LFA-1, ICAM-1 and VLA-4, *Immunology* 81:366.

Martiney, J.A., Litwak, M., Berman, J.W., Arezzo, J.C., and Brosnan, C.F., 1990, Pathophysiologic effect of interleukin-1ß in the rabbit retina, *Am. J. Pathol.* 137:1411.

Martiney, J.A., Berman, J.W., and Brosnan, C.F., 1992, Chronic inflammatory effects of interleukin-1 on the blood-retina barrier, *J. Neuroimmunol.* 41:167.

McCarron, R.M., Racke, M., Spatz, M., and McFarlin, D.E., 1991, Cerebral vascular endothelial cells are effective targets for in vitro lysis by encephalitogenic T lymphocytes, *J. Immunol.* 147:503.

McCarron, R.M., Wang, L., Cowan, E.P., and Spatz, M., 1991, Class II MHC antigen expression by cultured human cerebral vascular endothelial cells, *Brain Res.* 566:325.

McCarron, R.M., Wang, L., Racke, M.K., McFarlin, D.E., and Spatz, M., 1993, Cytokine-regulated adhesion between encephalitogenic T lymphocytes and cerebrovascular endothelial cells, *J. Neuroimmunol.* 43:23.

McMenamin, P.G., Forrester, J.V., Steptoe, R.J., and Dua, H.S., 1992, Ultrastructural pathology of experimental autoimmune uveitis, *Lab. Invest.* 67:42.

McMenamin, P.G., Broekhuyse, R.M., and Forrester, J.V., 1993, Ultrastructural pathology of experimental autoimmune uveitis: a review, *Micron* 24:521.

Neill, J.M., and Barnstable, C.J., 1990, Expression of the cell surface antigens RET-PE2 and N-CAM by rat retinal pigment epithelial cells during development and in tissue culture, *Exp. Eye Res.* 5:573.

Planck, S.R., Dang, T.T., Graves, D., Tara, D., Ansel, J.C., and Rosenbaum, J.T., 1992, Retinal pigment epithelial cells secrete interleukin-6 in response to interleukin-1, *Invest. Ophthalmol. Vis. Sci.* 33:78.

Pryce, G., Male, D., and Sedgwick, J., 1989, Antigen presentation in brain: brain endothelial cells are poor stimulators of T cell proliferation, *Immunology* 66:207.

Pryce, G., Male, D., and Sarkar, C., 1991, Control of lymphocyte migration into brain: selective interactions of lymphocyte subpopulations with brain endothelium, *Immunology* 72:393.

Raine, C.S., Cannella, B., Duijvestijn, A.M., and Cross, A.H., 1990, Homing to central nervous system vasculature by antigen-specific lymphocytes. II Lymphocyte/endothelial cell adhesion during the initial stages of autoimmune demyelination, *Lab. Invest.* 63:476.

Risau, W., Engelhardt, B., and Wekerle, H., 1990, Immune function of the blood-brain barrier: incomplete presentation of protein (auto-)antigens by rat brain microvascular endothelium in vitro, *J. Cell Biol.* 110:1751.

Robinson, P.J., and Rapoport, S.I., 1987, Size selectivity of blood-brain barrier permeability at various times after osmotic opening, *Am. J. Physiol.* 253:R459.

de Rosbo, N.K., Bernard, C.C.A., Simmands, R.D., and Carnegie, P.R., 1985, Concomitant detection of changes in myelin basic protein and permeability of blood spinal cord barrier in experimental allergic encephalomyelitis by electroimmunoblotting, *J. Neuroimmunol.* 6:349.

Sedgwick, J.D., Hughes, C.C., Male, D.K., MacPhee, I.A.M., and Ter Meulen, V., 1990, Antigen-specific damage to brain vascular endothelial cells mediated by encephalitogenic and nonencephalitogenic CD4+ T cell lines in vitro, *J. Immunol.* 145:2474.

Stoul, W., Kaplan, M.S., and Gonatas, N.K., 1979, A quantative assay for experimental allergic encephalomyelitis in the rat based on permeability of the spinal cord to $^{125}$I-human gamma-globulin, *J. Immunol.* 122:920.

Towler, H.M.A., Greenwood, J., and Butt, A.M., 1994, Electrical resistance across the inner blood-retinal barrier in the intact retina of the anaesthetized Lewis rat, *J. Physiol.* 480:10P.

Waldschmidt, M.M., Fabry, Z., Keiner, J., Love Homan, L., and Hart, M.N., 1991, Adhesion of splenocytes to brain microvascular endothelium in the BALB/c and SJL/j mouse systems, *J. Neuroimmunol.* 35:191.

Wang, Y.F., Calder, V.L., Greenwood, J., and Lightman, S.L., 1993, Lymphocyte adhesion to cultured endothelial cells of the blood-retinal barrier, *J. Neuroimmunol.* 48:161.

Wang, Y., Calder, V.L., Lightman, S.L., and Greenwood, J., 1995, Antigen presentation by rat brain and retinal endothelial cells, *J. Neuroimmunol.* in press.

# INDEX

MIX
Papier aus verantwortungsvollen Quellen
Paper from responsible sources
FSC® C105338
FSC
www.fsc.org

If you have any concerns about our products,
you can contact us on
**ProductSafety@springernature.com**

In case Publisher is established outside the EU,
the EU authorized representative is:
**Springer Nature Customer Service Center GmbH**
**Europaplatz 3, 69115 Heidelberg, Germany**

Printed by Libri Plureos GmbH
in Hamburg, Germany